BARRON'S

Regents Exams and Answers

Earth Science— Physical Setting

Revised Edition

EDWARD J. DENECKE, JR., M.A.
Formerly, Science Department
William H. Carr J.H.S. 194Q
Whitestone, N.Y.

Published by Kaplan North America, LLC, dba Barron's Educational Series
1515 West Cypress Creek Road
Fort Lauderdale, Florida 33309
www.barronseduc.com

ISBN: 978-1-5062-6465-3

10 9 8 7 6 5 4 3

Kaplan North America, LLC, dba Barron's Educational Series print books are available at special quantity discounts to use for sales promotions, employee premiums, or educational purposes. For more information or to purchase books, please call the Simon & Schuster special sales department at 866-506-1949.

Contents

Preface

This book is designed to strengthen your understanding and mastery of the material in the New York State Physical Setting/Earth Science Core Curriculum, a comprehensive course of study in earth science on the secondary level. This book has been specifically written to assist you in preparing for the Regents examination covering this course. It contains the following special features that will help you in studying for this exam:

(1) *Complete sets of questions from previous Physical Setting/Earth Science Regents examinations are provided.* Attempting to answer these questions will make you familiar with the topics tested on the examination and the degree of difficulty to which you are expected to master each topic. It will also increase your confidence as you become familiar with the nature and language of the various question types that will appear on the examination.

(2) *Correct answers to all Regents questions are fully explained.* Explanations of correct answers are detailed, yet simply and clearly stated. Each explanation has been written to help you grasp the reasoning used to arrive at the correct answer rather than just stating the correct choice. Careful study of the step-by-step explanations of how the correct answer is arrived at for each question will improve your mastery of the subject. Even if you have answered the question correctly, you should read the explanation to check your reasoning and gain insight into the topic. This insight may be valuable when answering a more difficult question on the same topic.

(3) *Wrong choices are explained.* If you've ever answered a test question incorrectly and not understood *why* your answer was wrong, you will find this feature helpful. Understanding why a choice is incorrect will help you to clear up misconceptions you may have and to see errors in your reasoning.

(4) *Reference Tables for Physical Setting/Earth Science.* These tables are provided to all students for use while taking the Physical Setting/Earth Science Regents examination. Some questions *require* the use of these tables, and that is

usually indicated in the question. When answering other questions, you may find information in the tables *useful*. These tables are included in this book so that you may become *thoroughly familiar* with all of the information contained in them.

(5) *Topic Outline and Self-Analysis Charts.* The New York State Physical Setting/Earth Science Core Curriculum has been cross-referenced to questions from past Physical Setting/Earth Science Regents examinations. This will help you to identify your areas of weakness and to locate questions to review particular earth science concepts. Refer to "How to Use the Topic Outline" and Self-Analysis Charts for more detailed instructions.

(6) *Test-Taking Tips.* These helpful tips and strategies will help you to increase your grade on the Physical Setting/Earth Science Regents examination.

The author would like to acknowledge his editor, Linda Turner, for her invaluable assistance in writing and producing this book; his wife, Gerry, for her infinite patience; and his children, Merry, Abby, and Ben, for being a constant source of joy.

How to Use This Book

FORMAT OF THE PHYSICAL SETTING/EARTH SCIENCE REGENTS EXAMINATION

The examination consists of a performance portion containing three tasks to measure laboratory skills and a written portion containing multiple-choice, constructed-response, and extended constructed-response questions. You will be required to answer ALL of the questions on the Physical Setting/Earth Science Regents examination.

THE LABORATORY PERFORMANCE TEST

Since laboratory experiences are an essential part of a science course, a portion of the Physical Setting/Earth Science Regents examination is devoted to assessing laboratory skills. Tasks have been identified from laboratory experiments that you will have performed during the school year. These tasks, which represent skills that you are expected to have mastered, change only slightly, if at all, from year to year.

The performance portion of the examination is administered separately from the written portion, normally two weeks earlier. Arrangements for administering the performance portion are made at each school in accordance with guidelines set by the New York State Education Department.

The scoring for each task is based upon accuracy. Values within a certain range are granted the full point value allotted to each task.

Additional information regarding the performance portion, including an indication of the three tasks to be completed, will be provided by your teacher when this portion of the examination is given. The following is an outline of these tasks. The time allowed for completing the tasks at each station is 9 minutes.

Note: The following description represents the information that the State Education Department has stated may be shared with students before taking the performance portion of the examination. You should be familiar with the skills being assessed because you have used them in laboratory activities throughout the year. However, you will not be allowed to practice the entire test or any of the individual stations before this performance component is administered.

Station 1 . . *Mineral and Rock Identification*

The student determines the properties of a mineral and identifies that mineral using a flowchart. Then the student classifies two different rock samples and states a reason for each classification based on observed characteristics.

Station 2 . . *Locating an Epicenter*

The student determines the location of an earthquake epicenter using various types of data that were recorded at three seismic stations.

Station 3 . . *Constructing and Analyzing an Asteroid's Elliptical Orbit*

The student constructs a model of an asteroid's elliptical orbit and compares the eccentricity of the orbit with that of a given planet.

THE WRITTEN TEST

The written portion of the Physical Setting/Earth Science Regents examination represents 85 points of the total score and has three parts: A, B, and C. You should be prepared to answer questions in multiple-choice, constructed-response, and extended constructed-response formats. Questions will be content- and skills-based and may require you to graph data, complete a data table, label or draw diagrams, design experiments, make calculations, or write short or extended responses. In addition, you may be required to hypothesize, to interpret, analyze, or evaluate data, or to apply your scientific knowledge and skills to real-world situations. Some of the questions will require use of the 2011 Edition of the *Reference Tables for Physical Setting/Earth Science*.

> You will be required to answer ALL of the questions in the Physical Setting/Earth Science Regents examination!

The following is a brief description of the types of questions comprising each part of the examination:

Part A—Multiple Choice A multiple-choice question offers several answers from which you choose the one that best answers the question or completes the statement. Part A of the exam focuses on earth science content from Standard 4 (see Topic Outline) and represents 30–40 percent of the examination. Many practice questions of this type from previous Regents examinations are included at the end of each chapter.

Part B—Multiple Choice (B-1) and Constructed Response (B-2) In a constructed-response question there is no list of choices from which to choose an answer; rather you are required to provide the answer. Constructed-response questions test skills ranging from constructing graphs or topographic maps to formulating hypotheses, evaluating experimental designs, and drawing conclusions based upon data. In Part B, which represents 25–35 percent of the examination, you are asked to demonstrate skills identified in Standards 1, 2, 6, and 7 in the context of earth science. Practice questions of this type are also included at the end of each chapter.

Part C—Extended Constructed Response These are constructed-response questions that require more time (15–20 minutes per item) and effort on your part to answer. Questions in Part C require you to apply your earth science knowledge and skills to real-world problems and applications. You may be asked to write short essays, design controlled experiments, predict outcomes, or analyze the risks and benefits of various solutions to a problem. Part C represents 15–25 percent of the examination. Again, practice questions are provided at the end of most chapters.

You should review constructed-response and extended constructed-response questions from previous examinations to familiarize yourself with these types of questions. Parts B and C are recent additions, and each examination may contain questions unlike those in previous exams. Keep these points in mind:

- Write complete sentences whenever you are asked to answer in your own words.
- Show all work, and use correct units in mathematical problems.
- Be able to interpret map scales and legends, to construct isolines, and to draw profiles.

Although the questions on each written test are unique, they are similar to questions on past Regents examinations. By studying questions that have been asked in the past, therefore, you can strengthen your skills and knowledge in preparation for the examination you will take.

CONTENTS OF THIS BOOK

THE TOPIC OUTLINE AND THE REFERENCE TABLES FOR PHYSICAL SETTING/EARTH SCIENCE

The first part of this book consists of two special features, the Topic Outline and the Earth Science Reference Tables. For detailed instructions on how to use the *Topic Outline*, see page 28. The *Reference Tables for Physical Setting/Earth Science* in this book are the same ones that are provided to students for use while taking the Regents examination. You should be familiar with these tables at the time of the exam, since you will have used them during the school year. Some questions require the use of these tables, and this fact is usually indicated in the question. For other questions you may find information in the tables helpful. It is therefore a good idea to become generally familiar with what is contained in the tables. For the Physical Setting/Earth Science exam, use the tables beginning on page 41.

REGENTS EXAMINATIONS AND ANSWERS

In the second, much longer part of the book there are five Physical Setting/ Earth Science Regents examinations with Answer Keys. Every correct answer is fully explained. Also, to help clear up misconceptions that may lead students astray, in many cases explanations for wrong answers are provided as well.

THE SELF-ANALYSIS CHART

At the end of the *Answers* section for each examination, a *Self-Analysis Chart* is provided to aid you in pinpointing your areas of weakness. The upper portion of the chart, labeled "Standards 1, 2, 6, and 7: Skills and Applications" lists the numbers of questions requiring the use of specific skills, or the ability to apply knowledge. These skills and abilities are described more completely in the *Major Understandings* of Standards 1, 2, 6, and 7 listed in the Topic Outline. The lower portion of the chart, labeled "Standard 4: The Physical Setting/Earth Science" lists the numbers of questions requiring knowledge of specific earth science concepts and is subdivided into sections labeled *"Astronomy," "Geology,"* and *"Meteorology."* These concepts are described more completely in the *Major Understandings* of Standard 4 listed in the Topic Outline.

Many question numbers may appear in both portions of the chart because many questions involve using a skill, such as deductive reasoning, together with your knowledge of specific earth science concepts to answer the question. Therefore, you should consider these two sections of the chart independently, using the upper section of the Self-Analysis Chart to pinpoint areas of weakness in your skills, and the lower section of the chart to pinpoint areas of weakness in your knowledge of earth science concepts.

Test-Taking Tips

The following pages contain several tips to help you achieve a good grade on the Physical Setting/Earth Science Regents exam. They are divided into GENERAL HELPFUL TIPS and SPECIFIC HELPFUL TIPS.

GENERAL HELPFUL TIPS

TIP 1

Be Confident and Prepared

SUGGESTIONS

- Review previous tests.
- Use a clock or watch, and take previous exams at home under examination conditions (i.e., don't have the radio or television on).
- Get a review book. (The preferred book is Barron's *Let's Review: Earth Science.*)
- Talk over the answers to questions on these tests with someone else, such as another student in your class or someone at home.
- Finish all your homework assignments.
- Look over classroom exams that your teacher gave during the term.
- Take class notes carefully.
- Practice good study habits.
- Know that there are answers for every question.
- Be aware that the people who made up the Regents exam want you to pass.
- Remember that thousands of students over the last few years have taken and passed an Earth Science Regents. You can pass too!
- On the night prior to the exam day, lay out all the things you will need, such as clothing, pens, and admission cards.

- Go to bed early; eat wisely.
- Bring at least two pens to the exam room.
- Bring your favorite good luck charm/jewelry to the exam.
- Once you are in the exam room, arrange things, get comfortable, be relaxed, attend to personal needs (the bathroom).
- Keep your eyes on your own paper; do not let them wander over to anyone else's paper.
- Be polite in making any reasonable requests of the exam room proctor, such as changing your seat or having window shades raised or lowered.

TIP 2

Read Test Instructions and Questions Carefully

SUGGESTIONS

- Be familiar with the test directions ahead of time.
- Decide upon the task(s) that you have to complete.
- Know how the test will be graded.
- Know which question or questions are worth the most points.
- Give only the information that is requested.
- Where a choice of questions exists, read all of them and answer *only* the number requested.
- Underline important words and phrases.
- Ask for assistance from the exam room proctor if you do not understand the directions.

TIP 3

Budget Your Test Time in a Balanced Manner

SUGGESTIONS

- Bring a watch or clock to the test.
- Know how much time is allowed.
- Arrive on time; leave your home earlier than usual.
- Prepare a time schedule and try to stick to it. Remember that Regents exams are longer than classroom tests, so you will need to pace yourself accordingly.
- Answer the easier questions first.
- Devote more time to the harder questions and to those worth more credit.
- Don't get "hung up" on a question that is proving to be very difficult; go on to another question and return later to the difficult one.
- Ask the exam room proctor for permission to go to the lavatory, if necessary, or if only to "take a break" from sitting in the room.
- Plan to stay in the room for the entire three hours. If you finish early, read over your work—there may be some things that you omitted or that you may wish to add. You also may wish to refine your grammar, spelling, and penmanship.

TIP 4

Be "Kind" to the Exam Grader/Evaluator

SUGGESTIONS

- Assume that you are the teacher grading/evaluating your test paper.
- Answer questions in an orderly sequence.
- Write legibly.
- Answer Part C questions with complete sentences.
- Proofread your answers prior to submitting your exam paper.

TIP 5

Use Your Reasoning Skills

SUGGESTIONS

- Answer *all* questions.
- Relate (connect) the question to anything that you studied, wrote in your notebook, or heard your teacher say in class.
- Relate (connect) the question to any film you saw in class, any project you did, or to anything you may have learned from newspapers, magazines, or television.
- Decide whether your answers would be approved by your teacher.
- Look over the entire test to see whether one part of it can help you answer another part.
- Be cautious when changing an answer. Try to remember why you selected the first answer to be sure that the new answer is better.

TIP 6

Don't Be Afraid to Guess

SUGGESTIONS

- In general, go with your first answer choice.
- Eliminate obvious incorrect choices.
- If you are still unsure of an answer, make an educated guess.
- There is no penalty for guessing; therefore, answer ALL questions. An omitted answer gets no credit.

Let's now review the six GENERAL HELPFUL TIPS for short-answer questions:

SUMMARY OF TIPS

1. Be confident and prepared.
2. Read test instructions and questions carefully.
3. Budget your test time in a balanced manner.
4. Be "kind" to the exam grader/evaluator.
5. Use your reasoning skills.
6. Don't be afraid to guess.

SPECIFIC HELPFUL TIPS FOR THE MULTIPLE-CHOICE QUESTIONS

TIP 1

Read the Question and Try to Answer It Before Looking at the Possible Choices

Thinking out the answer to a question before looking at the possible choices stimulates your memory. Key facts, concepts, and relationships come to mind. If the answer you anticipated is among the possible choices, it is likely to be the correct one. But ***keep reading*** through all the choices! By thinking of the answer first, you are less likely to be fooled by a wrong answer. But if you don't read all of the choices, you might miss a more correct or more complete choice farther down the list.

Example

Earth's hydrosphere is best described as the

(1) solid outer layer of Earth
(2) liquid outer layer of Earth
(3) magma layer located below Earth's stiffer mantle
(4) gaseous layer extending several hundred kilometers from Earth into space

As you read the question, you will probably think of the terms *lithosphere, hydrosphere*, and *atmosphere*. You may recall that "hydro" refers to water and remember that the hydrosphere is all the liquid water resting on the lithosphere. Therefore, you anticipate an answer referring to Earth's layer of liquid water. With this in mind, you recognize that the correct answer is choice 2, because choice 1 refers to a solid layer, choice 4 refers to a gaseous layer, and choice 3 refers to a layer of magma, which is liquid rock—not liquid water.

TIP 2

Always Read *All* of the Choices Before Choosing an Answer

Read the question and all of the choices carefully. Make sure you are reading exactly what has been written and not just skimming for words that you are anticipating. Continue reading even if you find your anticipated answer because there may be a more complete or more correct answer farther down the list.

Example

The length of an Earth year is based on Earth's

(1) rotation of 15°/hr
(2) revolution of 15°/hr
(3) rotation of approximately 1°/day
(4) revolution of approximately 1°/day

When reading this question you might anticipate that Earth's year is based on its revolution around the Sun. If you rushed through the answers and chose the first one to mention revolution—choice 2—you would miss this question. If you carefully read *all* of the choices, you realize that both choice 2 and choice 4 refer to revolution, and that choice 4 also correctly states the rate at which Earth revolves.

TIP 3

Answer the Easy Questions First; Do Not Spend Too Much Time on Any One Question

In this part of the test, all of the questions have the same value. You get as much credit for an easy question as a hard one. Don't waste a lot of time on questions you can't answer quickly. Answer all of the easy questions first, then go back and figure out the hard ones in the time left over. Easy questions usually contain short sentences and answers with few words. The question can usually be answered quickly from the information presented.

Example

An air mass classified as mT usually forms over which type of Earth surface?

(1) cool land
(2) cool water
(3) warm land
(4) warm water

The correct answer, choice 4, is obvious if you know that mT stands for maritime tropical. Since tropical means warm and maritime refers to the sea, such an air mass usually forms over warm water. But if you don't know what mT means, and don't realize you can find its meaning in the *Reference Tables for Physical Setting/Earth Science*, don't waste a lot of time on the question.

TIP 4

Look for and Underline Key Words or Phrases in the Question That Tell You What the Questioner Wants You to Do

Read every word of the question. Make sure you understand what is being asked before you move to the possible answers. Sometimes there are key words or phrases that will help you pick the correct answer. Some examples of key words are *all*, *none*, *some*, *most*, *never*, *always*, *least*, *greatest*, *less*, *more*, *maximum*, *minimum*, *highest*, *lowest*, *average*, *most nearly*, and *best*. Pay careful attention to these words. If you overlook a key word, you may miss a question to which you really know the answer.

Example

A student read in a newspaper that the maximum length of the daylight period for the year in Syracuse, New York, had just been reached. What was the date of the newspaper?

(1) March 22
(2) June 22
(3) September 22
(4) December 22

The key phrase in the question is "the maximum length of the daylight period for the year" and the question asks for the "date" on which this was reported in a newspaper. The questioner wants you to choose that *date* on which the *daylight period* is at a *maximum*. Since the longest period of daylight in New York State occurs at the summer solstice, June 21, the newspaper probably reported it the next day, June 22—choice 2.

TIP 5

Use Partial Knowledge to Eliminate Wrong Choices

When you have been through all the questions once, go back and find questions about which you have some knowledge. Read all of the possible choices and use your partial knowledge to eliminate one or two of the answers. Cross off incorrect choices to narrow possible answers. Then choose between the remaining answers. If you can eliminate two wrong answers, your chance of guessing the right answer is greater.

Example

In which list are celestial features correctly shown in order of increasing size?

(1) galaxy → solar system → universe → planet
(2) solar system → galaxy → planet → universe
(3) planet → solar system → galaxy → universe
(4) universe → galaxy → solar system → planet

Let's suppose you *know* that the universe is the biggest feature because it contains all the others. You *think* a planet might be the smallest of the features, but you are not absolutely sure. You have no idea of the sizes of the other features. You can use this partial knowledge to greatly increase your chances of guessing correctly. Because the question asks which list shows the features in *order of increasing size*, the list should end with the largest feature—the universe. Therefore, you can eliminate choices 1 and 4. Of the remaining two choices, you would then guess the answer to be choice 3, since you think a planet might be the smallest feature. And choice 3 is the correct answer.

TIP 6

Look for Clues Among the Choices in the Question, and Use Information from One Part of the Test to Help You with Others

By reading questions and choices carefully, you may often find words or phrases that provide clues to an answer. This tip is important because, if you are alert, you may find links and connections between various questions on the exam. One question may give you a clue (or even an answer) to another question.

Example

The diagram below represents part of Earth's latitude-longitude system.

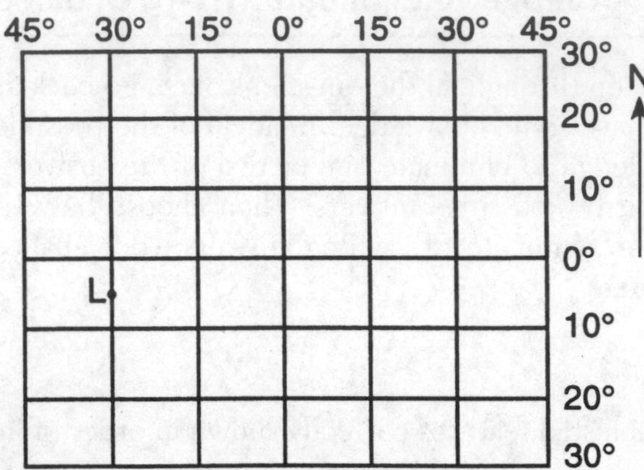

What is the latitude and longitude of point ***L***?

(1) 5° E 30° N
(2) 5° W 30° S
(3) 5° N 30° E
(4) 5° S 30° W

The first clue is that all of the possible answers give 5° and 30° as the coordinates of point *L*. If you look at the axes, you see that 5° is the coordinate of point *L* on the vertical axis to the right. Next you should realize that the vertical axis is a North-South axis, and you can eliminate choices 1 and 2 because they give the 5° coordinates as East or West. If the vertical axis is a North-South axis, coordinates above the 0° mark (the Equator) are N; those below it are S. Since point *L* is 5° below the 0° mark, it is an S coordinate, as shown in only one of the remaining choices—choice 4.

TIP 7

Know What Is in the *Reference Tables for Physical Setting/ Earth Science,* and Carefully Examine Any Diagrams, Graphs, or Tables Given with a Question

Many questions expect you to apply what these graphs, tables, and diagrams contain. If you can't answer a question with the information given, there is a good chance that you will find the information you need in the *Reference Tables for Physical Setting/Earth Science*.

Example 1

You need information given in the *Reference Tables for Physical Setting/ Earth Science* to answer the question.

The air outside a classroom has a dry-bulb temperature of 10°C and a wet-bulb temperature of 4°C. What is the relative humidity of the air?

(1) 1%
(2) 14%
(3) 33%
(4) 54%

In this question, you must know that there is a chart in the *Reference Tables for Physical Setting/Earth Science* that can be used to determine relative humidity given wet- and dry-bulb temperatures. If you know how to use this chart, you look for the intersection of the dry-bulb temperature and the difference between the wet- and dry-bulb temperatures, that is, the intersection of the row for 10°C and the column for 6°C, where you find the value for the % Relative Humidity—33—choice 3.

Example 2

You need to apply information in a chart, graph, or diagram given with the question.

The table below shows the noontime data for air pressure and air temperature at a location over a period of 1 week.

Date	Nov. 9	Nov. 10	Nov. 11	Nov. 12	Nov. 13	Nov. 14	Nov. 15
Air Temperature (°C)	1	6	0	−2	−4	5	10
Air pressure (mb)	1,024	998	1,015	1,021	1,030	1,013	?

Based on the data provided, which air pressure would most likely occur at noon on November 15?

(1) 987 mb
(2) 1,015 mb
(3) 1,017 mb
(4) 1,022 mb

In this question, you need to apply the information in the chart. By carefully examining the data in the table, you see that as air temperature increases, air pressure decreases. Note that as temperature goes from 1°C to 5°C to 6°C, the air pressure goes from 1,024 mb to 1,013 mb to 998 mb. Therefore, if the temperature increases to 10°C, the air pressure should drop even lower than 998 mb. Now read the choices and note that only one is lower than 998 mb, choice 1—987 mb.

TIP 8

Do Not Choose an Answer That Contains an Element of Truth but That Is Incorrect as It Relates to the Question

It is very important to remember that just because a statement is true, or contains an element of truth, does not mean that it is the correct answer to the question. Statements that are true, or partially true, make attractive distractors (wrong answers). Several answers may contain an element of truth, but you must decide which one is related most directly to the question.

Example

Major ocean and air currents appear to curve to the right in the Northern Hemisphere because

(1) Earth has seasons
(2) Earth's axis is tilted
(3) Earth rotates on its axis
(4) Earth revolves around the Sun

Each of the answer choices is a true statement. However, the question focuses on the curving of major ocean and air currents, that is, the Coriolis Effect. If you understand the Coriolis Effect, you know that its primary cause is Earth's rotation on its axis—choice 3.

TIP 9

Beware of Broad, Sweeping Generalizations in the Answer Choices—They Are Generally Incorrect

Pay careful attention to the use of words such as *always*, *never*, *only*, *must*, and so on. Such words usually make general statements improbable. This tip is particularly helpful when two or more choices may be correct. Often, the broadest, most inclusive choice is correct because it may include the other choice(s).

Example

According to the Rock Cycle in Earth's Crust diagram in the *Reference Tables for Physical Setting/Earth Science*, which type(s) of rock can be the source of deposited sediments?

(1) igneous and metamorphic rocks, only
(2) metamorphic and sedimentary rocks, only
(3) sedimentary rocks, only
(4) igneous, metamorphic, and sedimentary rocks

In this question, the best answer is choice 4—it is the least restrictive, includes all the other choices, and gives the broadest possibilities.

TIP 10

Don't Be Afraid to Guess If You're Not Sure of the Correct Answer, but Do Not Keep Changing Your Answer

If you are not sure of an answer, don't be afraid to guess. You have nothing to lose if you guess wrong. A wrong answer and no answer are both worth zero points. However, the correct answer is right there on the page, and if you guess correctly, you can increase your point score. If you can eliminate one or more wrong answers before guessing, your chance of choosing the correct answer jumps from one in four to one in three, or one in two. But don't think that you can pass a Regents exam by guessing; the odds are that if you guessed on every question, you would get one in four correct—and score a 25 on the exam! There is no substitute for careful, diligent preparation for the exam.

If you do have to guess, make an educated guess. An educated guess is made by considering the options and focusing on what you *do* know. Use what you *do* know to eliminate improbable answers. Then read the question and answer as a complete sentence and ask yourself if it is true, or sounds true. Often, when reading the choices to yourself, one of the answers will "sound right."

Once you have made your choice, be slow to change your answer without a good reason, because research shows that a person's first choices are usually correct. Most people who change their answers change from a correct one to a wrong one. Choose the answer that seems best to you and move on to the next question. Change your answer only if you are absolutely sure you made a mistake (for example, if another question on the test reminds you of the correct answer).

Example

In Earth's geologic past there were long periods that were much warmer than the present climate. What is the primary evidence that these long warm periods existed?

(1) United States National Weather Service records
(2) polar magnetic directions preserved in the rock record
(3) radioactive decay rates
(4) plant and animal fossils

Suppose you had no idea how to answer this question. You could use what you *do* know to make an educated guess. You do know that Earth's geologic past covers billions of years. You do know that the United States didn't exist until the late 1700s. So you can eliminate choice 1 because it is unlikely that USNWS records go back far enough to tell you about long warm periods in Earth's geologic past. You also know that compasses work in both winter and summer, so you have the idea that Earth's magnetic field doesn't really change with temperature. So now you eliminate choice 2. You remember reading something about radioactive decay and heat, and you know that the plants and animals that live in places with warm climates are very different from those that live in cold climates. At least, though, you've narrowed it down to two choices. Because you are more sure of the connection between climate, plants, and animals, you select choice 4 as your answer—and get the question right!

SUMMARY OF TIPS

1. Read the question and try to answer it before looking at the possible choices.
2. Always read *all* of the choices before choosing an answer.
3. Answer the easy questions first; do not spend too much time on any one question.
4. Look for and underline key words or phrases in the question that tell you what the questioner wants you to do.
5. Use partial knowledge to eliminate wrong choices.
6. Look for clues among the choices in the question, and use information from one part of the test to help you with others.
7. Know what is in the *Reference Tables for Physical Setting/ Earth Science*, and carefully examine any diagrams, graphs, or tables given with a question.
8. Do not choose an answer that contains an element of truth but that is incorrect as it relates to the question.
9. Beware of broad, sweeping generalizations in the answer choices—they are generally incorrect.
10. Don't be afraid to guess if you're not sure of the correct answer, but do not keep changing your answer.

HELPFUL TIPS FOR QUESTIONS ON PARTS B AND C

Part B includes both multiple-choice and constructed-response questions and represents 25–35 percent of the examination. Part C consists of extended constructed-response questions and represents 15–25 percent of the examination. The precise number of questions in each part may vary from year to year. Some questions may require the use of the *Reference Tables for Physical Setting/Earth Science*.

TIP 1

Know What a Constructed-Response Question Is

"Constructed-response" means that you must provide, in writing, your own response to a question, rather than choosing your answer from a list of possible choices. There are two types of constructed-response questions: open-response and free-response. For an open-response question, there is only one correct answer, while a free-response question has many alternative answers.

Example of an open-response question:

Base your answers to questions 80 and 81 on the diagram below and on your knowledge of earth science. The diagram represents the apparent path of the Sun on the dates indicated for an observer in New York State. The diagram also shows the angle of Polaris above the horizon.

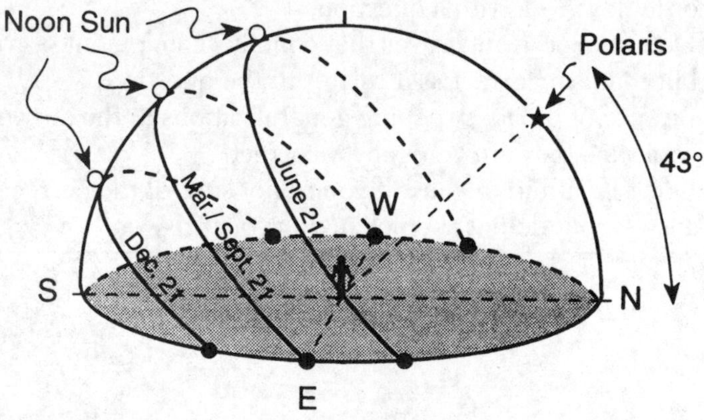

80 State the latitude of the location represented by the diagram to the *nearest degree.* Include the latitude direction in your answer. [2]

This is an open-response question; it has only one possible answer: 43°N.

Example of a free-response question.

81 The climate of most locations near the Equator is warm and moist.

(a) Explain why the climate is warm. [1]

___

___

___

(b) Explain why the climate is moist. [1]

___

___

___

Although an acceptable response must be scientifically correct, there are many alternative ways of expressing your answer to this question. Therefore, it is a free-response question.

Acceptable answers to (a) might include statements such as these: *"Locations near the Equator receive more direct sunlight"* and *"The intensity of sunlight is greater near the Equator."* Both responses, although different, would receive full credit.

Tip 2

Know What an Extended Constructed-Response Question Is

Extended constructed-response questions are constructed-response questions that require more time and effort on your part to answer. These questions often have multiple parts and higher point values and may require you to graph data, complete a data table, label or draw diagrams, design experiments, make calculations, or write short or extended responses. In addition, you may be required to hypothesize, interpret, analyze, or evaluate data, or apply your scientific knowledge and skills to real-world situations.

Tip 3

Understand Why These Kinds of Questions Are on the Examination

Constructed-response and extended constructed-response questions are designed to:

- Assess your *understanding* of earth science concepts rather than straight recall of factual information.
- Require you to demonstrate that you can "pull all the pieces together," that is, integrate data, to solve a problem.
- Provide insight into how you think by requiring you to defend your answer in writing or to show all of your work.
- Assess knowledge and skills that cannot easily be assessed using a multiple-choice format.

Tip 4

Know What Will Be Expected of You for These Parts of the Examination

Parts B-2 and C of the examination will require you to demonstrate in a variety of ways that you have certain knowledge, skills, and abilities. You may be asked to construct a graph or draw a diagram. You may be asked to solve a mathematical problem, showing all of your work. You may be asked to provide explanations for physical phenomena or to write an essay on a topic in earth science.

Some of the skills and abilities that may be assessed in Parts B-2 and C of the examination are:

- Apply the concepts of eccentricity, rate, gradient, standard error of measurement, and density in context.
- Determine the relationships among velocity, slope, sediment size, channel shape, and volume of a stream.
- Understand the relationships among the planets' distances from the Sun, gravitational forces, periods of revolution, and speeds of revolution.
- In a field, use isolines to determine a source of pollution.
- Show how observations of celestial motions supports the early idea that stars moved around a stationary Earth (the geocentric model), but further investigation led scientists to understand that most of these motions are a result of Earth's revolution around the Sun (the heliocentric model).

- Design an experiment to test sediment properties and the rate of deposition of particles.
- Graph the changing length of a shadow based on the motion of the Sun.
- Analyze patterns of topography and drainage, and design solutions to deal effectively with runoff.
- Analyze weather maps to predict future weather events.
- Critique printed or electronic materials that exemplify miscommunication and/or misconceptions of current commonly accepted scientific knowledge.
- Discuss how early warning systems can protect society and the environment from natural disasters such as hurricanes, tornadoes, earthquakes, tsunamis, floods, and volcanoes.
- Analyze a depositional-erosional system of a stream.
- Draw a simple contour map of a model landform.
- Construct and interpret a profile based on an isoline map.
- Use flowcharts to identify rocks and minerals.
- Develop a scale model to represent planet size and/or distance from the Sun.
- Develop a scale model of units of geologic time.
- Use topographical maps to determine distances and elevations.
- Analyze the interrelationship between gravity and inertia and its effects on the orbits of planets or satellites.
- Graph and interpret the nature of cyclic changes such as sunspots, tides, and atmospheric carbon dioxide concentrations.
- Determine, based on current data of plate movements, past and future positions of land masses.
- Using given weather data, identify the interface between air masses, such as cold fronts, warm fronts, and stationary fronts.
- Compare and contrast the effects of human activities as they relate to quality of life on Earth systems (global warming, land use, preservation of natural resources, pollution).
- Analyze the issues related to local energy needs and develop a viable energy-generation plan for a community.
- Evaluate the political, economic, and environmental impacts of global distribution and use of mineral resources and fossil fuels.
- Consider the environmental and social implications of various solutions to a specific environmental Earth resources problem.
- Using a topographic map, determine the safest and most efficient route for rescue purposes.

A complete list of skills and abilities is described in greater detail in Standards 1, 2, 6, and 7 of the Topic Outline and Index.

Tip 5

Work to Improve Your Score on These Types of Questions

1. **Determine what the question is asking you to do. Underline key words in the question.**

   Most questions of this type have an action word that identifies what you must do to earn points.

   Below, the action words in a question you looked at earlier are bold and underlined. Other key words are underlined.

<u>State</u> the <u>latitude</u> of the location represented by the diagram to the <u>nearest degree</u>. **<u>Include</u>** the <u>latitude direction</u> in your answer. [2]

2. **Check to see what points are being awarded for.**

   The points are listed in brackets after each question. If more than one point is being awarded, you can usually make a good guess about what each point will be given for if you read the question again carefully. In the example above, you have probably guessed that one point is given for stating the latitude to the nearest degree, and one point is given for including the correct direction. The answer is 43° North. If you wrote just 43°, you would get only one point. For full credit you must state both 43° and North (or N).

3. **Always include units in your answers, and check to make sure that the form of your answer matches the form asked for in the question.**

   Consider this example:

Air Temperature	21°C
Barometric Pressure	993.1 mb
Wind Direction	From the east
Windspeed	25 knots

<u>State</u> the <u>barometric pressure</u> in its proper form, <u>as used on a station model.</u>

What you are asked to do is state the barometric pressure. However, if you write 993.1 mb, your answer will be marked incorrect because the form asked for is "as used on a station model." On a station model, the decimal point is dropped and the last three digits only are listed. Your answer should be given as 931 because that is what would appear on a station model.

4. When asked to respond in the form of a statement or essay, use complete sentences and organize your answer carefully.

A complete sentence begins with a capital letter, has a subject and a verb, and ends with a period. Very often, a point is awarded solely for answering in a complete sentence—even if the answer is wrong! Therefore, read over your statements or essays carefully.

5. Examine the answer sheet for Part B-2 or C for clues to how the question is to be answered.

The answer sheet for Part B-2 or C will often give you a format for answering the question. For example, the answer sheet for the question below provides several blank lines, so you are probably being asked to write a short essay.

82 Using one or more complete sentences, state one reason that ultraviolet rays are dangerous. [2]

If the answer sheet has a blank graph, you are probably being asked to construct a graph. Be sure to include all elements asked for in the question. For example, on the graph below you must not only plot points but also plot them with specific shapes, label axes, and choose an appropriate scale for each axis.

83 Plot the data for rock sample *A* for the 20 minutes of the investigation. Surround each point with a small circle and connect the points. [1]

Example:

84. Plot the data for rock sample *B* for the 20 minutes of the investigation. Surround each point with a small triangle and connect the points. [1]

Example:

85. Plot the data for rock sample *C* for the 20 minutes of the investigation. Surround each point with a small square and connect the points. [1]

Example:

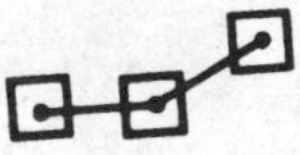

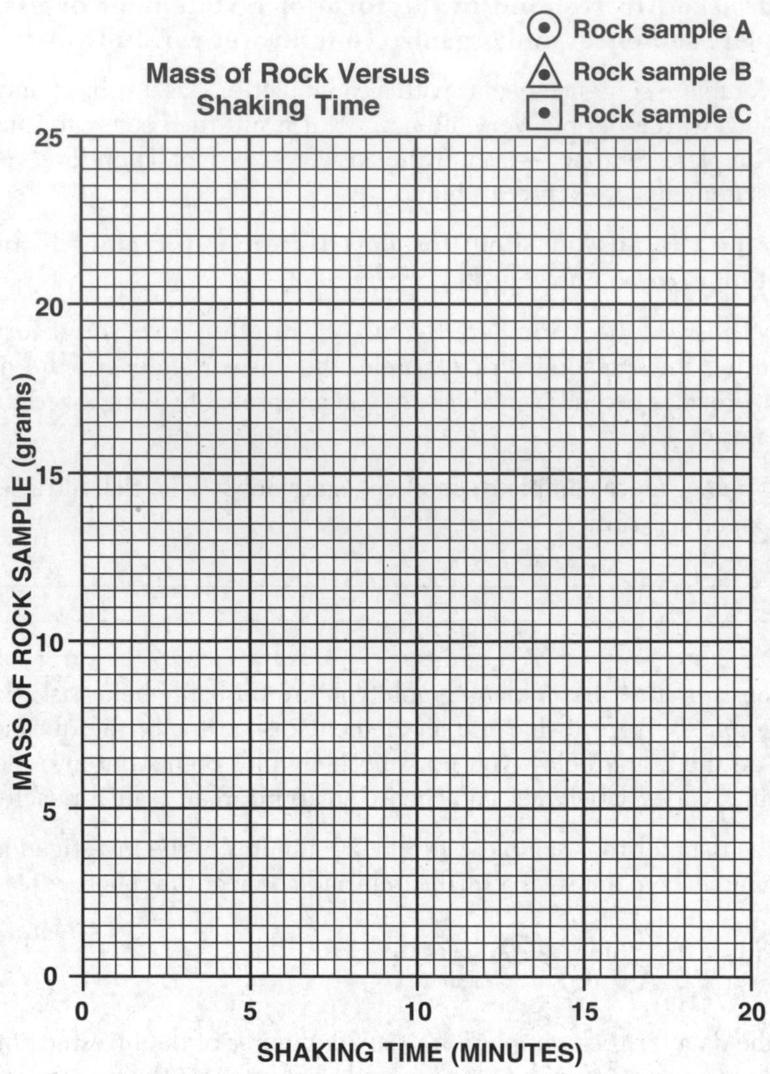

6. Review questions from Parts B-2 and C of the sample exams.

Use the Answers Explained sections of this book to review how questions of these types are answered and how point values are awarded. Focus on the kinds of questions being asked rather than on specific questions.

Topic Outline and Question Index

HOW TO USE THE TOPIC OUTLINE FOR THE PHYSICAL SETTING/EARTH SCIENCE EXAMINATION

The Topic Outline follows the organization of the Physical Setting/Earth Science Core Curriculum, which was written to accompany the New York State Mathematics, Science, and Technology Learning Standards, which are organized in the following way:

STANDARDS—statements of what you are expected to know and be able to do

1. **KEY IDEAS**—broad, unifying, general statements of the most important ideas relating to the Standard
2. **PERFORMANCE INDICATORS**—statements describing how you should be able to demonstrate that you understand the Key Ideas
3. **MAJOR UNDERSTANDINGS**—specific statements of underlying concepts needed to do the things described in the Performance Indicators

The Topic Outline is divided into the five Standards (Standards 1, 2, 4, 6, and 7) that will be addressed on the Physical Setting/Earth Science Regents examination. Each Standard is subdivided into Key Ideas. Each Key Idea is further subdivided into Performance Indicators and their associated Major Understandings, so that you may locate questions for individual concepts. Suppose, for example, you wish to check your mastery of concepts related to tectonic plate boundaries. If you look in the Topic Outline under Standard 4, Key Idea 2, Performance Indicator 2.1, Major Understanding 2.1*l*, you will find *"The lithosphere consists of separate plates that ride on the more fluid asthenosphere and move slowly in relationship to one another, creating convergent, divergent, and transform plate boundaries."* By following a parallel line across the page, you can locate appropriate questions from past examinations. On the June 2012 exam, for example, question 85 tests concepts related to plate boundaries. In a similar manner, you can locate questions on this topic from other past examinations.

When you have identified questions from the Topic Outline, turn to the appropriate examination in the book for the first keyed question. Try to answer it, and then check your response, reviewing explanations where desirable. Continue in this manner for other keyed questions.

You will notice that for some headings in the Topic Outline there is more than one test item on a particular examination. This usually occurs because more than one concept is involved. To find a question on a particular concept, you may find it necessary to check the keyed questions on several examinations.

TOPIC OUTLINE AND QUESTION INDEX—PHYSICAL SETTING/EARTH SCIENCE EXAMINATION

STANDARD 1—Analysis, Inquiry, and Design: Students will use mathematical analysis, scientific inquiry, and engineering design, as appropriate, to pose questions, seek answers, and develop solutions.

	EXAMINATION QUESTIONS				
	August 2017	**June 2018**	**August 2018**	**June 2019**	**August 2019**
MATHEMATICAL ANALYSIS **Key Idea 1:** Abstraction and symbolic representation	60, 69, 70	5, 49, 60, 64, 69, 74, 80	30, 63, 74, 76	11, 53, 60, 83	46, 53, 57, 66, 67, 72, 84
Key Idea 2: Deductive and inductive reasoning	28, 49, 51, 63, 70, 76	13, 16, 25, 27, 36, 43, 44, 65	20, 43, 45, 47, 66, 77, 78, 83	26, 36, 37, 51	14, 42, 43, 66, 68, 79, 80
Key Idea 3: Critical thinking skills	28, 67, 68	48, 60, 66	62	66, 81	
SCIENTIFIC INQUIRY **Key Idea 1:** The central purpose of scientific inquiry is to develop explanations of natural phenomena in a continuing, creative process.	2, 10, 12, 26, 36, 39, 47, 48, 50, 55, 56, 65, 67, 71, 74, 79, 81	4, 22, 30, 38, 42, 46, 54, 56, 57, 63, 71, 76, 77, 84	5, 8, 33, 34, 46, 58–60, 66, 68, 79, 84, 85	9, 32, 43, 49, 57, 58, 61–65, 67, 71, 73, 74, 76, 77, 79	8, 10, 12, 21 39, 61, 62, 71, 72, 73, 76, 77, 81
Key Idea 2: Beyond the use of reasoning and consensus, scientific inquiry involves the testing of proposed explanations involving the use of conventional techniques and procedures and usually requiring considerable ingenuity.			59	71	2
Key Idea 3: The observations made while testing proposed explanations, when analyzed using conventional and invented methods, provide new insights into phenomena.	6, 8–10, 13, 15, 17–19, 24, 25, 29, 32–35, 37, 38, 40, 50, 55, 57, 60, 62–66, 73, 77, 80, 82–85	2, 3, 5, 6, 8, 10, 13, 15, 16, 21, 24, 25, 27–29, 31, 35, 39, 41, 45, 47, 49, 51–54, 59, 61, 62, 64, 65, 67, 68, 70, 71, 78–80, 85	2, 3, 7, 10, 11, 15, 17, 18, 19, 20, 21, 23, 24, 25, 26, 28, 30, 35, 37, 39, 42, 44, 50, 55, 56, 57, 61, 65, 72, 75, 76, 80, 81, 82	4, 5, 7, 8, 10, 11, 13, 14, 19, 21, 25–27, 30, 33–35, 40, 45–47, 50, 52, 53, 55, 56, 59, 63, 66–68, 72, 75, 77, 79, 80, 83, 85	1, 3, 4, 5, 14, 18, 20, 21, 22, 23, 24, 25, 28, 33, 34, 35, 36, 40, 41, 42,43, 44, 47, 48,49, 50, 52, 54, 55, 58, 59, 60, 64, 65, 66, 69, 70, 71, 74, 78
ENGINEERING DESIGN **Key Idea 1:** Engineering design is an iterative process involving modeling and optimization (finding the best solution within given constraints); this process is used to develop technological solutions to problems within given constraints.					

	EXAMINATION QUESTIONS				
	August 2017	**June 2018**	**August 2018**	**June 2019**	**August 2019**
STANDARD 2—Students will access, generate, process, and transfer information, using appropriate technologies.					
INFORMATION SYSTEMS **Key Idea 1:** Information technology is used to retrieve, process, and communicate information as a tool to enhance learning.		68			
Key Idea 2: Knowledge of the impacts and limitations of information systems is essential to its effective and ethical use.					
Key Idea 3: Information technology can have positive and negative impacts on society, depending upon how it is used.	72	7			
STANDARD 6—Interconnectedness: Common Themes: Students will understand the relationships and common themes that connect mathematics, science, and technology and apply the themes to these and other areas of learning.					
INTERCONNECTEDNESS COMMON THEMES **Key Idea 1: Systems Thinking** Through systems thinking, people can recognize the commonalities that exist among all systems and how parts of a system interrelate and combine to perform specific functions.	7, 30, 45, 78	20, 32, 52, 54, 56, 83, 84	8, 41, 67, 68, 85	15, 29, 70	10, 30, 38
Key Idea 2: Models Models are simplified representations of objects, structures, or systems used in analysis, explanation, interpretation, or design.	6, 9, 17, 20, 23, 26, 27, 30–32, 34–36, 39, 41–46, 52–55, 57–59, 61, 64, 66, 67, 69, 73, 75, 76, 81	1, 8–12, 18–20, 22–24, 30, 32–37, 40, 41, 44, 45, 47–50, 55, 58, 59, 61, 66, 67, 70, 71, 73–75, 80–84	1, 4, 13, 14, 21, 24, 27–29, 31-38, 40, 42, 43, 45, 47–49, 51–54, 59-63, 65, 69–74, 80-82	1–4, 7, 12, 14–16, 19, 21–23, 27–29, 32, 38–46, 50, 52, 55, 56, 62–64, 66–69, 73–78, 81–85	7, 9, 15, 17, 20, 25, 26, 27, 28, 29, 30, 36, 37, 39, 44, 45, 46, 47, 48, 49, 50, 51, 52, 54, 55, 56, 60, 61, 62, 63, 64, 67, 69, 70, 75, 80, 82, 83, 84, 85
Key Idea 3: Magnitude and Scale The grouping of magnitudes of size, time, frequency, and pressures or other units of measurement into a series of relative order provides a useful way to deal with the immense range and the changes in scale that affect the behavior and design of systems.	1, 76	48, 50, 75	64	59, 84	
Key Idea 4: Equilibrium and Stability Equilibrium is a state of stability due either to a lack of change (static equilibrium) or a balance between opposing forces (dynamic equilibrium).				80	
Key Idea 5: Patterns of Change Identifying patterns of change is necessary for making predictions about future behavior and conditions.	7, 16, 30, 40, 43, 45, 52, 53, 62, 79	5, 9, 12, 19, 23, 35, 37, 41–43, 55, 57, 68, 84	6, 41, 45, 47, 53, 59, 78	19, 36, 37, 51, 54, 62, 71, 75, 77, 79, 80	2, 5, 23, 36, 51, 61, 71, 83
Key Idea 6: Optimization In order to arrive at the best solution that meets criteria within constraints, it is often necessary to make trade-offs.		14, 33			
STANDARD 7—Interdisciplinary Problem Solving: Students will apply the knowledge and thinking skills of mathematics, science, and technology to address real-life problems and make informed decisions.					
INTERDISCIPLINARY PROBLEM SOLVING **Key Idea 1: Connections** The knowledge and skills of mathematics, science, and technology are used together to make informed decisions and solve problems, especially those relating to issues of science/technology/society, consumer decision making, design, and inquiry into phenomena.			12		

	EXAMINATION QUESTIONS				
	August 2017	**June 2018**	**August 2018**	**June 2019**	**August 2019**
Key Idea 2: Strategies Solving interdisciplinary problems involves a variety of skills and strategies, including effective work habits; gathering and processing information; generating and analyzing ideas; realizing ideas; making connections among the common themes of mathematics, science, and technology; and presenting results.	72	7		48	
STANDARD 4—Students will understand and apply scientific concepts, principles, and theories pertaining to the physical setting and living environment and recognize the historical development of ideas in science.					
Physical Setting Key Idea 1: **EARTH AND CELESTIAL PHENOMENA CAN BE DESCRIBED BY PRINCIPLES OF RELATIVE MOTION AND PERSPECTIVE.**					
Intro 1: People have observed the stars for thousands of years, using them to find direction, note the passage of time, and to express their values and traditions. As our technology has progressed, so has understanding of celestial objects and events.					
Intro 2: Theories of the universe have developed over many centuries. Although to a casual observer celestial bodies appeared to orbit a stationary Earth, scientific discoveries led us to the understanding that Earth is one planet that orbits the Sun, a typical star in a vast and ancient universe. We now infer an origin and an age and evolution of the universe, as we speculate about its future.					
Intro 3: As we look at Earth, we find clues to its origin and how it has changed through nearly five billion years, as well as the evolution of life on Earth.					
Physical Setting Performance Indicator 1.1: Explain complex phenomena, such as tides, variations in day length, solar insolation, apparent motion of the planets, and annual traverse of the constellations.					
MAJOR UNDERSTANDINGS 1.1a Most objects in the solar system are in regular and predictable motion. • These motions explain such phenomena as the day, the year, seasons, phases of the moon, eclipses, and tides. • Gravity influences the motions of celestial objects. The force of gravity between two objects in the universe depends on their masses and the distance between them. *See Standard 6, Key Idea 4*	6, 40, 51–53, 66	1, 9, 81	39, 40, 41	3, 38–40, 60, 61, 64	2, 82, 83, 84, 85
1.1b Nine planets move around the Sun in nearly circular orbits. • The orbit of each planet is an ellipse with the Sun located at one of the foci. • Earth is orbited by one Moon and many artificial satellites. *See Standard 1: Mathematical Analysis, Key Idea 2, and Standard 6, Key Ideas 3 and 4*			67, 68	59	3
1.1c Earth's coordinate system of latitude and longitude, with the Equator and Prime Meridian as reference lines, is based upon Earth's rotation and our observation of the Sun and stars.	69	39			37, 38, 39

	EXAMINATION QUESTIONS				
	August 2017	June 2018	August 2018	June 2019	August 2019
1.1d Earth rotates on an imaginary axis at a rate of 15 degrees per hour. To people on Earth, this turning of the planet makes it seem as though the Sun, the Moon, and the stars are moving around Earth once a day. Rotation provides a basis for our system of local time; meridians of longitude are the basis for time zones. *See Standard 1: Scientific Inquiry, Key Idea 3*	39, 41, 68	56	1, 29, 30	1, 62	7, 36
1.1e The Foucault pendulum and the Coriolis effect provide evidence of Earth's rotation.	3	4	6	9, 65	45, 46
1.1f Earth's changing position with regard to the Sun and the Moon has noticeable effects. • Earth revolves around the Sun with its rotational axis tilted at 23.5 degrees to a line perpendicular to the plane of its orbit, with the North Pole aligned with Polaris. • During Earth's one-year period of revolution, the tilt of its axis results in changes in the angle of incidence of the Sun's rays at a given latitude; these changes cause variation in the heating of the surface. This produces seasonal variation in weather.	26, 67	5, 11, 57	5, 51, 52, 53, 54	41, 63	9
1.1g Seasonal changes in the apparent positions of constellations provide evidence of Earth's revolution.	5	12	4	6	5
1.1h The Sun's apparent path through the sky varies with latitude and season.		55			79, 81
1.1i Approximately 70 percent of Earth's surface is covered by a relatively thin layer of water, which responds to the gravitational attraction of the Moon and the Sun with a daily cycle of high and low tides.	4	36–38		36, 37	
Physical Setting Performance Indicator 1.2: Describe current theories about the origin of the universe and solar system.					
1.2a The universe is vast and estimated to be over ten billion years old. The current theory is that the universe was created from an explosion called the big bang. Evidence for this theory includes: • cosmic background radiation; • a redshift (the Doppler effect) in the light from very distant galaxies.	2	3, 10		2, 5	6, 75, 76, 77
1.2b Stars form when gravity causes clouds of molecules to contract until nuclear fusion of light elements into heavier ones occurs. Fusion releases great amounts of energy over millions of years. • The stars differ from each other in size, temperature, and age. • Our Sun is a medium-sized star within a spiral galaxy of stars known as the Milky Way. Our galaxy contains billions of stars, and the universe contains billions of such galaxies.	1, 54–57	40–42, 79	45, 46, 47		78
1.2c Our solar system formed about five billion years ago from a giant cloud of gas and debris. Gravity caused Earth and the other planets to become layered according to density differences in their materials. • The characteristics of the planets of the solar system are affected by each planet's location in relationship to the Sun. • The terrestrial planets are small, rocky, and dense. The Jovian planets are large and gaseous, and have low density.	36, 37	2, 80, 82	2, 3, 66	66–68	4

	EXAMINATION QUESTIONS				
	August 2017	**June 2018**	**August 2018**	**June 2019**	**August 2019**
1.2d Asteroids, comets, and meteors are components of our solar system. • Impact events have been correlated with mass extinction and global climatic change. • Impact craters can be identified in Earth's crust.				58	1, 8
1.2e Earth's early atmosphere formed as a result of the out-gassing of water vapor, carbon dioxide, nitrogen, and lesser amounts of other gases from its interior.					
1.2f Earth's oceans formed as a result of precipitation over millions of years. The presence of an early ocean is indicated by sedimentary rocks of marine origin, dating back about four billion years.					
1.2g Earth has continuously been recycling water since the outgassing of water early in its history. This constant recirculation of water at and near Earth's surface is described by the hydrologic (water) cycle. • Water is returned from the atmosphere to Earth's surface by precipitation. Water returns to the atmosphere by evaporation or transpiration from plants. A portion of the precipitation becomes runoff over the land or infiltrates into the ground to become stored in the soil or groundwater below the water table. Soil capillarity influences these processes. • The amount of precipitation that seeps into the ground or runs off is influenced by climate, slope of the land, soil, rock type, vegetation, land use, and degree of saturation. • Porosity, permeability, and water retention affect runoff and infiltration.	7, 42, 43	20, 63, 64	8, 9	11, 12	10, 11, 12
1.2h The evolution of life caused dramatic changes in the composition of Earth's atmosphere. Free oxygen did not form in the atmosphere until oxygen-producing organisms evolved.					21
1.2i The pattern of evolution of life-forms on Earth is at least partially preserved in the rock record. • Fossil evidence indicates that a wide variety of life-forms has existed in the past and that most of these forms have become extinct. • Human existence has been very brief compared to the expanse of geologic time.	29	16, 69, 72	17	4, 21	22
1.2j Geologic history can be reconstructed by observing sequences of rock types and fossils to correlate bed-rock at various locations. • The characteristics of rocks indicate the processes by which they formed and the environments in which 69 these processes took place. • Fossils preserved in rocks provide information about past environmental conditions. • Geologists have divided Earth history into time units based upon the fossil record. • Age relationships among bodies of rocks can be determined using principles of original horizontality, superposition, inclusions, cross-cutting relationships, contact metamorphism, and unconformities. The presence of volcanic ash layers, index fossils, and meteoritic debris can provide additional information. • The regular rate of nuclear decay (half-life time period) of radioactive isotopes allows geologists to determine the absolute age of materials found in some rocks.	31, 47–50, 74, 75, 77	17, 19, 23, 43, 44, 70, 71, 85	18, 19, 21, 55, 56, 59, 74, 75, 76	13, 22, 32, 47, 51–53, 78	23, 26, 27, 57, 58, 59, 60, 61, 62, 74

	EXAMINATION QUESTIONS				
	August 2017	June 2018	August 2018	June 2019	August 2019
Physical Setting Key Idea 2: **MANY OF THE PHENOMENA THAT WE OBSERVE ON EARTH INVOLVE INTERACTIONS AMONG COMPONENTS OF AIR, WATER, AND LAND.**					
Intro 1: Earth may be considered a huge machine driven by two engines, one internal and one external. These heat engines convert heat energy into mechanical energy.					
Intro 2: Earth's external heat engine is powered primarily by solar energy and influenced by gravity. Nearly all the energy for circulating the atmosphere and oceans is supplied by the Sun. As insolation strikes the atmosphere, a small percentage is directly absorbed, especially by gases such as ozone, carbon dioxide, and water vapor. Clouds and Earth's surface reflect some energy back to space, and Earth's surface absorbs some energy. Energy is transferred between Earth's surface and the atmosphere by radiation, conduction, evaporation, and convection. Temperature variations within the atmosphere cause differences in density that cause atmospheric circulation, which is affected by Earth's rotation. The interaction of these processes results in the complex atmospheric occurrence known as weather.					
Intro 3: Average temperatures on Earth are the result of the total amount of insolation absorbed by Earth's surface and its atmosphere and the amount of long-wave energy radiated back into space. However, throughout geologic time, ice ages occurred in the middle latitudes. In addition, average temperatures may have been significantly warmer at times in the geologic past. This suggests that Earth had climate changes that were most likely associated with long periods of imbalances of its heat budget.					
Intro 4: Earth's internal heat engine is powered by heat from the decay of radioactive materials and residual heat from Earth's formation. Differences in density resulting from heat flow within Earth's interior caused the changes explained by the theory of plate tectonics: movement of the lithospheric plates, earthquakes, volcanoes, and the deformation and metamorphism of rocks during the formation of young mountains.					
Intro 5: Precipitation resulting from the external heat engine's weather systems supplies moisture to Earth's surface that contributes to the weathering of rocks. Running water erodes mountains that were originally uplifted by Earth's internal heat engine and transports sediments to other locations, where they are deposited and may undergo the processes that transform them into sedimentary rocks.					
Intro 6: Global climate is determined by the interaction of solar energy with Earth's surface and atmosphere. This energy transfer is influenced by dynamic processes such as cloud cover and Earth rotation, and the positions of mountain ranges and oceans.					
Physical Setting Performance Indicator 2.1: **Use the concepts of density and heat energy to explain observations of weather patterns, seasonal changes, and the movements of Earth's plates.**					
MAJOR UNDERSTANDINGS 2.1a Earth systems have internal and external sources of energy, both of which create heat.					
2.1b The transfer of heat energy within the atmosphere, the the hydrosphere, and Earth's interior results in the formation of regions of different densities. These densitydifferences result in motion.		45, 46	8		

	EXAMINATION QUESTIONS				
	August 2017	June 2018	August 2018	June 2019	August 2019
2.1c Weather patterns become evident when weather variables are observed, measured, and recorded. These variables include air temperature, air pressure, moisture (relative humidity and dewpoint), precipitation (rain, snow, hail, sleet, etc.), wind speed and direction, and cloud cover.		6	10	10	43
2.1d Weather variables are measured using instruments such as thermometers, barometers, psychrometers, precipitation gauges, anemometers, and wind vanes.	28, 34, 71		73	57	42, 44
2.1e Weather variables are interrelated. For example: • Temperature and humidity affect air pressure and probability of precipitation. • Air pressure gradient controls wind velocity.	27, 70	66	36	43	13
2.1f Air temperature, dewpoint, cloud formation, and precipitation are affected by the expansion andcontraction of air due to vertical atmospheric movement.					71
2.1g Weather variables can be represented in a variety of formats including radar and satellite images, weather maps (including station models, isobars, and fronts), atmospheric cross sections, and computer models. *See Standard 2, Information Systems, Key Idea 1*	8, 9	47	37, 38, 72	14, 55, 56	69
2.1h Atmospheric moisture, temperature, and pressure distributions; jet streams; wind; air masses and frontal boundaries; and the movement of cyclonic systems and associated tornadoes, thunderstorms, and hurricanes occur in observable patterns. Loss of property, personal injury, and loss of life can be reduced by effective emergency preparedness. *Also see Standard 6, Key Idea 5, and Standard 7, Key Idea 2*	10, 11, 72	7, 8, 67, 68	31, 32, 69, 70, 71	42, 44, 54	14, 15, 70
2.1i Seasonal changes can be explained using concepts of density and heat energy. These changes include the shifting of global temperature zones, the shifting of planetary wind and ocean current patterns, the occurrence of monsoons, hurricanes, flooding, and severe weather.			34		16
2.1j Properties of Earth's internal structure (crust, mantle, inner core, and outer core) can be inferred from the analysis of the behavior of seismic waves, including velocity and refraction. • Analysis of seismic waves allows the determination of the location of earthquake epicenters, and the measurement of earthquake magnitude; this analysis leads to the inference that Earth's interior is composed of layers that differ in composition and states of matter.	38, 62, 63, 65	13, 25	20, 42, 43, 44	26	24, 25, 50
2.1k The outward transfer of Earth's internal heat drives convective circulation in the mantle that moves the lithospheric plates comprising Earth's surface.					

	EXAMINATION QUESTIONS				
	August 2017	**June 2018**	**August 2018**	**June 2019**	**August 2019**
2.1l The lithosphere consists of separate plates that ride on the more fluid asthenosphere and move slowly in relationship to one another, creating convergent, divergent, and transform plate boundaries. These motions indicate Earth is a dynamic geologic system. • These plate boundaries are the sites of most earthquakes, volcanoes, and young mountain ranges. • Compared to continental crust, ocean crust is thinner and denser. New ocean crust continues to form at mid-ocean ridges. • Earthquakes and volcanoes present geologic hazards to humans. Loss of property, personal injury, and loss of life can be reduced by effective emergency preparedness. *See Standard 7, Key Idea 2*	18, 35	24, 59, 60	80, 81	25, 27, 45, 46, 48	48, 49
2.1m Many processes of the rock cycle are consequences of plate dynamics. These include the production of magma (and subsequent igneous rock formation and contact metamorphism) at both subduction and rifting regions, regional metamorphism within subduction zones, and the creation of major depositional basins through downwarping of the crust.				15	
2.1n Many of Earth's surface features, such as mid-ocean ridges/rifts, trenches/subduction zones/island arcs, mountain ranges (folded, faulted, and volcanic), hot spots, and the magnetic and age patterns in surface bedrock, are a consequence of forces associated with plate motion and interaction.	17, 19	61, 62	82	7	51, 52, 53
2.1o Plate motions have resulted in global changes in geography, climate, and the patterns of organic evolution. *See Standard 6, Key Idea 5*	32			19	
2.1p Landforms are the result of the interaction of tectonic forces and the processes of weathering, erosion, and deposition.					
2.1q Topographic maps represent landforms through the use of contour lines that are isolines connecting points of equal elevation. Gradients and profiles can be determined from changes in elevation over a given distance. *See Standard 6, Key Ideas 2 and 3, and Standard 7, Key Idea 2*	58–61	48–50, 73–75	24, 62, 63, 64	81–84	66, 67, 68
2.1r Climate variations, structure, and characteristics of bedrock influence the development of landscape features including mountains, plateaus, plains, valleys, ridges, escarpments, and stream drainage patterns.	20, 21	26, 32	33, 65	23, 24	
2.1s Weathering is the physical and chemical breakdown of rocks at or near Earth's surface. Soils are the result of weathering and biological activity over long periods of time.		30	58	31	31
2.1t Natural agents of erosion, generally driven by gravity, remove, transport, and deposit weathered rock particles. Each agent of erosion produces distinctive changes in the material that it transports and creates characteristic surface features and landscapes. In certain erosional situations, loss of property, personal injury, and loss of life can be reduced by effective emergency preparedness. *See Standard 7, Key Idea 2*			60		

	EXAMINATION QUESTIONS				
	August 2017	June 2018	August 2018	June 2019	August 2019
2.1u The natural agents of erosion include:					
• **Streams (running water):** Gradient, discharge, and channel shape influence a stream's velocity and the erosion and deposition of sediments. Sediments transported by streams tend to become rounded as a result of abrasion. Stream features include V-shaped valleys, deltas, flood plains, and meanders. A watershed is the area drained by a stream and its tributaries. *See Standard 1: Engineering Design, Key Idea 1; Standard 1: Mathematical Analysis, Key Idea 2; and Standard 6, Key Idea 1*	44, 45, 79	27	22, 48, 49, 50	69–72	29, 30, 55, 56, 63, 64, 65
• **Glaciers (moving ice):** Glacial erosional processes include the formation of U-shaped valleys, parallel scratches, and grooves in bedrock. Glacial features include moraines, drumlins, kettle lakes, finger lakes, and outwash plains.	76	83, 84	83, 84	28, 29	32
• **Wave Action:** Erosion and deposition cause changes in shoreline features, including beaches, sandbars, and barrier islands. Wave action rounds sediments as a result of abrasion. Waves approaching a shoreline move sand parallel to the shore within the zone of breaking waves.		22, 34	27		
Other • **Wind:** Erosion of sediments by wind is most common in arid climates and along shorelines. Wind-generated features include dunes and sand-blasted bedrock. • **Mass Movement:** Earth materials move downslope under the influence of gravity.	22, 23			49	
2.1v Patterns of deposition result from a loss of energy within the transporting system and are influenced by the size, shape, and density of the transported particles. Sediment deposits may be sorted or unsorted. *See Standard 1: Scientific Inquiry, Key Idea 2*	46, 78	65	85		
2.1w Sediments of inorganic and organic origin often accumulate in depositional environments. Sedimentary rocks form when sediments are compacted and/or cemented after burial or as the result of chemical precipitation from seawater.		28			
Physical Setting Performance Indicator 2.2: Explain how incoming solar radiation, ocean currents, and land masses affect weather and climate.					
2.2a Insolation (solar radiation) heats Earth's surface and atmosphere unequally due to variations in: • the intensity caused by differences in atmospheric transparency and angle of incidence, which vary with time of day, latitude, and season; • characteristics of the materials absorbing the energy, such as color, texture, transparency, state of matter, and specific heat; • duration, which varies with seasons and latitude.	13–16	58, 76–78	7, 14, 16, 77, 78, 79	17, 74	15, 19, 80
2.2b The transfer of heat energy within the atmosphere, the hydrosphere, and Earth's surface occurs as the result of radiation, convection, and conduction. • Heating of Earth's surface and atmosphere by the Sun drives convection within the atmosphere and oceans, producing winds and ocean currents.			11		18, 20, 47

	EXAMINATION QUESTIONS				
	August 2017	**June 2018**	**August 2018**	**June 2019**	**August 2019**
2.2c A location's climate is influenced by latitude, proximity to large bodies of water, ocean currents, prevailing winds, vegetative cover, elevation, and mountain ranges.	12, 80, 81	15, 18, 21, 33	13, 15	16, 73, 75, 76	17, 72, 73
2.2d Temperature and precipitation patterns are altered by: • natural events, such as El Niño and volcanic eruptions; • human influences, including deforestation, urbanization, and the production of greenhouse gases such as carbon dioxide and methane.	30	14	12	18, 20	
Physical Setting Key Idea 3: **MATTER IS MADE UP OF PARTICLES WHOSE PROPERTIES DETERMINE THE OBSERVABLE CHARACTERISTICS OF MATTER AND ITS REACTIVITY.**					
Intro 1: Observation and classification have helped us understand the great variety and complexity of Earth materials. Minerals are the naturally occurring inorganic solid elements, compounds, and mixtures from which rocks are made. We classify minerals on the basis of their chemical composition and observable properties. Rocks are generally classified by their origin (igneous, metamorphic, and sedimentary), texture, and mineral content.					
Intro 2: Rocks and minerals help us understand Earth's historical development and its dynamics. They are important to us because of their availability and properties. The use and distribution of mineral resources and fossil fuels have important economic and environmental impacts. As limited resources, they must be used wisely.					
Physical Setting Performance Indicator 3.1: Explain the properties of materials in terms of the arrangement and properties of the atoms that compose them.					
3.1a Minerals have physical properties determined by their chemical composition and crystal structure. • Minerals can be identified by well-defined physical and chemical properties, such as cleavage, fracture, color, density, hardness, streak, luster, crystal shape, and reaction with acid. • Chemical composition and physical properties determine how minerals are used by humans. *See Standard 6, Key Idea 2*	24	51, 52	57	34, 35	34, 35
3.1b Minerals are formed inorganically by the process of crystallization as a result of specific environmental conditions. These include: • cooling and solidification of magma; • precipitation from water caused by such processes as evaporation, chemical reactions, and temperature changes; • rearrangement of atoms in existing minerals subjected to conditions of high temperature and pressure.		35		79, 80	
3.1c Rocks are usually composed of one or more minerals. • Rocks are classified by their origin, mineral content, and texture. • Conditions that existed when a rock formed can be inferred from the rock's mineral content and texture. • The properties of rocks determine how they are used and also influence land usage by humans.	25, 33, 64, 73, 82–85	29, 31, 53, 54	23, 25, 26, 28, 35, 61	30, 33, 50, 77, 85	33, 40, 41, 54

	EXAMINATION QUESTIONS				
	August 2017	June 2018	August 2018	June 2019	August 2019
Reference Tables for Physical Settings/Earth Science 2011 Edition (Revised)	8–11, 13, 15, 18, 19, 24, 25, 29, 32–35, 37, 38, 48, 50, 55, 57, 60, 62–65, 73, 77, 80, 82–85	2, 3, 5, 6, 8, 10, 13, 15, 16, 21, 24, 25, 27–29, 31, 35, 39, 41, 45, 47, 49, 51–54, 59, 61, 62, 64, 65, 67, 68, 70, 71, 78, 80, 85	2, 3, 7, 10, 11, 15, 17–21, 23–26, 28, 30, 34, 35, 37, 42, 44, 50, 55–57, 61, 65, 72, 75, 76, 80–82	4, 5, 7, 8, 10, 11, 13, 14, 19, 21, 25–27, 30, 33–35, 40, 45–47, 50, 52, 53, 55, 56, 59, 63, 66, 67, 68, 72, 75–77, 79, 80, 83, 85	1, 3, 4, 14, 18, 20, 21, 22, 23, 24, 25, 28, 33, 34, 35, 36, 40, 41, 42, 43, 44, 47, 48, 49, 50, 52, 54, 55, 58, 59, 60, 64, 65 66, 69, 70, 71, 74, 78

Earth Science Reference Tables and Charts

The University of the State of New York • THE STATE EDUCATION DEPARTMENT • Albany, New York 12234 • www.nysed.gov

Reference Tables for Physical Setting/EARTH SCIENCE

Radioactive Decay Data

RADIOACTIVE ISOTOPE	DISINTEGRATION	HALF-LIFE (years)
Carbon-14	$^{14}C \rightarrow {}^{14}N$	5.7×10^3
Potassium-40	$^{40}K \rightarrow {}^{40}Ar$, $^{40}K \rightarrow {}^{40}Ca$	1.3×10^9
Uranium-238	$^{238}U \rightarrow {}^{206}Pb$	4.5×10^9
Rubidium-87	$^{87}Rb \rightarrow {}^{87}Sr$	4.9×10^{10}

Specific Heats of Common Materials

MATERIAL	SPECIFIC HEAT (Joules/gram • °C)
Liquid water	4.18
Solid water (ice)	2.11
Water vapor	2.00
Dry air	1.01
Basalt	0.84
Granite	0.79
Iron	0.45
Copper	0.38
Lead	0.13

Equations

$$\text{Eccentricity} = \frac{\text{distance between foci}}{\text{length of major axis}}$$

$$\text{Gradient} = \frac{\text{change in field value}}{\text{distance}}$$

$$\text{Rate of change} = \frac{\text{change in value}}{\text{time}}$$

$$\text{Density} = \frac{\text{mass}}{\text{volume}}$$

Properties of Water

Heat energy gained during melting .... 334 J/g
Heat energy released during freezing .... 334 J/g
Heat energy gained during vaporization ..... 2260 J/g
Heat energy released during condensation ... 2260 J/g

Density at 3.98°C .... 1.0 g/mL

Average Chemical Composition of Earth's Crust, Hydrosphere, and Troposphere

ELEMENT (symbol)	CRUST		HYDROSPHERE	TROPOSPHERE
	Percent by mass	Percent by volume	Percent by volume	Percent by volume
Oxygen (O)	46.10	94.04	33.0	21.0
Silicon (Si)	28.20	0.88		
Aluminum (Al)	8.23	0.48		
Iron (Fe)	5.63	0.49		
Calcium (Ca)	4.15	1.18		
Sodium (Na)	2.36	1.11		
Magnesium (Mg)	2.33	0.33		
Potassium (K)	2.09	1.42		
Nitrogen (N)				78.0
Hydrogen (H)			66.0	
Other	0.91	0.07	1.0	1.0

2011 EDITION

This edition of the Earth Science Reference Tables should be used in the classroom beginning in the 2011–12 school year. The first examination for which these tables will be used is the January 2012 Regents Examination in Physical Setting / Earth Science.

Eurypterus remipes

New York State Fossil

Generalized Bedrock Geology of New York State

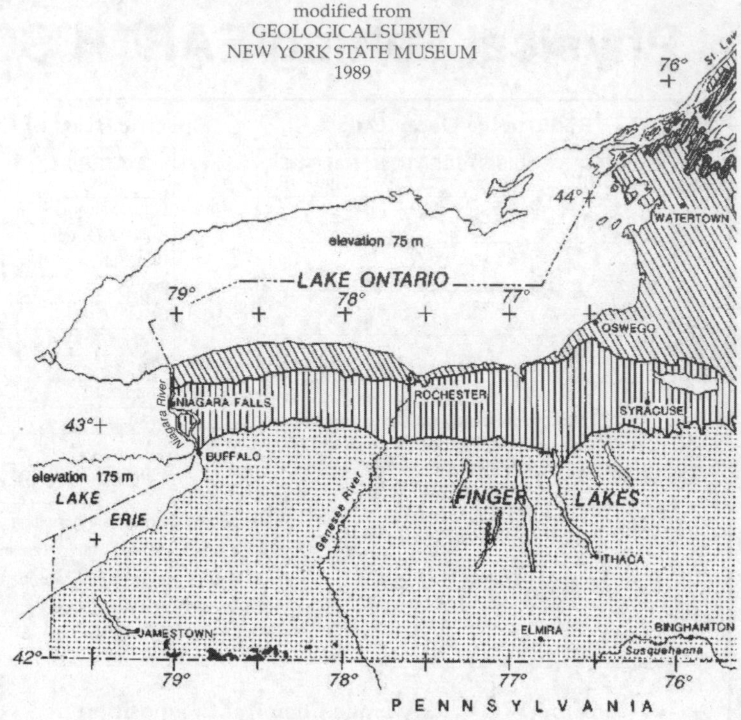
modified from
GEOLOGICAL SURVEY
NEW YORK STATE MUSEUM
1989
elevation 75 m
LAKE ONTARIO
WATERTOWN
OSWEGO
ROCHESTER
NIAGARA FALLS
Niagara River
SYRACUSE
BUFFALO
elevation 175 m
LAKE
ERIE
Genesee River
FINGER LAKES
ITHACA
JAMESTOWN
ELMIRA
BINGHAMTON
Susquehanna
St. Law
79°
78°
77°
76°
44°
43°
42°
PENNSYLVANIA

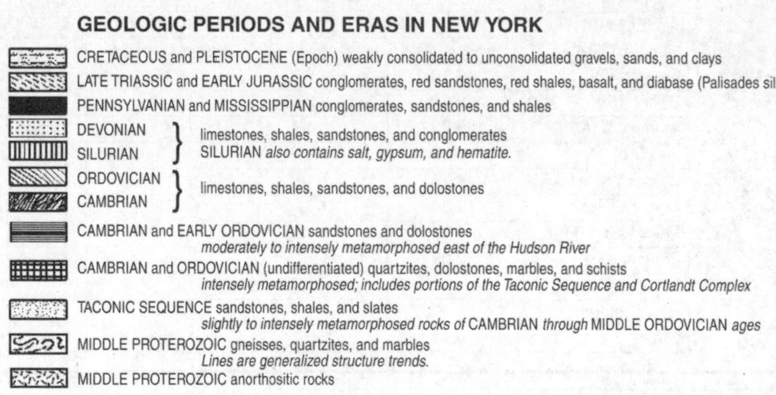
GEOLOGIC PERIODS AND ERAS IN NEW YORK
CRETACEOUS and PLEISTOCENE (Epoch) weakly consolidated to unconsolidated gravels, sands, and clays
LATE TRIASSIC and EARLY JURASSIC conglomerates, red sandstones, red shales, basalt, and diabase (Palisades sill)
PENNSYLVANIAN and MISSISSIPPIAN conglomerates, sandstones, and shales
DEVONIAN
SILURIAN
limestones, shales, sandstones, and conglomerates
SILURIAN also contains salt, gypsum, and hematite.
ORDOVICIAN
CAMBRIAN
limestones, shales, sandstones, and dolostones
CAMBRIAN and EARLY ORDOVICIAN sandstones and dolostones
moderately to intensely metamorphosed east of the Hudson River
CAMBRIAN and ORDOVICIAN (undifferentiated) quartzites, dolostones, marbles, and schists
intensely metamorphosed; includes portions of the Taconic Sequence and Cortlandt Complex
TACONIC SEQUENCE sandstones, shales, and slates
slightly to intensely metamorphosed rocks of CAMBRIAN through MIDDLE ORDOVICIAN ages
MIDDLE PROTEROZOIC gneisses, quartzites, and marbles
Lines are generalized structure trends.
MIDDLE PROTEROZOIC anorthositic rocks

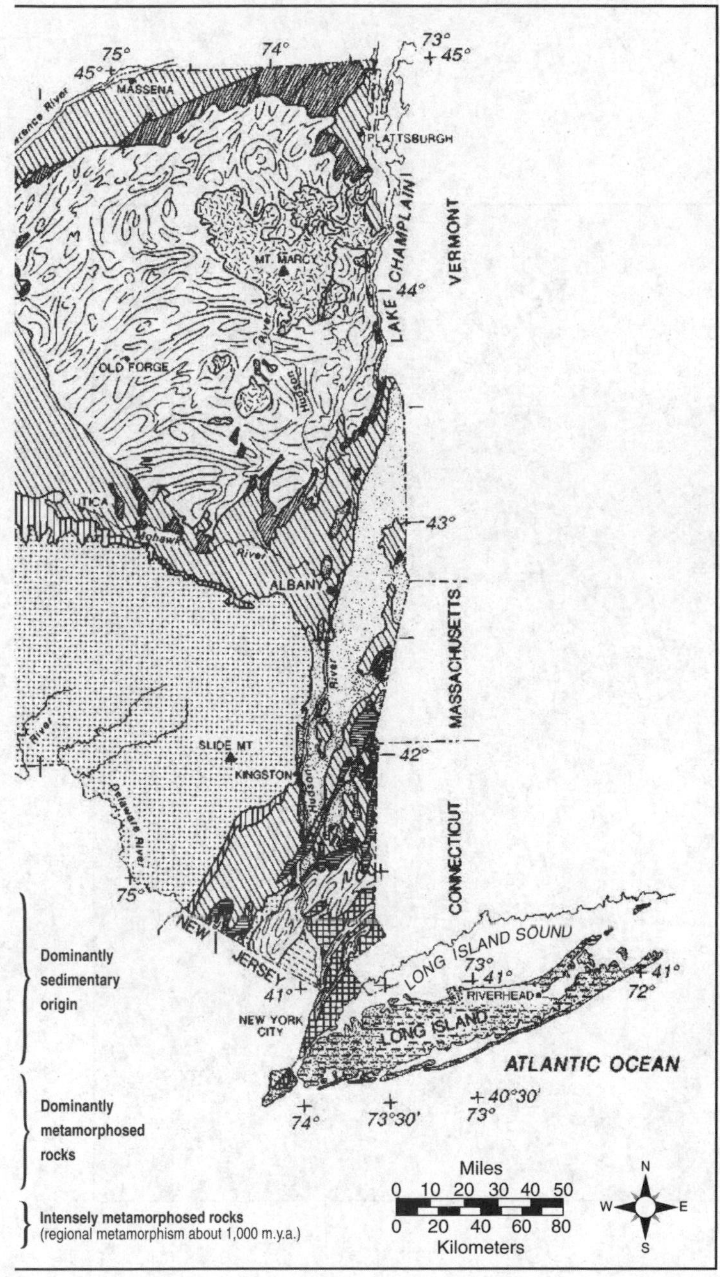
75°
74°
73°
45°
45°
MASSENA
PLATTSBURGH
LAKE CHAMPLAIN
VERMONT
MT. MARCY
44°
OLD FORGE
Hudson River
UTICA
Mohawk River
43°
ALBANY
MASSACHUSETTS
Hudson River
42°
SLIDE MT
KINGSTON
Delaware River
CONNECTICUT
75°
NEW JERSEY
41°
LONG ISLAND SOUND
73°
41°
RIVERHEAD
41°
72°
NEW YORK CITY
LONG ISLAND
ATLANTIC OCEAN
74°
73°30'
40°30'
73°
Dominantly sedimentary origin
Dominantly metamorphosed rocks
Intensely metamorphosed rocks
(regional metamorphism about 1,000 m.y.a.)
Miles
0 10 20 30 40 50
0 20 40 60 80
Kilometers
N
W
E
S

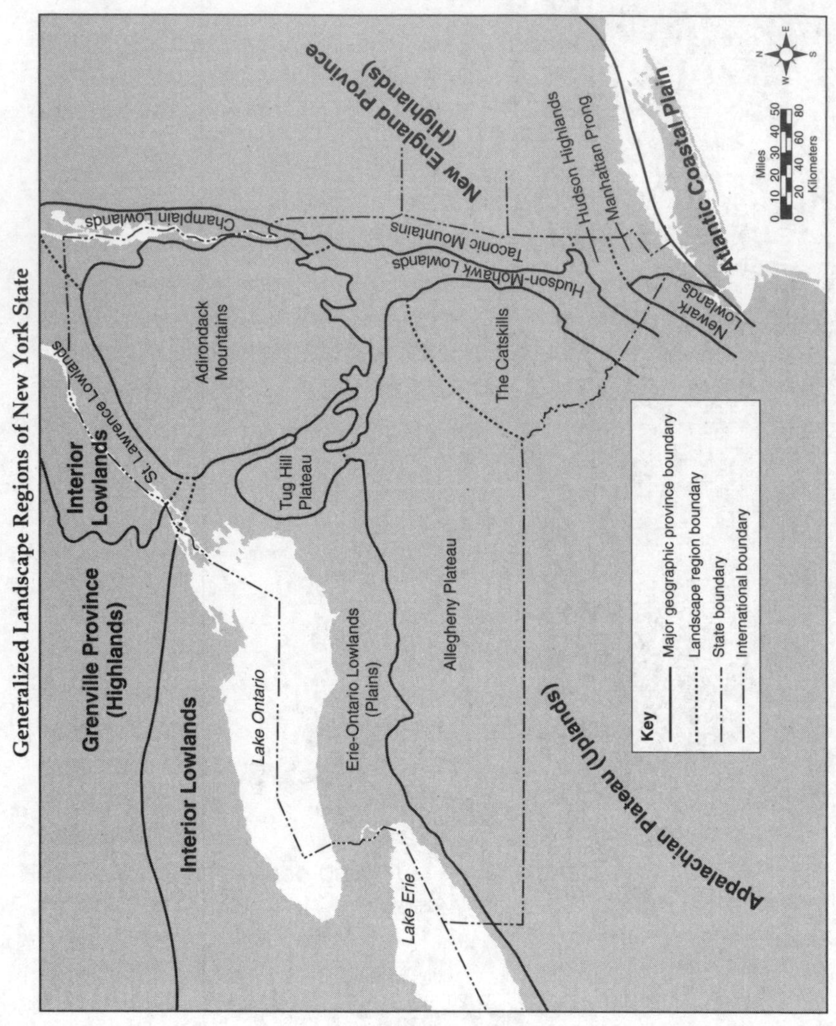
Generalized Landscape Regions of New York State
Grenville Province
(Highlands)
Interior Lowlands
Interior
Lowlands
St. Lawrence Lowlands
Champlain Lowlands
Adirondack
Mountains
Tug Hill
Plateau
Lake Ontario
Erie-Ontario Lowlands
(Plains)
Lake Erie
Allegheny Plateau
The Catskills
Hudson-Mohawk Lowlands
Taconic Mountains
New England Province
(Highlands)
Hudson Highlands
Manhattan Prong
Newark
Lowlands
Atlantic Coastal Plain
Appalachian Plateau (Uplands)
Key
Major geographic province boundary
Landscape region boundary
State boundary
International boundary
Miles
0 10 20 30 40 50
0 20 40 60 80
Kilometers
N
E
S
W

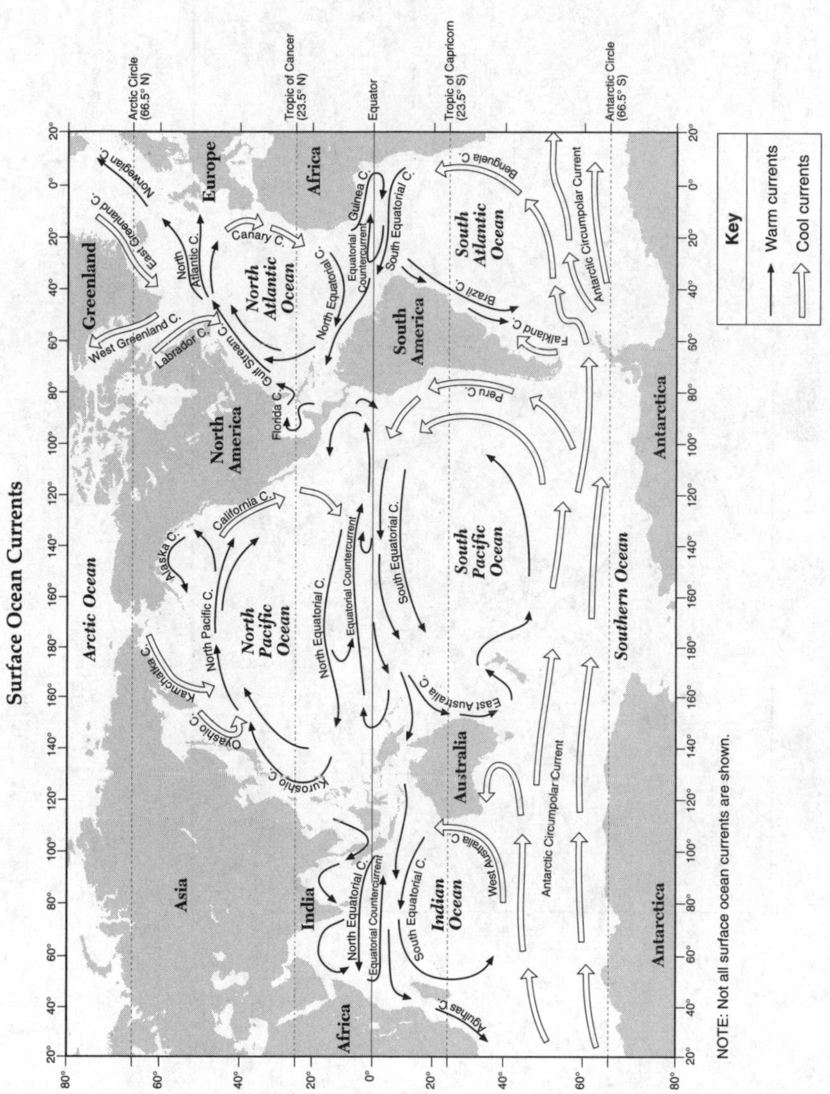
Surface Ocean Currents
Arctic Circle (66.5° N)
Tropic of Cancer (23.5° N)
Equator
Tropic of Capricorn (23.5° S)
Antarctic Circle (66.5° S)
Arctic Ocean
Greenland
Europe
Asia
India
Africa
North America
South America
Australia
Antarctica
North Pacific Ocean
South Pacific Ocean
North Atlantic Ocean
South Atlantic Ocean
Indian Ocean
Southern Ocean
Norwegian C.
East Greenland C.
West Greenland C.
Labrador C.
North Atlantic C.
Canary C.
Gulf Stream C.
Florida C.
North Equatorial C.
Equatorial Countercurrent
Guinea C.
South Equatorial C.
Brazil C.
Benguela C.
Falkland C.
Antarctic Circumpolar Current
Peru C.
California C.
Alaska C.
North Pacific C.
Kamchatka C.
Oyashio C.
Kuroshio C.
East Australia C.
West Australia C.
Agulhas C.
Key
Warm currents
Cool currents
NOTE: Not all surface ocean currents are shown.

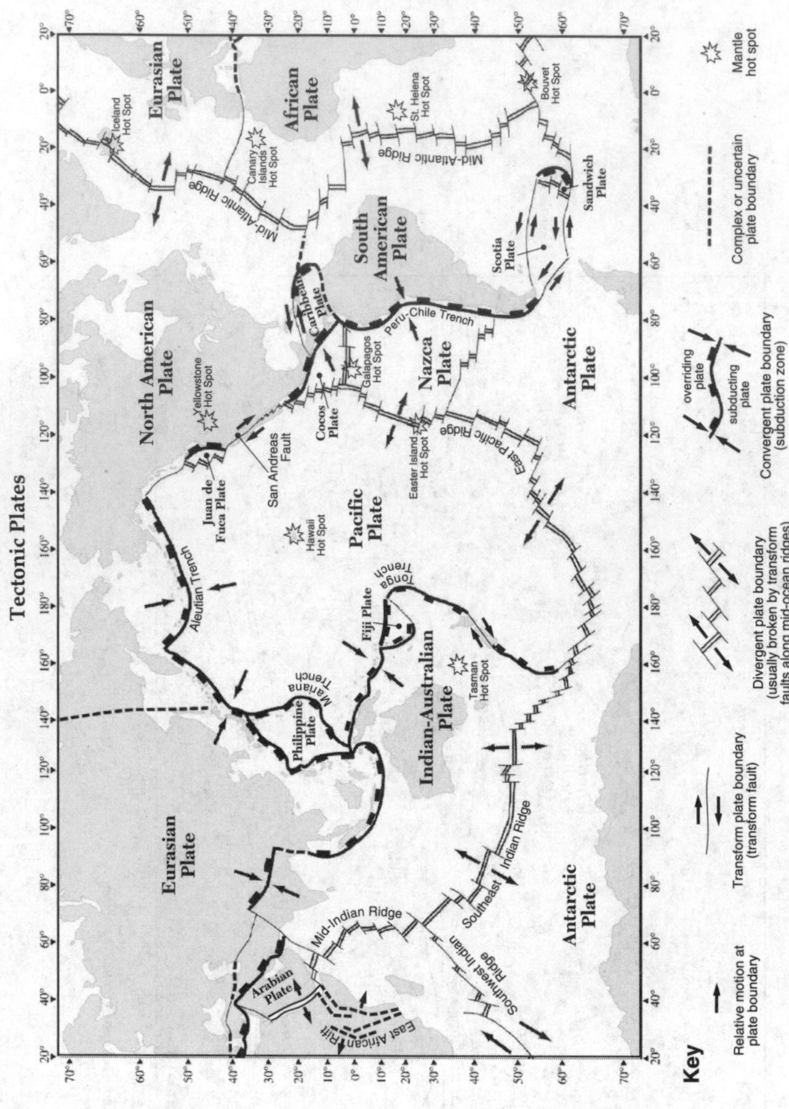
Tectonic Plates
Eurasian Plate
North American Plate
Eurasian Plate
African Plate
South American Plate
Caribbean Plate
Cocos Plate
Nazca Plate
Pacific Plate
Juan de Fuca Plate
Philippine Plate
Fiji Plate
Indian-Australian Plate
Arabian Plate
Antarctic Plate
Antarctic Plate
Scotia Plate
Sandwich Plate
Iceland Hot Spot
Canary Islands Hot Spot
St. Helena Hot Spot
Bouvet Hot Spot
Yellowstone Hot Spot
Galapagos Hot Spot
Easter Island Hot Spot
Hawaii Hot Spot
Tasman Hot Spot
Mid-Atlantic Ridge
Mid-Atlantic Ridge
San Andreas Fault
Peru-Chile Trench
East Pacific Ridge
Aleutian Trench
Mariana Trench
Tonga Trench
Mid-Indian Ridge
Southeast Indian Ridge
Southwest Indian Ridge
East African Rift
Key
Relative motion at plate boundary
Transform plate boundary (transform fault)
Divergent plate boundary (usually broken by transform faults along mid-ocean ridges)
overriding plate
subducting plate
Convergent plate boundary (subduction zone)
Complex or uncertain plate boundary
Mantle hot spot
NOTE: Not all mantle hot spots, plates, and boundaries are shown.

Rock Cycle in Earth's Crust

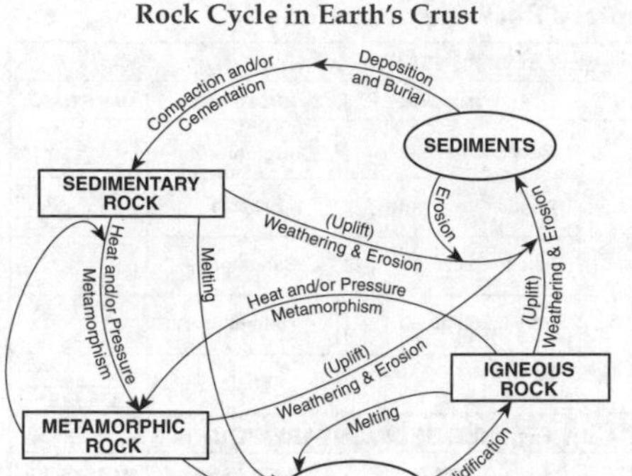

Relationship of Transported Particle Size to Water Velocity

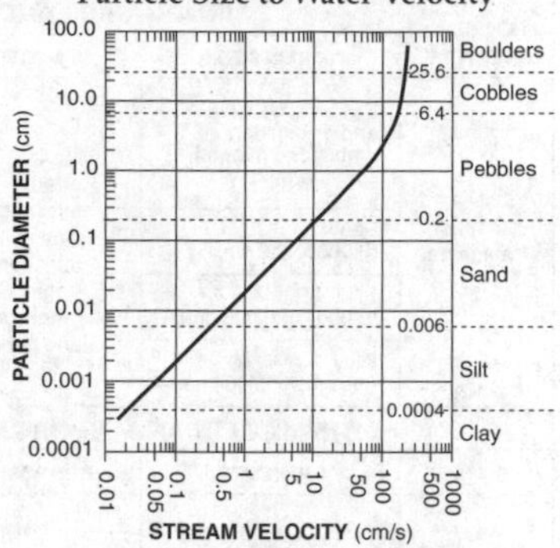

This generalized graph shows the water velocity needed to maintain, but not start, movement. Variations occur due to differences in particle density and shape.

Scheme for Igneous Rock Identification

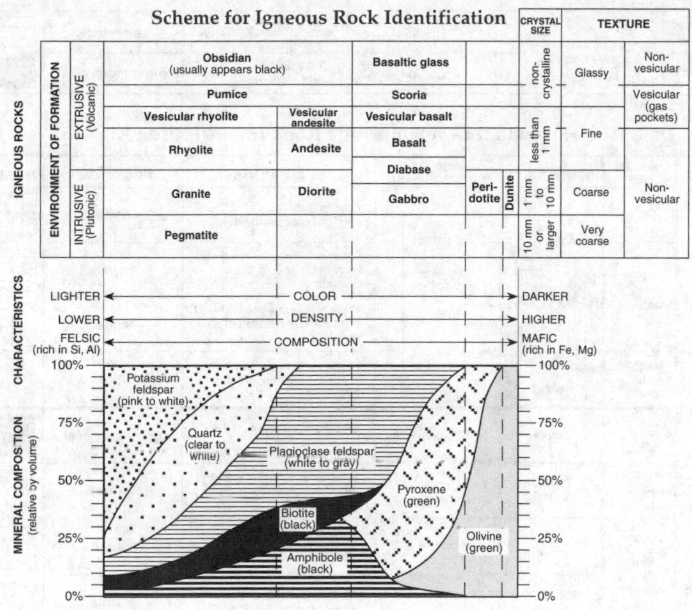

Scheme for Sedimentary Rock Identification

INORGANIC LAND-DERIVED SEDIMENTARY ROCKS					
TEXTURE	GRAIN SIZE	COMPOSITION	COMMENTS	ROCK NAME	MAP SYMBOL
Clastic (fragmental)	Pebbles, cobbles, and/or boulders embedded in sand, silt, and/or clay	Mostly quartz, feldspar, and clay minerals; may contain fragments of other rocks and minerals	Rounded fragments	**Conglomerate**	
			Angular fragments	**Breccia**	
	Sand (0.006 to 0.2 cm)		Fine to coarse	**Sandstone**	
	Silt (0.0004 to 0.006 cm)		Very fine grain	**Siltstone**	
	Clay (less than 0.0004 cm)		Compact; may split easily	**Shale**	

CHEMICALLY AND/OR ORGANICALLY FORMED SEDIMENTARY ROCKS					
TEXTURE	GRAIN SIZE	COMPOSITION	COMMENTS	ROCK NAME	MAP SYMBOL
Crystalline	Fine to coarse crystals	Halite	Crystals from chemical precipitates and evaporites	**Rock salt**	
		Gypsum		**Rock gypsum**	
		Dolomite		**Dolostone**	
Crystalline or bioclastic	Microscopic to very coarse	Calcite	Precipitates of biologic origin or cemented shell fragments	**Limestone**	
Bioclastic		Carbon	Compacted plant remains	**Bituminous coal**	

Scheme for Metamorphic Rock Identification

TEXTURE		GRAIN SIZE	COMPOSITION	TYPE OF METAMORPHISM	COMMENTS	ROCK NAME	MAP SYMBOL
FOLIATED	MINERAL ALIGNMENT	Fine	MICA	Regional (Heat and pressure increases)	Low-grade metamorphism of shale	**Slate**	
		Fine to medium	MICA, QUARTZ, FELDSPAR, AMPHIBOLE, GARNET		Foliation surfaces shiny from microscopic mica crystals	**Phyllite**	
			MICA, QUARTZ, FELDSPAR, AMPHIBOLE, GARNET, PYROXENE		Platy mica crystals visible from metamorphism of clay or feldspars	**Schist**	
	BAND-ING	Medium to coarse			High-grade metamorphism; mineral types segregated into bands	**Gneiss**	
NONFOLIATED		Fine	Carbon	Regional	Metamorphism of bituminous coal	**Anthracite coal**	
		Fine	Various minerals	Contact (heat)	Various rocks changed by heat from nearby magma/lava	**Hornfels**	
		Fine to coarse	Quartz	Regional or contact	Metamorphism of quartz sandstone	**Quartzite**	
			Calcite and/or dolomite		Metamorphism of limestone or dolostone	**Marble**	
		Coarse	Various minerals		Pebbles may be distorted or stretched	**Metaconglomerate**	

Inferred Properties of Earth's Interior

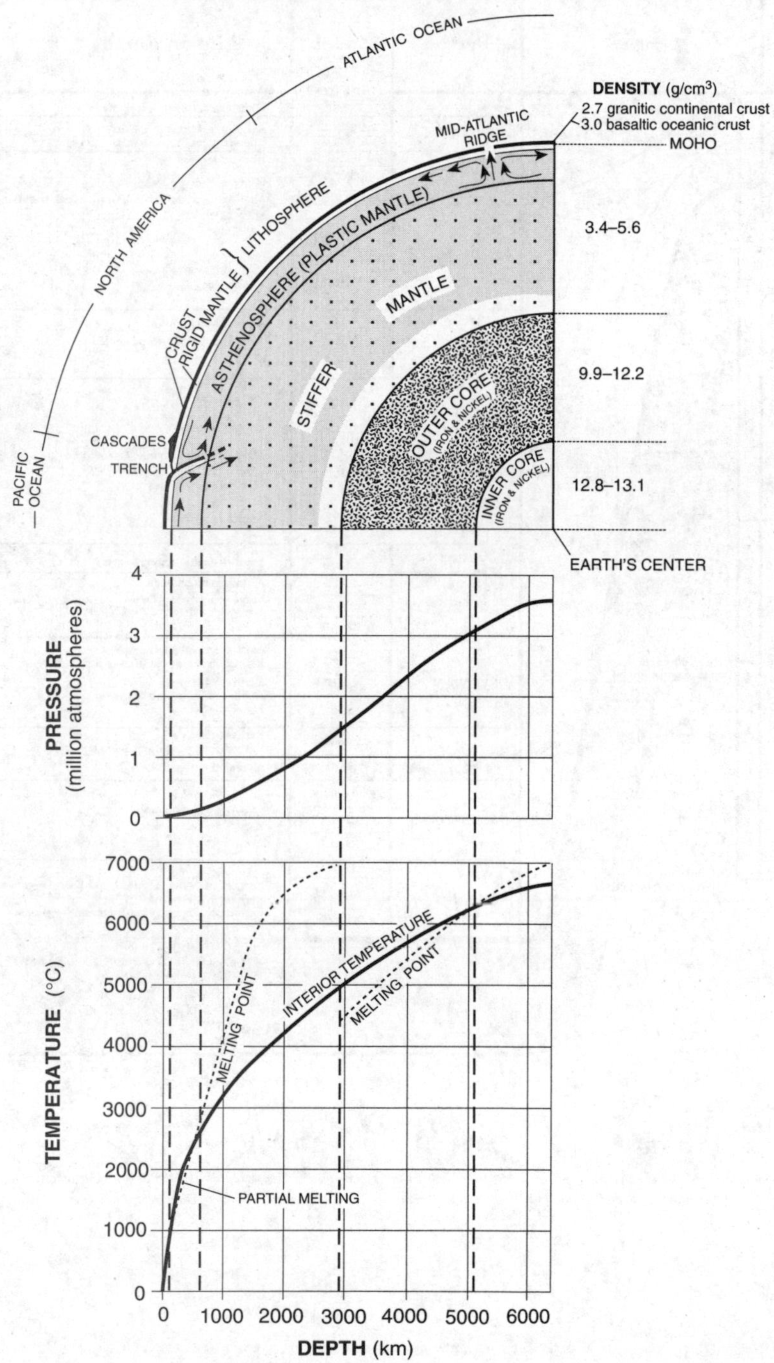

GEOLOGIC HISTORY

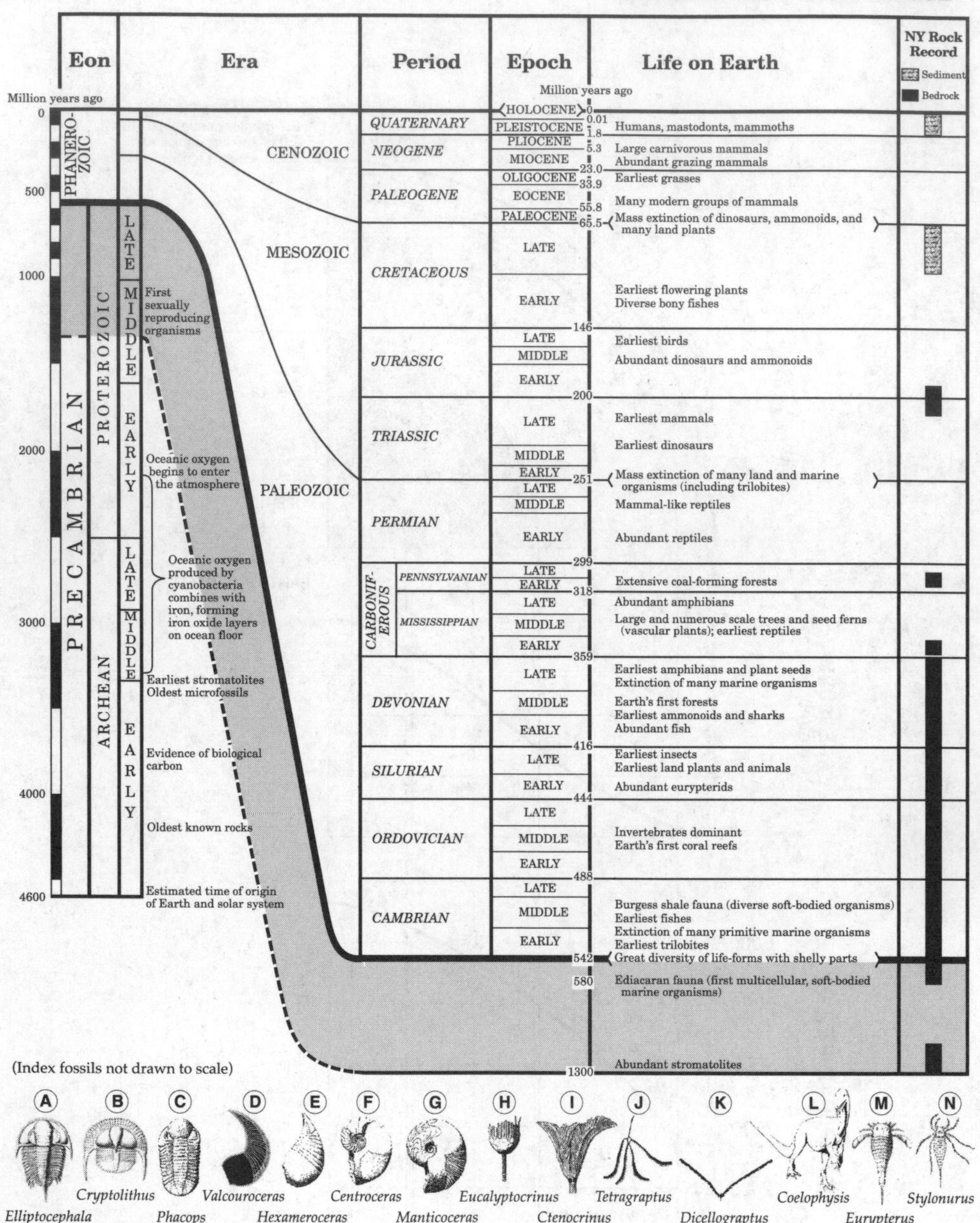

Eon
Era
Period
Epoch
Life on Earth
NY Rock Record
Sediment
Bedrock
Million years ago
0
500
1000
2000
3000
4000
4600
PHANERO-ZOIC
PRECAMBRIAN
PROTEROZOIC
ARCHEAN
LATE
MIDDLE
EARLY
First sexually reproducing organisms
Oceanic oxygen begins to enter the atmosphere
Oceanic oxygen produced by cyanobacteria combines with iron, forming iron oxide layers on ocean floor
Earliest stromatolites
Oldest microfossils
Evidence of biological carbon
Oldest known rocks
Estimated time of origin of Earth and solar system
CENOZOIC
MESOZOIC
PALEOZOIC
QUATERNARY
NEOGENE
PALEOGENE
CRETACEOUS
JURASSIC
TRIASSIC
PERMIAN
CARBONIFEROUS
PENNSYLVANIAN
MISSISSIPPIAN
DEVONIAN
SILURIAN
ORDOVICIAN
CAMBRIAN
HOLOCENE
PLEISTOCENE
PLIOCENE
MIOCENE
OLIGOCENE
EOCENE
PALEOCENE
0
0.01
1.8
5.3
23.0
33.9
55.8
65.5
146
200
251
299
318
359
416
444
488
542
580
1300
Humans, mastodonts, mammoths
Large carnivorous mammals
Abundant grazing mammals
Earliest grasses
Many modern groups of mammals
Mass extinction of dinosaurs, ammonoids, and many land plants
Earliest flowering plants
Diverse bony fishes
Earliest birds
Abundant dinosaurs and ammonoids
Earliest mammals
Earliest dinosaurs
Mass extinction of many land and marine organisms (including trilobites)
Mammal-like reptiles
Abundant reptiles
Extensive coal-forming forests
Abundant amphibians
Large and numerous scale trees and seed ferns (vascular plants); earliest reptiles
Earliest amphibians and plant seeds
Extinction of many marine organisms
Earth's first forests
Earliest ammonoids and sharks
Abundant fish
Earliest insects
Earliest land plants and animals
Abundant eurypterids
Invertebrates dominant
Earth's first coral reefs
Burgess shale fauna (diverse soft-bodied organisms)
Earliest fishes
Extinction of many primitive marine organisms
Earliest trilobites
Great diversity of life-forms with shelly parts
Ediacaran fauna (first multicellular, soft-bodied marine organisms)
Abundant stromatolites
(Index fossils not drawn to scale)
A Elliptocephala
B Cryptolithus
C Phacops
D Valcouroceras
E Hexameroceras
F Centroceras
G Manticoceras
H Eucalyptocrinus
I Ctenocrinus
J Tetragraptus
K Dicellograptus
L Coelophysis
M Eurypterus
N Stylonurus

OF NEW YORK STATE

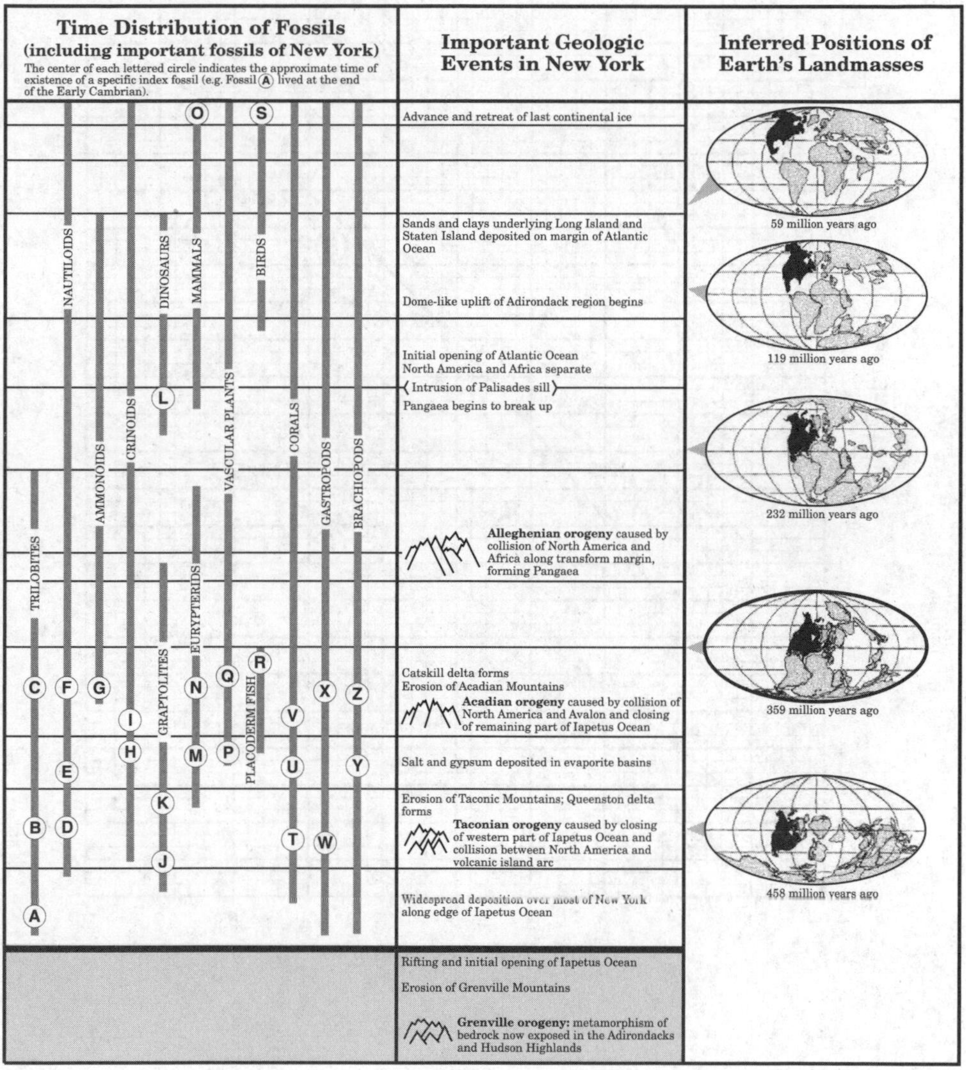

Time Distribution of Fossils (including important fossils of New York)

The center of each lettered circle indicates the approximate time of existence of a specific index fossil (e.g. Fossil (A) lived at the end of the Early Cambrian).

TRILOBITES, NAUTILOIDS, AMMONOIDS, CRINOIDS, GRAPTOLITES, DINOSAURS, EURYPTERIDS, MAMMALS, VASCULAR PLANTS, PLACODERM FISH, BIRDS, CORALS, GASTROPODS, BRACHIOPODS

O S L C F G I H E B D A N Q M P K J R V U T X W Z Y

Important Geologic Events in New York

Advance and retreat of last continental ice

Sands and clays underlying Long Island and Staten Island deposited on margin of Atlantic Ocean

Dome-like uplift of Adirondack region begins

Initial opening of Atlantic Ocean
North America and Africa separate

Intrusion of Palisades sill

Pangaea begins to break up

Alleghenian orogeny caused by collision of North America and Africa along transform margin, forming Pangaea

Catskill delta forms
Erosion of Acadian Mountains

Acadian orogeny caused by collision of North America and Avalon and closing of remaining part of Iapetus Ocean

Salt and gypsum deposited in evaporite basins

Erosion of Taconic Mountains; Queenston delta forms

Taconian orogeny caused by closing of western part of Iapetus Ocean and collision between North America and volcanic island arc

Widespread deposition over most of New York along edge of Iapetus Ocean

Rifting and initial opening of Iapetus Ocean

Erosion of Grenville Mountains

Grenville orogeny: metamorphism of bedrock now exposed in the Adirondacks and Hudson Highlands

Inferred Positions of Earth's Landmasses

59 million years ago

119 million years ago

232 million years ago

359 million years ago

458 million years ago

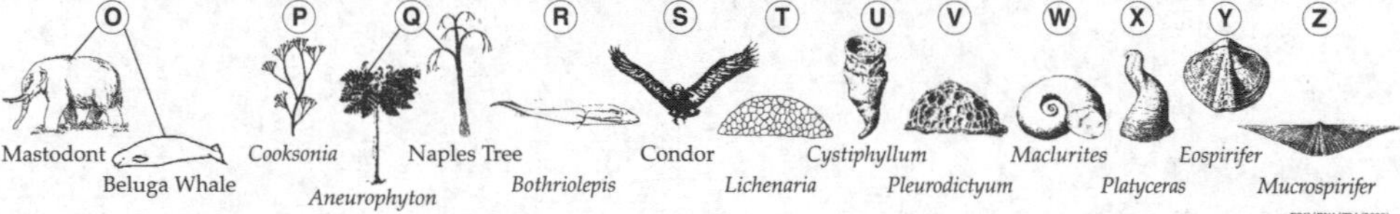

ESC/BW/TN (2009)

Earthquake P-Wave and S-Wave Travel Time

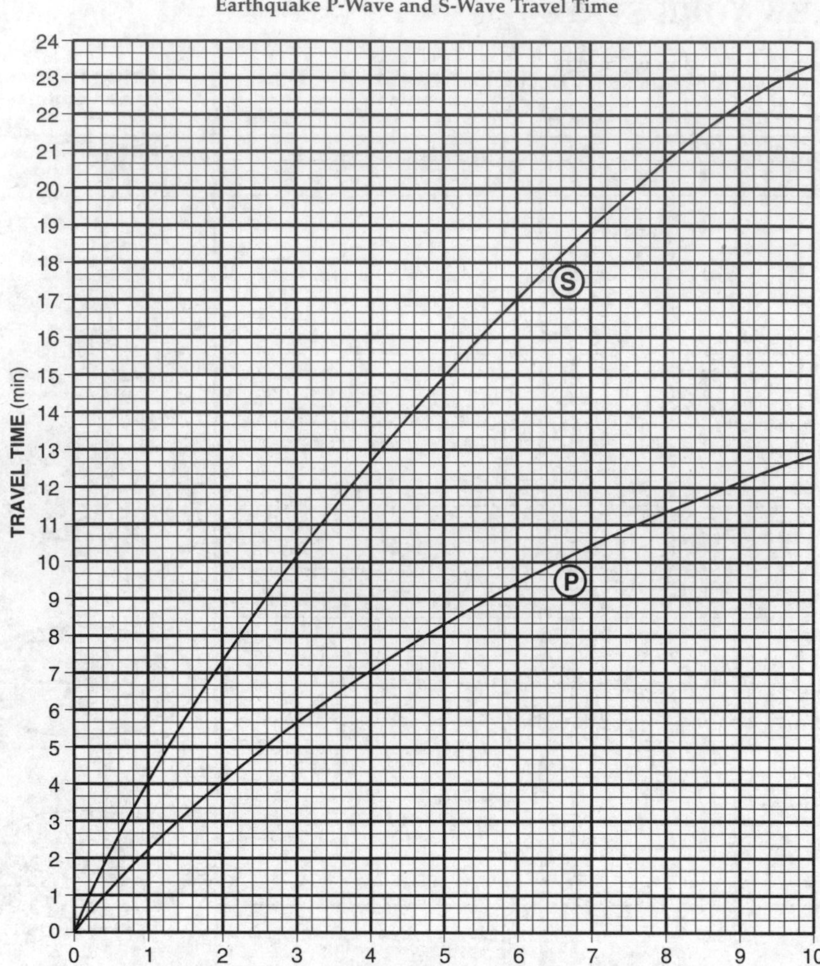

Dewpoint (°C)

Dry-Bulb Temperature (°C)	Difference Between Wet-Bulb and Dry-Bulb Temperatures (C°)															
	0	1	2	3	4	5	6	7	8	9	10	11	12	13	14	15
−20	−20	−33														
−18	−18	−28														
−16	−16	−24														
−14	−14	−21	−36													
−12	−12	−18	−28													
−10	−10	−14	−22													
−8	−8	−12	−18	−29												
−6	−6	−10	−14	−22												
−4	−4	−7	−12	−17	−29											
−2	−2	−5	−8	−13	−20											
0	0	−3	−6	−9	−15	−24										
2	2	−1	−3	−6	−11	−17										
4	4	1	−1	−4	−7	−11	−19									
6	6	4	1	−1	−4	−7	−13	−21								
8	8	6	3	1	−2	−5	−9	−14								
10	10	8	6	4	1	−2	−5	−9	−14	−28						
12	12	10	8	6	4	1	−2	−5	−9	−16						
14	14	12	11	9	6	4	1	−2	−5	−10	−17					
16	16	14	13	11	9	7	4	1	−1	−6	−10	−17				
18	18	16	15	13	11	9	7	4	2	−2	−5	−10	−19			
20	20	19	17	15	14	12	10	7	4	2	−2	−5	−10	−19		
22	22	21	19	17	16	14	12	10	8	5	3	−1	−5	−10	−19	
24	24	23	21	20	18	16	14	12	10	8	6	2	−1	−5	−10	−18
26	26	25	23	22	20	18	17	15	13	11	9	6	3	0	−4	−9
28	28	27	25	24	22	21	19	17	16	14	11	9	7	4	1	−3
30	30	29	27	26	24	23	21	19	18	16	14	12	10	8	5	1

Relative Humidity (%)

Dry-Bulb Temperature (°C)	Difference Between Wet-Bulb and Dry-Bulb Temperatures (C°)															
	0	1	2	3	4	5	6	7	8	9	10	11	12	13	14	15
−20	100	28														
−18	100	40														
−16	100	48														
−14	100	55	11													
−12	100	61	23													
−10	100	66	33													
−8	100	71	41	13												
−6	100	73	48	20												
−4	100	77	54	32	11											
−2	100	79	58	37	20	1										
0	100	81	63	45	28	11										
2	100	83	67	51	36	20	6									
4	100	85	70	56	42	27	14									
6	100	86	72	59	46	35	22	10								
8	100	87	74	62	51	39	28	17	6							
10	100	88	76	65	54	43	33	24	13	4						
12	100	88	78	67	57	48	38	28	19	10	2					
14	100	89	79	69	60	50	41	33	25	16	8	1				
16	100	90	80	71	62	54	45	37	29	21	14	7	1			
18	100	91	81	72	64	56	48	40	33	26	19	12	6			
20	100	91	82	74	66	58	51	44	36	30	23	17	11	5		
22	100	92	83	75	68	60	53	46	40	33	27	21	15	10	4	
24	100	92	84	76	69	62	55	49	42	36	30	25	20	14	9	4
26	100	92	85	77	70	64	57	51	45	39	34	28	23	18	13	9
28	100	93	86	78	71	65	59	53	47	42	36	31	26	21	17	12
30	100	93	86	79	72	66	61	55	49	44	39	34	29	25	20	16

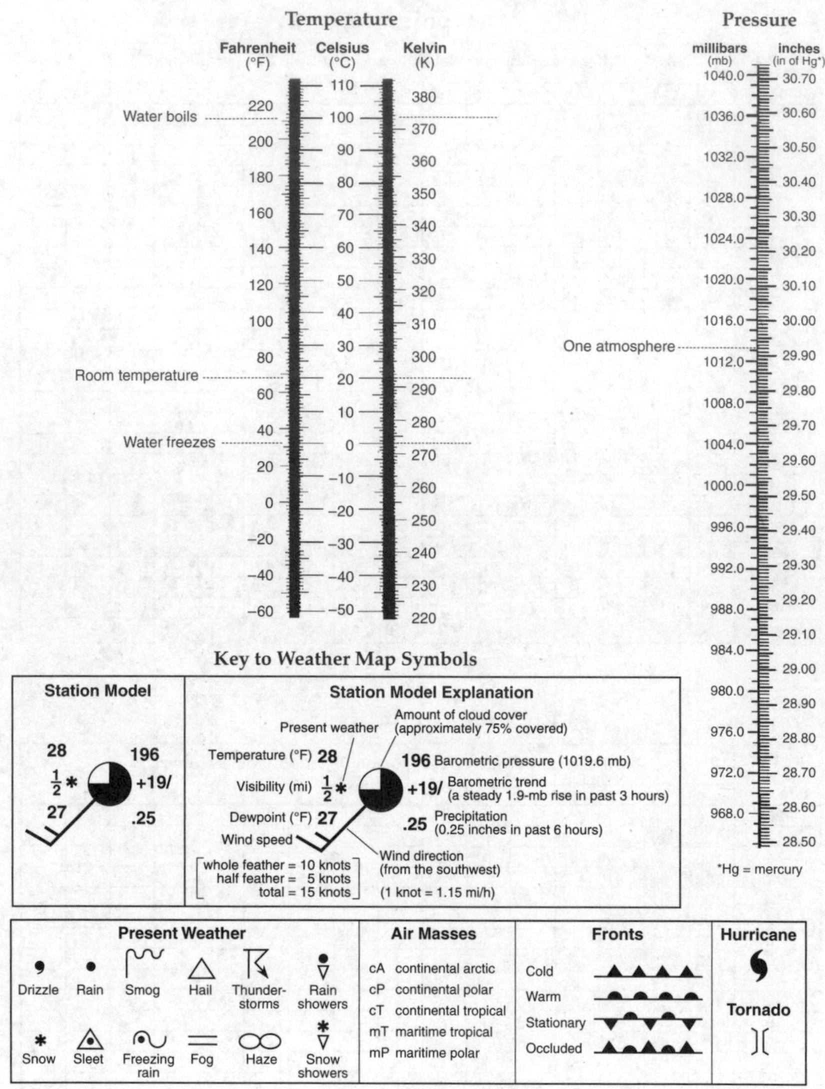
Temperature
Fahrenheit (°F)
Celsius (°C)
Kelvin (K)
Water boils
Room temperature
Water freezes
220
200
180
160
140
120
100
80
60
40
20
0
−20
−40
−60
110
100
90
80
70
60
50
40
30
20
10
0
−10
−20
−30
−40
−50
380
370
360
350
340
330
320
310
300
290
280
270
260
250
240
230
220
Pressure
millibars (mb)
inches (in of Hg*)
One atmosphere
1040.0
1036.0
1032.0
1028.0
1024.0
1020.0
1016.0
1012.0
1008.0
1004.0
1000.0
996.0
992.0
988.0
984.0
980.0
976.0
972.0
968.0
30.70
30.60
30.50
30.40
30.30
30.20
30.10
30.00
29.90
29.80
29.70
29.60
29.50
29.40
29.30
29.20
29.10
29.00
28.90
28.80
28.70
28.60
28.50
*Hg = mercury
Key to Weather Map Symbols
Station Model
28
196
½ *
+19/
27
.25
Station Model Explanation
Present weather
Amount of cloud cover (approximately 75% covered)
Temperature (°F) 28
196 Barometric pressure (1019.6 mb)
Visibility (mi) ½ *
+19/ Barometric trend (a steady 1.9-mb rise in past 3 hours)
Dewpoint (°F) 27
.25 Precipitation (0.25 inches in past 6 hours)
Wind speed
whole feather = 10 knots
half feather = 5 knots
total = 15 knots
Wind direction (from the southwest)
(1 knot = 1.15 mi/h)
Present Weather
Drizzle
Rain
Smog
Hail
Thunder-storms
Rain showers
Snow
Sleet
Freezing rain
Fog
Haze
Snow showers
Air Masses
cA continental arctic
cP continental polar
cT continental tropical
mT maritime tropical
mP maritime polar
Fronts
Cold
Warm
Stationary
Occluded
Hurricane
Tornado

Selected Properties of Earth's Atmosphere

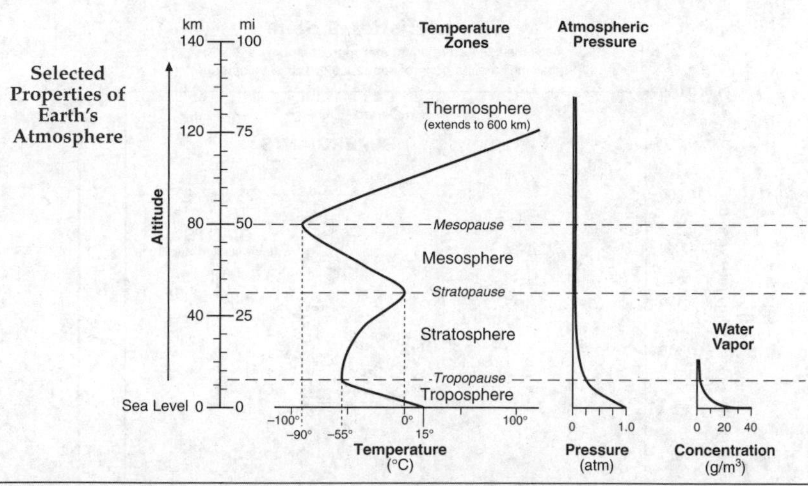

Planetary Wind and Moisture Belts in the Troposphere

The drawing on the right shows the locations of the belts near the time of an equinox. The locations shift somewhat with the changing latitude of the Sun's vertical ray. In the Northern Hemisphere, the belts shift northward in the summer and southward in the winter.

(Not drawn to scale)

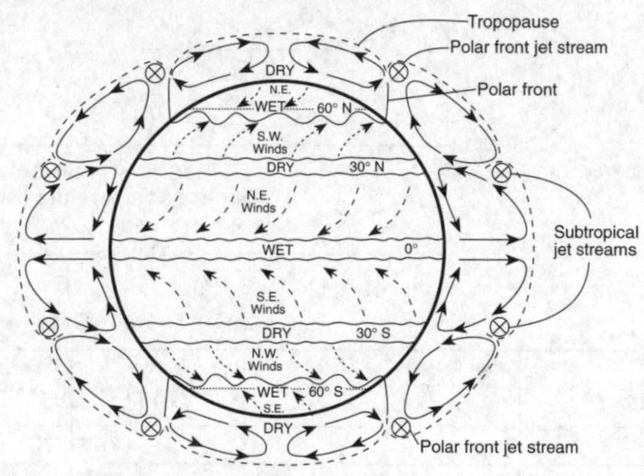

Electromagnetic Spectrum

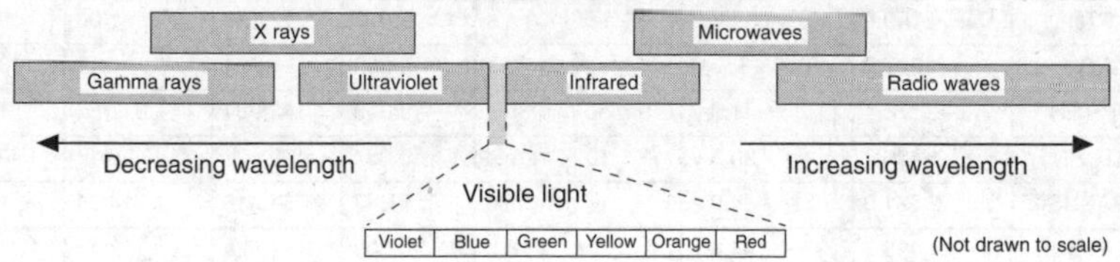

Characteristics of Stars

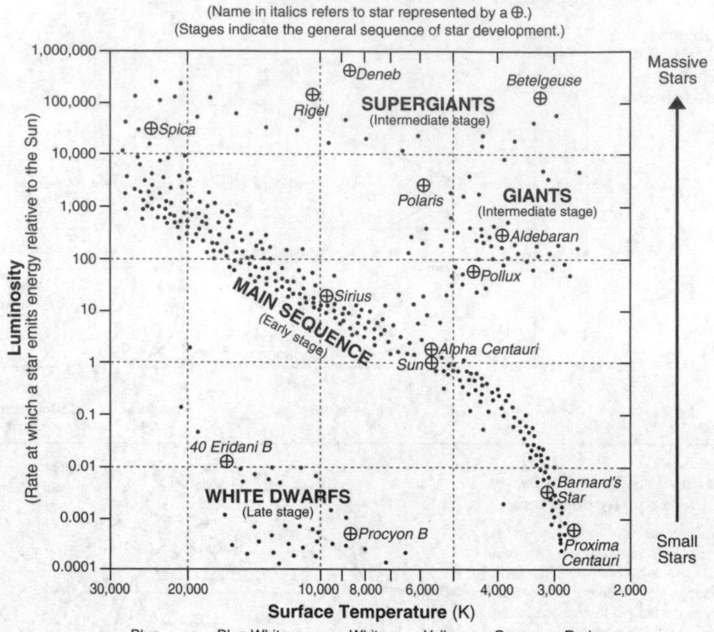

Solar System Data

Celestial Object	Mean Distance from Sun (million km)	Period of Revolution (d=days) (y=years)	Period of Rotation at Equator	Eccentricity of Orbit	Equatorial Diameter (km)	Mass (Earth = 1)	Density (g/cm³)
SUN	—	—	27 d	—	1,392,000	333,000.00	1.4
MERCURY	57.9	88 d	59 d	0.206	4,879	0.06	5.4
VENUS	108.2	224.7 d	243 d	0.007	12,104	0.82	5.2
EARTH	149.6	365.26 d	23 h 56 min 4 s	0.017	12,756	1.00	5.5
MARS	227.9	687 d	24 h 37 min 23 s	0.093	6,794	0.11	3.9
JUPITER	778.4	11.9 y	9 h 50 min 30 s	0.048	142,984	317.83	1.3
SATURN	1,426.7	29.5 y	10 h 14 min	0.054	120,536	95.16	0.7
URANUS	2,871.0	84.0 y	17 h 14 min	0.047	51,118	14.54	1.3
NEPTUNE	4,498.3	164.8 y	16 h	0.009	49,528	17.15	1.8
EARTH'S MOON	149.6 (0.386 from Earth)	27.3 d	27.3 d	0.055	3,476	0.01	3.3

Properties of Common Minerals

LUSTER	HARDNESS	CLEAVAGE	FRACTURE	COMMON COLORS	DISTINGUISHING CHARACTERISTICS	USE(S)	COMPOSITION*	MINERAL NAME
Metallic luster	1–2	✔		silver to gray	black streak, greasy feel	pencil lead, lubricants	C	**Graphite**
	2.5	✔		metallic silver	gray-black streak, cubic cleavage, density = 7.6 g/cm³	ore of lead, batteries	PbS	**Galena**
	5.5–6.5		✔	black to silver	black streak, magnetic	ore of iron, steel	Fe_3O_4	**Magnetite**
	6.5		✔	brassy yellow	green-black streak, (fool's gold)	ore of sulfur	FeS_2	**Pyrite**
Either	5.5–6.5 or 1		✔	metallic silver or earthy red	red-brown streak	ore of iron, jewelry	Fe_2O_3	**Hematite**
Nonmetallic luster	1	✔		white to green	greasy feel	ceramics, paper	$Mg_3Si_4O_{10}(OH)_2$	**Talc**
	2		✔	yellow to amber	white-yellow streak	sulfuric acid	S	**Sulfur**
	2	✔		white to pink or gray	easily scratched by fingernail	plaster of paris, drywall	$CaSO_4 \cdot 2H_2O$	**Selenite gypsum**
	2–2.5	✔		colorless to yellow	flexible in thin sheets	paint, roofing	$KAl_3Si_3O_{10}(OH)_2$	**Muscovite mica**
	2.5	✔		colorless to white	cubic cleavage, salty taste	food additive, melts ice	NaCl	**Halite**
	2.5–3	✔		black to dark brown	flexible in thin sheets	construction materials	$K(Mg,Fe)_3 AlSi_3O_{10}(OH)_2$	**Biotite mica**
	3	✔		colorless or variable	bubbles with acid, rhombohedral cleavage	cement, lime	$CaCO_3$	**Calcite**
	3.5	✔		colorless or variable	bubbles with acid when powdered	building stones	$CaMg(CO_3)_2$	**Dolomite**
	4	✔		colorless or variable	cleaves in 4 directions	hydrofluoric acid	CaF_2	**Fluorite**
	5–6	✔		black to dark green	cleaves in 2 directions at 90°	mineral collections, jewelry	$(Ca,Na)(Mg,Fe,Al)(Si,Al)_2O_6$	**Pyroxene** (commonly augite)
	5.5	✔		black to dark green	cleaves at 56° and 124°	mineral collections, jewelry	$CaNa(Mg,Fe)_4(Al,Fe,Ti)_3Si_6O_{22}(O,OH)_2$	**Amphibole** (commonly hornblende)
	6	✔		white to pink	cleaves in 2 directions at 90°	ceramics, glass	$KAlSi_3O_8$	**Potassium feldspar** (commonly orthoclase)
	6	✔		white to gray	cleaves in 2 directions, striations visible	ceramics, glass	$(Na,Ca)AlSi_3O_8$	**Plagioclase feldspar**
	6.5		✔	green to gray or brown	commonly light green and granular	furnace bricks, jewelry	$(Fe,Mg)_2SiO_4$	**Olivine**
	7		✔	colorless or variable	glassy luster, may form hexagonal crystals	glass, jewelry, electronics	SiO_2	**Quartz**
	6.5–7.5		✔	dark red to green	often seen as red glassy grains in NYS metamorphic rocks	jewelry (NYS gem), abrasives	$Fe_3Al_2Si_3O_{12}$	**Garnet**

*Chemical symbols:

Al = aluminum	Cl = chlorine	H = hydrogen	Na = sodium	S = sulfur
C = carbon	F = fluorine	K = potassium	O = oxygen	Si = silicon
Ca = calcium	Fe = iron	Mg = magnesium	Pb = lead	Ti = titanium

✔ = dominant form of breakage

Glossary of Earth Science Terms

abrasion the wearing away of rock material that results when one rock particle strikes another. In stream abrasion, particles carried by the stream may strike other particles, causing pieces to break off.

adiabatic change the change in temperature of a gas caused by expansion or compression. When a gas expands, for example, it cools and its temperature drops.

air mass a large area in the atmosphere in which the temperature and moisture conditions are similar.

altitude the angle between the line of sight to a star and the horizon; also, the elevation above sea level.

angle of insolation the angle at which the Sun's rays strike the Earth's surface.

anticline a series of folded rock layers that bends upward near the center. *See also* **syncline**.

apparent diameter the size that an object appears to an observer. When objects are close by, they appear larger than when they are farther away.

asthenosphere a region of the upper mantle between 100 and 350 kilometers in depth that behaves like a fluid.

atmosphere the envelope of air that encircles Earth. It has been divided into zones based largely on temperature differences.

axis an imaginary line around which an object rotates. Earth's axis extends from the North Pole to the South Pole.

barometer an instrument for measuring atmospheric pressure.

barometric pressure the amount of force exerted by the air per square inch or centimeter at a particular location.

bedrock the rock layer nearest Earth's surface, lying directly below any soil layers.

Big Bang theory the idea that the universe started out with all of its matter in a small volume and then expanded outward in all directions.

bioclastic a sedimentary rock consisting of fragmental or broken remains of organisms, such as limestone composed of shell fragments.

boiling point the temperature at which a liquid changes to a vapor.

capillarity (capillary action) the rising of water against gravity, as when water rises above the water table in soil because of the attraction between the water and the soil.

cementation the process by which sediments are bonded together by material dissolved in water when the water evaporates.

chemical weathering weathering that occurs because of the chemical reaction between material dissolved in water and local rock material. *See also* **physical weathering**.

cleavage the splitting of a mineral in distinctive directions caused by the arrangement of the atoms in a mineral. For example, the mineral mica cleaves in layers because the atoms in the crystal structure of mica are arranged in layers.

climate the average weather conditions, in terms of temperature and moisture, of an area over a long period of time.

clouds masses of water droplets suspended in the air. They form when air is cooled and moisture condenses.

cold front the boundary between two air masses of different temperatures, the point where the colder air mass moves under and pushes up the warmer air mass. *See also* **warm front**.

compaction the process that results when buried sediments are subjected to pressure, which packs them together. Together with cementation, this process causes sediments to be converted to rock.

condensation the process whereby, when moisture in the atmosphere is cooled, it changes from a vapor to a liquid.

conduction the method of heat transfer in solids in which faster moving molecules strike other molecules, causing them to speed up.

contact metamorphism changes in rock that result from the extreme heat produced by contact with magma or lava.

continental air mass an air mass that forms over land and therefore is relatively dry.

continental drift a hypothetical slow movement of the continents, which forms the basis of the theory that the present continents were once part of one large landmass that broke up. Since that time, the continents have been drifting apart.

continental glacier a large sheet of ice that covered much of a continent during a period of Earth's past geologic history.

contour interval the differences in elevation between contour lines on a contour or topographic map.

contour line a line on a topographic map that connects points having the same elevation.

contour map *See* **topographic map**.

convection the method of energy transfer in fluids in which the fluid expands when it is heated and rises. When the fluid cools, it contracts and sinks.

convection cell the circular pattern of movement in a fluid caused by the rising and sinking of the fluid due to differences in density caused by differences in temperature.

convergent boundary places where edges of adjacent plates are colliding.

core the innermost zone of Earth's interior. A solid inner core is surrounded by a molten outer core.

coriolis effect an apparent force, due to the rotation of the Earth, that deflects winds toward the right in the Northern Hemisphere and toward the left in the Southern Hemisphere.

correlation the matching of rock layers at different locations, based on composition, thickness, and in some cases fossil content, for the purpose of establishing that they represent the same rock layer.

cosmic background radiation a remnant of radiation left over from the original Big Bang that fills the universe.

crust the outermost solid layer of Earth, extending across the continents and under the oceans. It is thicker under the continents than under the oceans.

density the mass per unit volume of a substance.

deposition the process by which material carried by running water, glaciers, or wind settles out when the velocity of the carrier slows down.

dew moisture that condenses at Earth's surface when moist air touches a cool area.

dewpoint temperature the temperature at which the air becomes saturated and excess moisture begins to condense.

dike a rock layer that forms when molten rock material flows through breaks in rock layers and then cools and hardens.

discharge the total volume of water flowing in a river or stream per unit of time.

divergent boundary places where edges of adjacent plates are spreading apart.

Doppler effect the shift in wavelength of a spectrum line from its normal position due to relative motion between the source and the observer.

duration of insolation the number of hours that the Sun's rays strike Earth's surface over a 24-hour period; the number of hours of daylight.

earthquake the large-scale and rapid motion of rock layers that occurs when pressure is released.

electromagnetic spectrum the range of wavelengths of energy released by the Sun. A small range of wavelengths represents visible light. Other bands within the spectrum include ultraviolet radiation, cosmic rays, infrared radiation, and radio waves.

elevation the height above sea level.

ellipse a curve that has two centers, or foci. The sum of the distance between either focus and any point on the ellipse and the distance between the other focus and that point is a constant. The shape of Earth's orbit around the Sun is an ellipse with the Sun at one focus.

epicenter the location on Earth's surface directly above the point of origin or focus of an earthquake.

equator an imaginary circle around Earth lying halfway between the poles and dividing the Earth's surface into the Northern and Southern hemispheres.

equinox the two times a year when the Sun is directly overhead at the equator at noon. At these times, about March 21 and September 23, the number of hours of day and of night is the same.

erosion the process by which weathered rock material is carried away by agents such as running water, ice, wind, and gravity.

evaporation the process by which water is converted from a liquid to a vapor.

evapotranspiration the combined processes of evaporation and transpiration, representing the method by which water vapor returns to the atmosphere.

extinct referring to a plant or animal species that lived in the past but is no longer found alive on Earth.

fault movement within rock layers where pressure has caused the layers to break. The layers on one side of the break move up, while the opposite layers move down.

felsic referring to igneous rocks composed of minerals with a high content of aluminum. These rocks tend to be lighter in color than others.

focus the point of origin within Earth's crust or mantle of an earthquake.

foliated a metamorphic rock texture characterized by thin, leaflike layers caused by the flattening of mineral grains under heat and pressure.

formation (rock) a sequence of rock layers that cover a large area and were formed over a period of time.

fossil the preserved remains or traces of an animal or plant that lived in the past.

fossil record groups of fossils found in a rock layer that are used to interpret how and when the rock layer was formed and what the environment was like at that time.

Foucault pendulum a freely swinging pendulum whose path appears to change direction relative to Earth's surface in a predictable manner due to Earth's rotation.

fracture a term used to describe the irregular way in which some minerals break.
freezing point the temperature at which a liquid changes to a solid.
front the boundary between two different air masses. *See also* **cold front; warm front**.
frost moisture that condenses directly from a vapor to a solid when moist air touches a cold surface. When the air temperature is below freezing, frost forms instead of dew.

galaxy a system consisting of hundreds of billions of stars.
gradient the slope of the land or of a river or stream.
greenhouse effect the process by which longer wavelength radiation emitted from Earth's surface is absorbed by carbon dioxide in the atmosphere, causing a rise in air temperature.

half-life the amount of time it takes for half the mass of a radioactive substance to decay. For example, the half-life of carbon-14 is 5.6×10^3 years.
high-pressure center a location on a weather map where winds blow outward, in a clockwise direction, away from the center. This occurs because the air pressure is lower at the center than over the surrounding area. *See also* **low-pressure center**.
humidity the amount of moisture in the air.
hydrosphere the outer zone of Earth, which consists of the oceans and seas.
hypothesis the attempt to explain a scientific phenomenon on the basis of observations and other relevant information.

igneous referring to rocks that form when molten rock material cools and solidifies.
impermeable referring to a layer of material through which water cannot pass. Most soil layers are permeable, while most rock layers are not.
index fossil a fossil that is found over widespread areas and that formed from an organism that existed for a relatively short geologic period of time.
inference an interpretation or explanation of a natural phenomenon based on observations.
infiltration the process by which water at Earth's surface penetrates and filters down through porous soil and rock layers.
insolation the radiation reaching Earth's surface from the Sun. This term is a contraction of "incoming solar radiation."
intrusion the forcible entry of solidified molten rock material between rock layers or into breaks within a rock layer.
isobar a line on a weather map that connects points having the same air or barometric pressure.

land breeze a wind blowing from over the land to over a lake or ocean. It occurs when the water temperature is warmer than the land temperature. *See also* **sea breeze**.

landscape region a grouping of landscapes with similar relief, stream patterns, and soil associations.

latent heat of fusion the amount of energy required to convert one gram of ice at 0°C to water at 0°C.

latent heat of vaporization the amount of energy required to convert one gram of water at 100°C to vapor at 100°C.

latitude imaginary circles around Earth parallel to the equator and between the North and South poles.

lava molten rock material that reaches Earth's surface through volcanoes.

light-year the distance light travels in one year; roughly 9.5 trillion kilometers, or 6 trillion miles.

lithosphere the solid outer portion of Earth.

longitude imaginary circles around Earth that pass through the North and South poles.

low-pressure center a location on a weather map where winds blow inward, in a counterclockwise direction, into a center. This occurs because the air pressure is lower at the center than in the surrounding area. *See also* **high-pressure center**.

luminosity the total amount of energy radiated by a star in 1 second.

lunar eclipse the phenomenon that occurs when Earth passes between the Moon and the Sun, causing light from the Sun to be blocked from reaching the surface of the Moon. *See also* **solar eclipse**.

luster the type of shine exhibited by a mineral, based on the way its surface reflects light. Examples are metallic, glassy, and dull lusters.

mafic referring to igneous rocks composed of minerals with a high content of magnesium and iron. They tend to be darker in color than other rocks.

magma molten rock material beneath the Earth's surface.

magnetic north (pole) the point near the geographic North Pole toward which the needle on a compass points.

mantle the zone within Earth that lies between the crust and the outer core.

maritime air mass an air mass that forms over water and therefore has a relatively high moisture content.

mass the measure of the amount of matter contained in a sample.

meander a curve in a river or stream.

meridian a line of longitude measured in degrees. The prime meridian (0°) passes through Greenwich, England.

metamorphic referring to rocks that form when existing rocks are subjected to enough heat and pressure to cause partial melting of the minerals present.

meter the standard unit of length in the metric system; 1 meter = 39.37 inches.

mid-ocean ridge a chain of undersea mountains running through the center of an ocean.

millibar a unit of air pressure in the atmosphere, commonly used on weather maps.

mineral a naturally occurring substance that is always made of the same elements in a fixed proportion.

modified Mercalli scale a system that measures the strength of an earthquake based upon perception of motion by human observers and damage to structures built by humans.

Moho the boundary between the dense rock of the mantle and the less dense rock of the crust. It was named for the seismologist Andrija Mohorovicic, who recognized its existence after analyzing seismic wave behavior inside Earth.

moraine a large deposit formed when a glacier melts and leaves behind the material it is carrying.

mountain any part of Earth's crust that projects at least 300 m (1000 ft) above the surrounding land, has a limited summit area (as opposed to a plateau), steep sides, and considerable bare-rock surface.

observation a description of what is perceived by the senses. It can often be made more accurate by using instruments.

occluded front a weather front that forms when a cold front moves in behind a warm front, lifting the warm front off the ground.

orbit the path followed by one object as it revolves around another; for example, the orbit of Earth around the Sun.

outgassing the release of gases and water vapor from molten rocks, leading to the formation of Earth's atmosphere and oceans.

permeability the ability of water to penetrate through a material. The permeability of a soil depends upon the size of the soil grains and the amount of space between the grains.

physical weathering the breakdown of rocks due to physical changes, such as changes in temperature or the freezing and thawing of water, that fills cracks in the rock. *See also* **chemical weathering**.

plain a flat area at low elevation.

plateau a large, flat region elevated more than 150–300 m (500–1000 ft) above the surrounding land, or above sea level.

polar air mass an air mass that forms over the polar regions and therefore contains relatively cold air.

pollution a condition of the air, water, or land in which there is a surplus of materials present that may be harmful to living things.
porosity the percentage of open space in a soil sample.
precipitation all forms of moisture that reach the Earth's surface from the atmosphere, including rain, snow, hail, and sleet.
prevailing westerlies the wind belt stretching across the United States in which the general direction of wind movement is from southwest to northeast.
***P*-wave (primary wave)** the compression type of wave emitted by an earthquake that travels through both liquids and solids. *See also* **S-wave**.

radiation the method of energy transfer by which energy from the Sun reaches the Earth.
radioactive dating the use of the radioactive isotope to determine the age of a fossil or rock layer. This technique is possible because each radioactive isotope decays at a unique and fixed rate.
radioactive decay the process by which a radioactive element breaks down to emit particles and radiation that form a new element.
radioactive isotope an unstable form of an element that breaks down by radioactive decay. For example, carbon-14 is a radioactive isotope of carbon that breaks down by emitting electrons to form nitrogen-14.
rain gauge an instrument that collects atmospheric precipitation so that the amount of rainfall can be measured.
regional metamorphism changes in rock over an extensive area that occur due to the pressure and high temperatures associated with either deep burial or movements of Earth's crust.
relative humidity the percentage of moisture present in the air as compared to the maximum amount of moisture the air can hold at the prevailing temperature.
residual soil soil that is formed by the weathering of local rock material.
Richter scale a scale with a range of 1 to 10 that indicates the magnitude of an earthquake. For each successive number, the magnitude increases by a factor of 10.
rock naturally occurring materials that are composed of one or more minerals.
rock cycle the process by which each of the three rock types (igneous, metamorphic, and sedimentary) can be converted into each other type.
runoff excess precipitation reaching Earth's surface that flows into rivers and streams because the surface soil is saturated.

salinity a measure of the amount of salt dissolved in water.
saturation temperature *See* **dewpoint temperature**.

sea breeze a wind blowing from over a lake or ocean to over land. It occurs when the water temperature is cooler than the land temperature. *See also* **land breeze**.

sea floor spreading the concept that portions of the ocean floor are moving away from a central ridge because new material moving upward at the ridge is pushing the old material outward.

sedimentary referring to rocks that form by the compaction and cementation of sediments.

seismograph an instrument used to measure the disturbances caused by an earthquake.

sill a rock layer that forms when molten rock material forces its way between existing rock layers and then cools and solidifies.

sling psychrometer an instrument used to determine relative humidity.

solar eclipse the phenomenon that occurs when the Moon passes between Earth and the Sun, causing light from the Sun to be blocked from reaching the surface of Earth. *See also* **lunar eclipse**.

solar system the Sun and the various objects that orbit it, including the planets and their moons, comets, and asteroids.

solstice one of the two times a year, about June 22 and December 22, when the Sun is directly overhead at 23½° north and south latitude. The solstices represent the longest and shortest days of the year.

specific heat the relative amount of energy needed to raise the temperature of one gram of a material by one degree Celsius.

spectroscope an instrument used to break down the visible portion of the electromagnetic spectrum into individual wavelengths or colors.

spectrum the band of colors formed when a beam of white light is passed through a prism; each wavelength of light in the beam is refracted at a slightly different angle, causing the light to be arrayed in order of its constituent wavelengths.

stationary front a front that forms when the opposing warm and cold air masses are of equal energy.

station model a pattern of symbols on a weather map that is used to describe local weather conditions such as temperature, pressure, humidity, wind speed and direction, and cloud cover.

stratosphere the layer of the atmosphere directly above the troposphere.

streak a more accurate method of identifying the color of a mineral by rubbing it against a plate, causing powder to form.

stream load the total amount of material carried by a stream, including material that is carried in solution or suspension or is pushed along the bottom.

subduction the sliding of a denser ocean plate beneath a less dense continental plate resulting in the melting of the ocean plate as it plunges into the hot mantle.

subsoil the layer of soil just below the topsoil. It does not contain the organic matter found in the topsoil and has not been as extensively weathered.

S-wave (shear wave) the transverse type of wave emitted by an earthquake that travels through solids but is absorbed by liquids. *See also* **P-wave**.

syncline a series of folded rock layers that dips downward near the center. *See also* **anticline**.

terminal moraine the material deposited by the meltwater of a glacier at the point of its farthest advance.

texture the grain size of a rock. A course texture represents large grains; a fine texture, small grains.

theory an explanation for scientific observations. A theory is formed from a hypothesis when there is substantial evidence to support it.

till the material found in a glacial moraine and usually representing a wide range of particle sizes.

time zone a region in which the same time is used throughout, instead of the local time at each place within it.

topographic (contour) map a map of an area with contour lines to show the elevations at all locations, as well as other geologic features such as rivers, streams, and mountain peaks.

topsoil the uppermost layer of soil. It contains the most highly weathered rock fragments, as well as organic remains.

transform boundary places where edges of plates are sliding laterally past each other.

transpiration the process by which plants release moisture to the atmosphere.

transported soils soils formed from weathered rock material that have been carried from other locations and deposited.

trench a large valley on the ocean floor.

tropical air mass an air mass that has formed over the tropics and therefore contains relatively warm air.

Tropic of Cancer the line of latitude at 23½° North, which represents the farthest north that the noon Sun can be overhead.

Tropic of Capricorn the line of latitude at 23½° South, which represents the farthest south that the noon Sun can be overhead.

troposphere the layer of the atmosphere closest to Earth's surface. All weather phenomena occur within this zone.

unconformity a gap in a sequence of rock layers, resulting either from the removal of layers by erosion or failure of layers to form for long periods.
uniformitarianism the principle that the geologic processes acting today are the same as those that occurred in the past.

valley glacier a glacier that forms in a valley between mountain peaks.
vesicular a rock texture characterized by cavities formed by gas bubbles escaping from a lava as it cools and solidifies.
volcano a mountain formed when lava erupts from beneath the surface and then cools and solidifies.

warm front the boundary between two air masses of different temperatures; the point where the warmer air mass moves over the colder air mass and pushes it backward. *See also* **cold front**.
water cycle the cyclic process during which water leaves Earth's surface by evaporation and transpiration to enter the atmosphere. It returns to the surface as precipitation.
watershed the area drained by a stream and its tributaries.
water table the upper boundary of saturated rock or soil beneath Earth's surface.
weathering the processes in nature by which rock materials are broken down into smaller pieces to form sand, soil, and so on. *See also* **chemical weathering; physical weathering**.
wind belts zones around Earth in which the winds blow in the same general direction.

Regents Examinations, Answers, and Self-Analysis Charts

Examination August 2017

Physical Setting/Earth Science

PART A

Answer all questions in this part.

Directions (1–35): For *each* statement or question, choose the word or expression that, of those given, best completes the statement or answers the question. Some questions may require the use of the *2011 Edition Reference Tables for Physical Setting/Earth Science*. Record your answers in the space provided.

1 In which sequence are the celestial objects correctly listed in order from the smallest mass to the largest mass?

(1) Saturn, solar system, Milky Way, universe
(2) Saturn, universe, Milky Way, solar system
(3) Milky Way, Saturn, solar system, universe
(4) Milky Way, universe, solar system, Saturn 1 ____

2 The red shift of light from distant galaxies provides evidence that these galaxies are

(1) decreasing in size
(2) increasing in size
(3) decreasing in distance from Earth
(4) increasing in distance from Earth 2 ____

3 The best evidence of Earth's rotation is provided by the

(1) Foucault pendulum and global warming
(2) Foucault pendulum and Coriolis effect
(3) Moon phases and global warming
(4) Moon phases and Coriolis effect 3 ____

4 Which sphere of Earth covers approximately 70% of Earth's surface?

(1) atmosphere
(2) lithosphere
(3) hydrosphere
(4) asthenosphere

4 ____

5 Some of the constellations that are visible to New York State observers at midnight in December are different from the constellations that are visible at midnight in June because

(1) constellations rotate on an axis
(2) constellations revolve around Earth
(3) Earth rotates on its axis
(4) Earth revolves around the Sun

5 ____

6 The diagram below represents Earth and the Moon as viewed from above the North Pole. Points *A*, *B*, *C*, and *D* are locations on Earth's surface.

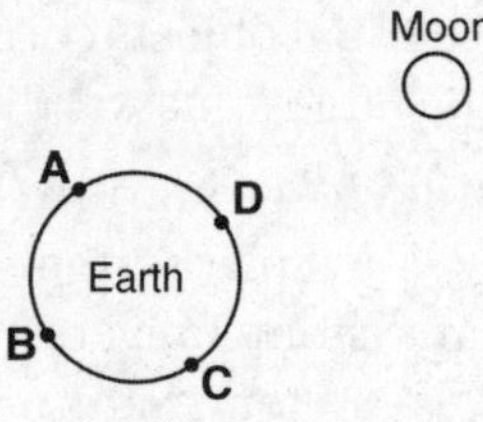

(Not drawn to scale)

According to the diagram, where will high ocean tides and low ocean tides most likely be located?

(1) high tides at *A* and *B*; low tides at *C* and *D*
(2) high tides at *B* and *D*; low tides at *A* and *C*
(3) high tides at *A* and *C*; low tides at *B* and *D*
(4) high tides at *C* and *D*; low tides at *A* and *B*

6 ____

7 Urbanization affects the amount of vegetation and runoff in an area by

(1) decreasing vegetation and decreasing runoff
(2) decreasing vegetation and increasing runoff
(3) increasing vegetation and decreasing runoff
(4) increasing vegetation and increasing runoff

7 ____

8 A severe thunderstorm warning was issued on a warm summer afternoon. Which present weather symbol represents the dangerous solid form of precipitation that is commonly associated with some of these severe thunderstorms?

(1) (2) (3) (4) 8 ____

9 The station model below represents the weather conditions for a location in New York State.

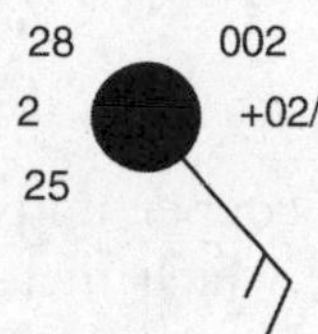

The barometric trend for the past three hours at this location indicates a steady increase of

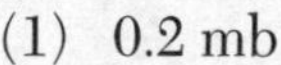

(1) 0.2 mb
(2) 2.0 mb
(3) 0.002 mb
(4) 0.02 mb 9 ____

10 Which New York State location is most often affected by lake-effect snow storms caused by winds blowing over Lake Ontario?

(1) Jamestown
(2) Plattsburgh
(3) Oswego
(4) Riverhead 10 ____

11 Which type of air mass would most likely form over the Pacific Ocean north of the Aleutian Trench?

(1) mP
(2) mT
(3) cP
(4) cT 11 ____

12 Mount Kilimanjaro is located in eastern Africa at 3° S. Which climate factor best explains the presence of permanent snow on its peak?

(1) latitude
(2) elevation
(3) prevailing winds
(4) ocean currents 12 ____

13 In which portion of the electromagnetic spectrum is the maximum intensity of Earth's outgoing radiation?

(1) visible light
(2) gamma rays
(3) infrared
(4) ultraviolet

13 ____

14 A solar water heater contains fluid-filled tubing that absorbs sunlight energy on its outside surface. Which tubing exterior will best absorb insolation?

(1) dark-colored and rough
(2) dark-colored and smooth
(3) light-colored and rough
(4) light-colored and smooth

14 ____

15 Equal masses of granite, iron, copper, and lead are placed in sunlight. Based on specific heat, which material will warm up the fastest?

(1) granite
(2) iron
(3) copper
(4) lead

15 ____

16 During explosive volcanic eruptions, large amounts of ash entering Earth's atmosphere often rise to an altitude of 20 kilometers. What is the most likely effect of this ash cloud?

(1) a decrease in the insolation reaching Earth's surface
(2) a decrease in the thickness of Earth's stratosphere layer
(3) an increase in the insolation reaching Earth's surface
(4) an increase in the thickness of Earth's stratosphere layer

16 ____

17 The map below shows some tectonic plate boundaries near South America and Africa. Letters *A*, *B*, *C*, and *D* represent locations on Earth's surface.

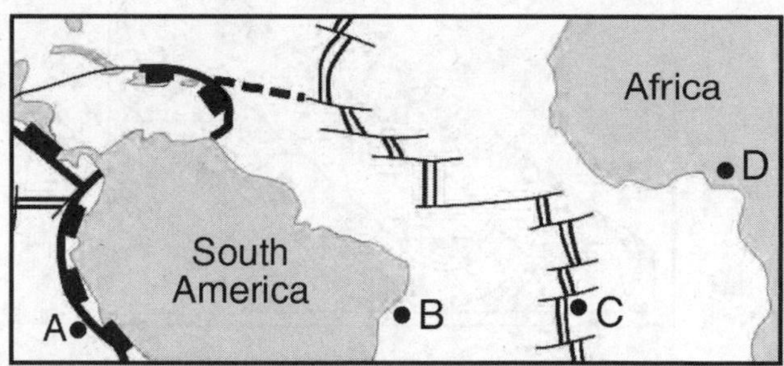

Which location most likely has the youngest bedrock?

(1) *A* (3) *C*
(2) *B* (4) *D* 17____

18 Compared to the average density and composition of oceanic crust, continental crust is

(1) less dense and more felsic
(2) less dense and more mafic
(3) more dense and more felsic
(4) more dense and more mafic 18____

19 The Hawaiian Islands were formed as a result of

(1) lava flowing over Earth's surface where two tectonic plates move apart
(2) an oceanic plate moving over a mantle hot spot
(3) two oceanic plates colliding to form an island arc
(4) tectonic plates sliding past each other 19____

20 The block diagram below represents the underlying bedrock structure of a landscape region.

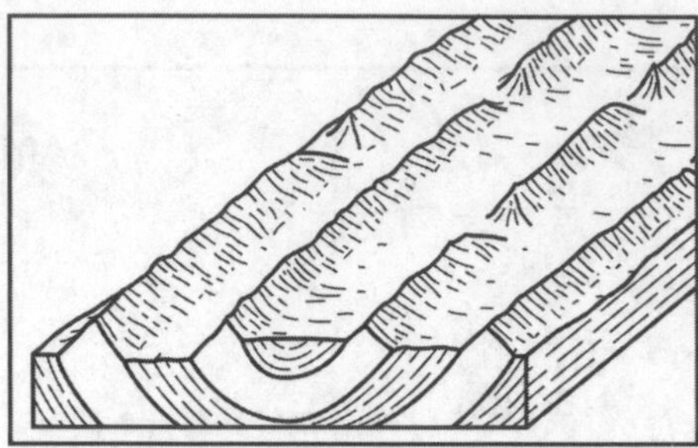

Which diagram represents the most likely stream drainage pattern on the surface of this landscape?

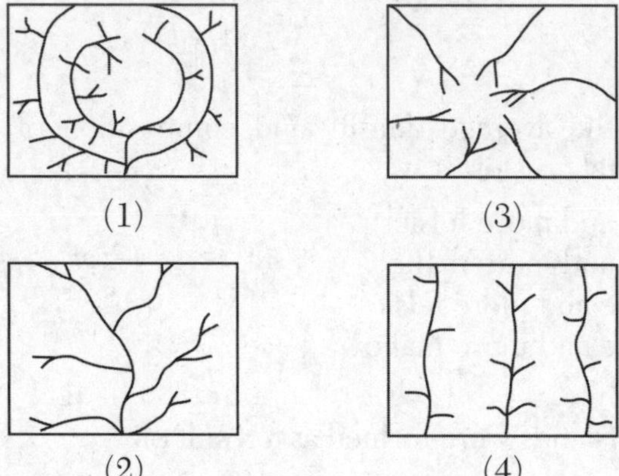

20 _____

21 Which characteristics identify mountain landscape regions?

(1) steep slopes with deformed bedrock
(2) steep slopes with horizontal bedrock
(3) gentle slopes with deformed bedrock
(4) gentle slopes with horizontal bedrock

21 _____

22 Which agent of erosion causes the sandblasting of bedrock?

(1) glaciers
(2) wind
(3) running water
(4) wave action

22 _____

23 The photograph below shows wire netting installed over a steep rock outcrop.

This wire netting has been installed to prevent loss of property or life resulting from

(1) crosscutting and downwarping
(2) folding and faulting
(3) weathering and erosion
(4) high winds and flooding 23 ____

24 Which two minerals are commercial sources of iron?

(1) galena and graphite
(2) muscovite mica and biotite mica
(3) garnet and fluorite
(4) hematite and magnetite 24 ____

25 Which mineral can be found in the rocks phyllite, sandstone, and granite?

(1) quartz
(2) pyroxene
(3) gypsum
(4) calcite 25 ____

26 Which diagram best represents Earth's axis position relative to Earth's orbital plane?

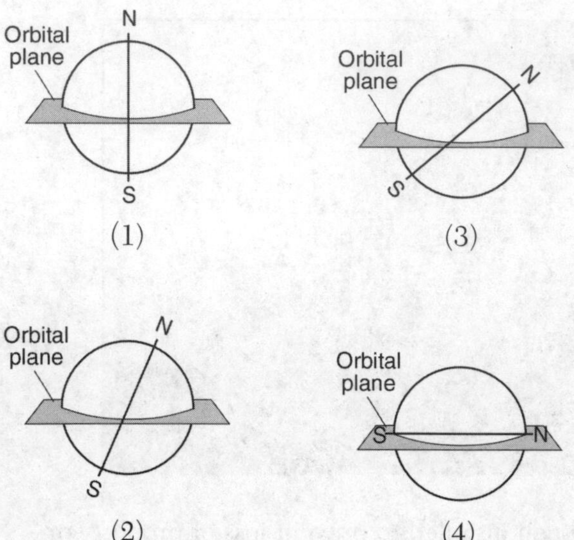

26 _____

27 In which cross section do the arrows represent the most probable direction of air movement over land and water surfaces at an ocean coast on a hot, sunny, summer afternoon?

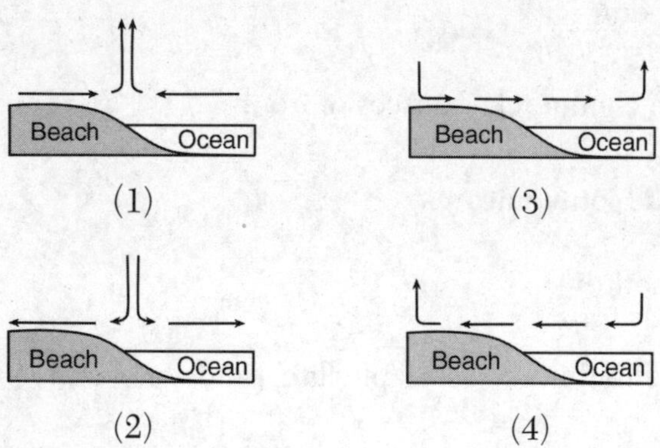

27 _____

28 The table below shows the average diameter and falling velocity of five types of precipitation. The data for moderate rain have been left blank.

Type of Precipitation	Average Diameter (millimeters)	Falling Velocity (meters/second)
drizzle	0.96	4.1
light rain	1.24	4.8
moderate rain		
heavy rain	2.05	6.7
excessive rain	2.40	7.3

What are the probable values for the average diameter and falling velocity of the raindrops found in moderate rain?

(1) average diameter = 1.20 mm
falling velocity = 4.6 m/sec
(2) average diameter = 1.20 mm
falling velocity = 5.7 m/sec
(3) average diameter = 1.60 mm
falling velocity = 4.6 m/sec
(4) average diameter = 1.60 mm
falling velocity = 5.7 m/sec

28 ____

29 Which New York State index fossil is classified as a coral?

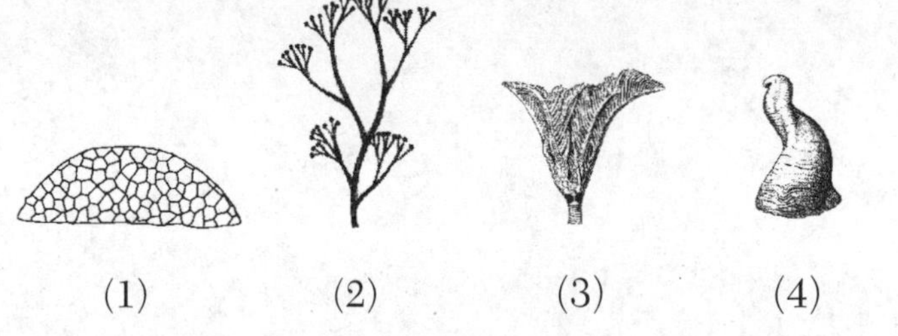

29 ____

30 The map below shows the weak trade winds and strong equatorial countercurrent in the Pacific Ocean during El Niño conditions. This causes warm surface ocean water to migrate eastward, lowering the atmospheric pressure above this warm water.

El Niño Conditions

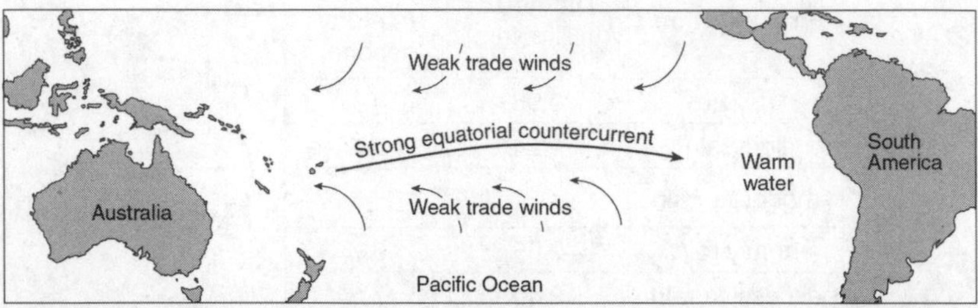

What are the most likely changes to atmospheric temperature and precipitation along the west coast of South America during El Niño conditions?

(1) lower temperatures and lower amounts of precipitation
(2) lower temperatures and higher amounts of precipitation
(3) higher temperatures and lower amounts of precipitation
(4) higher temperatures and higher amounts of precipitation 30 _____

31 The cross sections below represent two bedrock outcrops 15 kilometers apart. Numbers 1 through 9 indicate rock layers. Some layers contain index fossils. No overturning of rock layers has occurred.

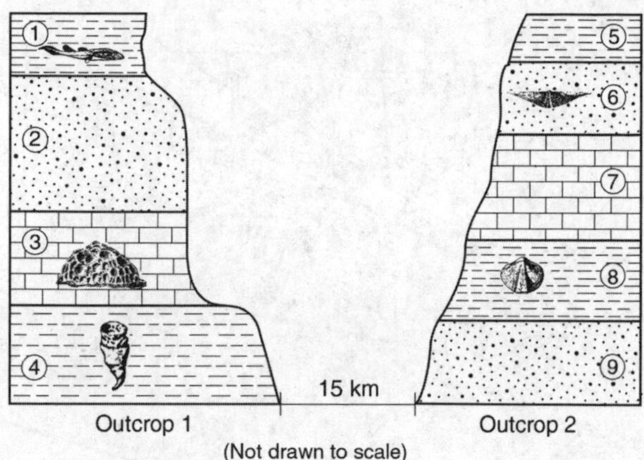

Which layers most likely formed during the same geologic time period?

(1) 1 and 8
(2) 2 and 9
(3) 3 and 7
(4) 4 and 5

31 ____

32 Labeled lines on the map below show the inferred location of Earth's equator during the middle of several geologic periods.

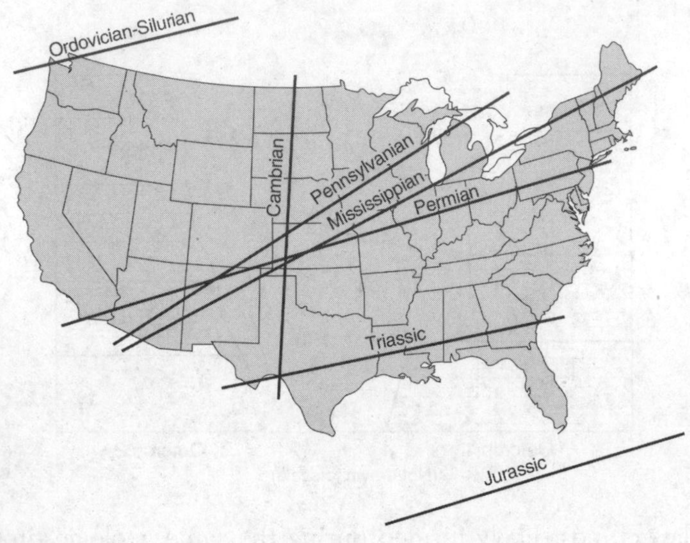

Approximately how many million years ago (mya) was the region around current-day Watertown, New York, located the nearest to the equator?

(1) 270 mya
(2) 300 mya
(3) 340 mya
(4) 450 mya

32 _____

33 Which table correctly matches rock textures with a sedimentary rock that exhibits each texture?

Rock Texture	Sedimentary Rock
clastic	sandstone
crystalline	breccia
bioclastic	rock gypsum

(1)

Rock Texture	Sedimentary Rock
clastic	shale
crystalline	dolostone
bioclastic	sandstone

(3)

Rock Texture	Sedimentary Rock
clastic	dolostone
crystalline	limestone
bioclastic	siltstone

(2)

Rock Texture	Sedimentary Rock
clastic	conglomerate
crystalline	rock salt
bioclastic	limestone

(4)

33 _____

34 The data table below shows the dry-bulb and wet-bulb temperatures measured with a psychrometer on four different days at the same location.

Temperatures Measured with a Psychrometer

Day	1	2	3	4
Dry-bulb temperature (°C)	0	5	10	15
Wet-bulb temperature (°C)	−5	0	5	10

According to the data shown in the table, which day had the highest relative humidity?

(1) 1 (3) 3
(2) 2 (4) 4 34 ____

35 The cross section below represents a lithospheric plate boundary.

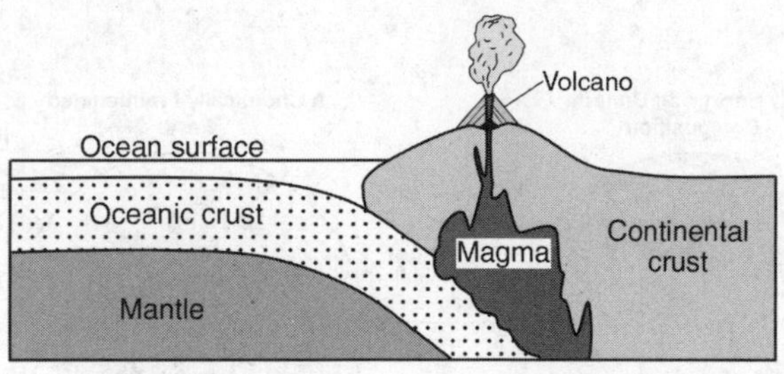

(Not drawn to scale)

In which diagram do the arrows show the relative directions of plate movement at this type of plate boundary?

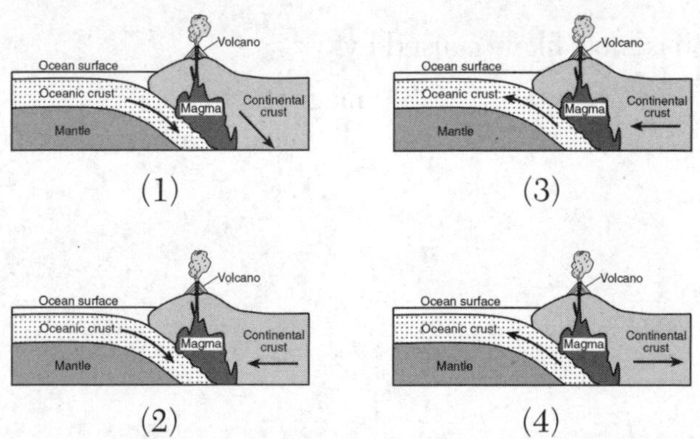

35 ____

PART B–1

Answer all questions in this part.

Directions (36–50): For *each* statement or question, choose the word or expression that, of those given, best completes the statement or answers the question. Some questions may require the use of the *2011 Edition Reference Tables for Physical Setting/Earth Science*. Record your answers in the space provided.

Base your answers to questions 36 through 38 on the information and diagram below and on your knowledge of Earth science. The diagram represents a simplified model of the early formation of Earth's interior.

> Early in its formation, Earth was a molten mass of evenly mixed composition. During the next few million years, the heavier and more dense elements sank to the center, while lighter and less dense elements rose toward the surface. This is called chemical fractionation.

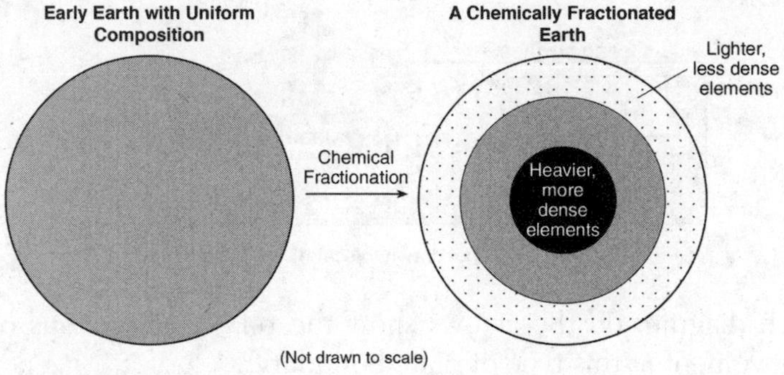

36 Chemical fractionation is most likely caused by

(1) solidification
(2) gravity
(3) magnetic force
(4) chemical weathering 36 ____

37 Approximately how many years ago did Earth and other planets in our solar system begin the process of chemical fractionation?

(1) 8.2 billion years ago
(2) 13.8 billion years ago
(3) 542 million years ago
(4) 4600 million years ago

37 ____

38 Which pair of elements sank to Earth's center during chemical fractionation?

(1) aluminum and silicon
(2) carbon and sulfur
(3) iron and nickel
(4) oxygen and potassium

38 ____

Base your answers to questions 39 through 41 on the diagram below and on your knowledge of Earth science. The diagram represents the apparent path of the Sun through the sky as viewed by an observer in the Northern Hemisphere. Points *A*, *B*, *C*, and *D* represent four positions of the Sun.

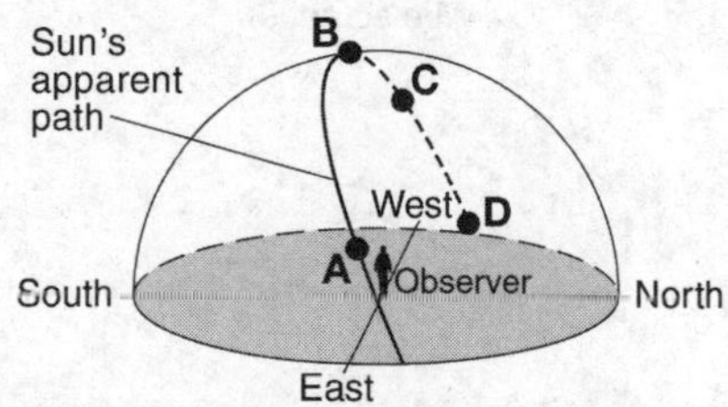

39 This apparent path of the Sun through the sky is caused by

(1) Earth's revolution around the Sun
(2) Earth's rotation on its axis
(3) the Sun's revolution around Earth
(4) the Sun's rotation on its axis

39 ____

40 The observer has the longest shadow when the Sun is at position

(1) *A*
(2) *B*
(3) *C*
(4) *D*

40 ____

41 What is the approximate time of day when the Sun is at position *C*?

(1) 6 a.m.
(2) 9 a.m.
(3) 3 p.m.
(4) 6 p.m.

41 ____

Base your answers to questions 42 and 43 on the diagram below and on your knowledge of Earth science. The diagram represents three tubes, *A*, *B*, and *C*, each containing an equal volume of uniform-sized spherical beads. The bottom of each tube is covered with a wire screen. *XY* is a reference line.

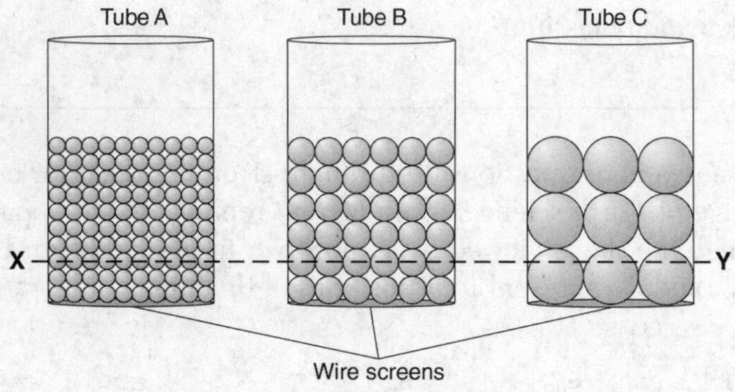

42 Which bar graph would best represent the rate of water infiltration through tubes *A*, *B*, and *C*?

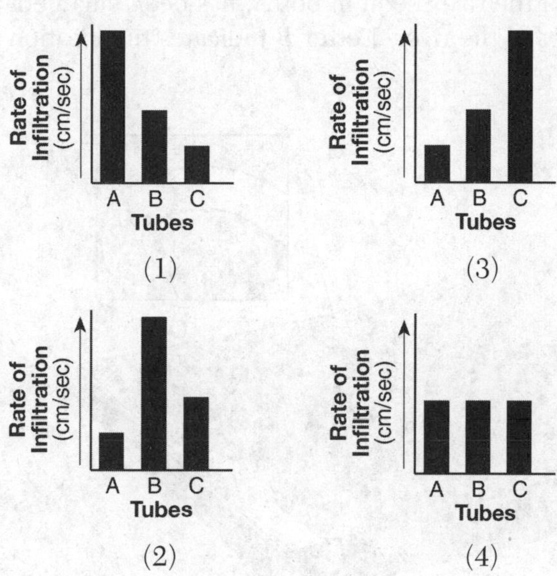

42 _____

43 These tubes are placed in water up to the level of line *XY* to demonstrate capillarity. After one hour, the height of the water above line *XY* will be

(1) highest in tube *A*
(2) highest in tube *B*
(3) highest in tube *C*
(4) the same height in all three tubes

43 _____

Base your answers to questions 44 through 46 on the block diagram below and on your knowledge of Earth science. The block diagram represents a river drainage system. A portion of the river, seen in box *A*, has been enlarged. Points *X* and *Y* are on opposite sides of the river. Letter *B* indicates the location where the river enters the ocean.

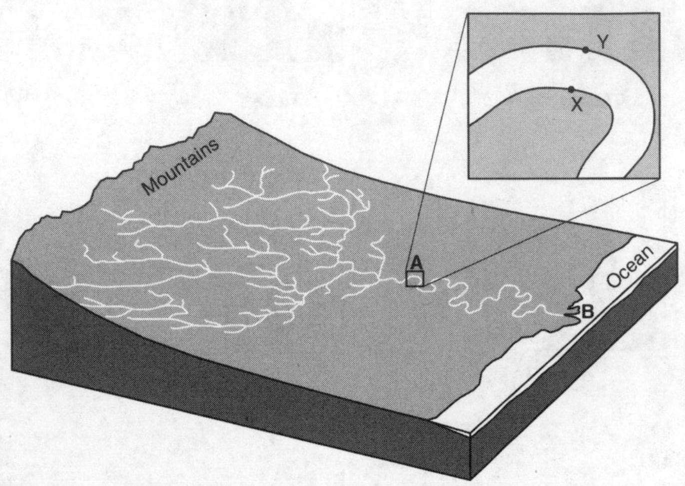

44 The area of land drained by this river and its tributaries is best described as the river's

(1) topography
(2) watershed
(3) water table
(4) floodplain

44 ____

45 Which cross section best represents the profile of the bottom of the river between points *X* and *Y* at location *A*?

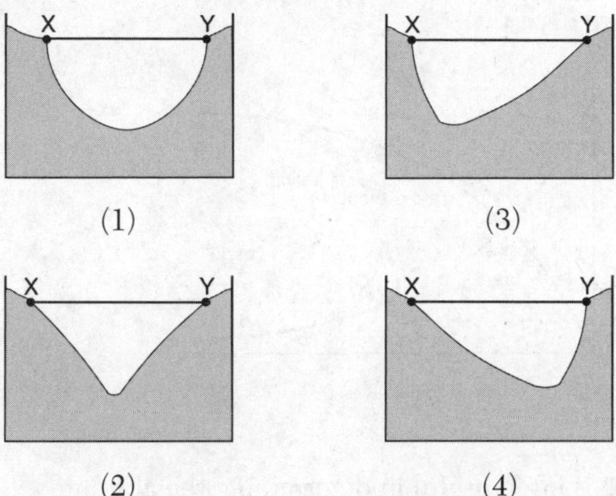

45 ____

46 Which cross section represents the most likely pattern of sediments deposited where the river enters the ocean at location *B*?

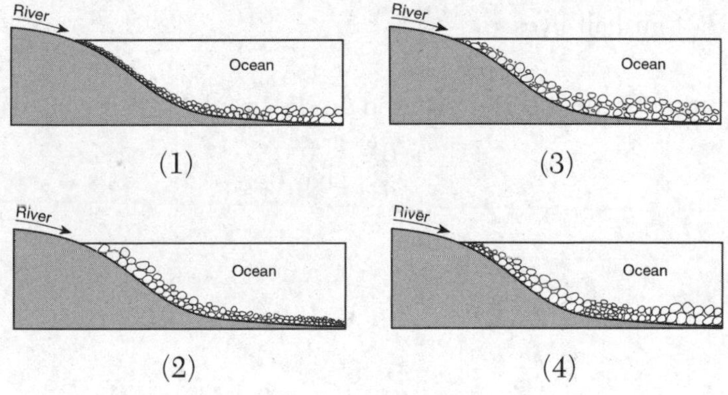

46 ____

Base your answers to questions 47 through 50 on the graph below and on your knowledge of Earth science. The graph shows the percentages of the radioactive isotope carbon-14 (^{14}C) and its disintegration product produced during four half-lives of radioactive decay.

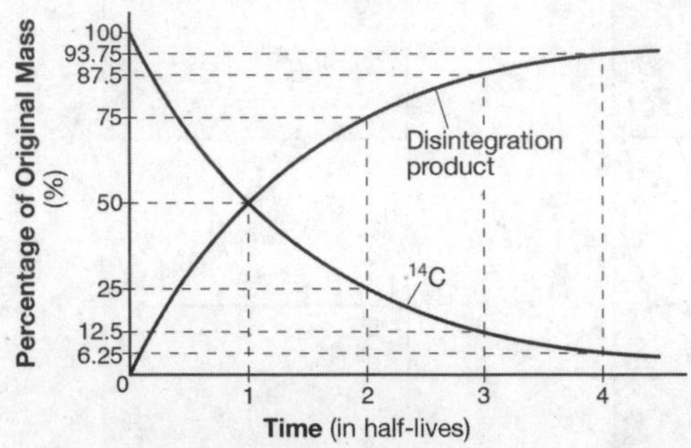

47 Radioactive carbon-14 is often useful in determining the absolute age of geologic samples because radioactive isotopes

(1) decay at a regular rate
(2) become less stable during decay
(3) remain unchanged over time
(4) stabilize after four half-lives

47 _____

48 Which disintegration product is represented on the graph?

(1) ^{206}Pb
(2) ^{87}Sr
(3) ^{40}Ar
(4) ^{14}N

48 _____

49 How many half-lives have passed if a sample contains 25% of its original carbon-14?

(1) 1 half-life
(2) 2 half-lives
(3) 3 half-lives
(4) 4 half-lives

49 ____

50 The age of which index fossil could be determined by using carbon-14?

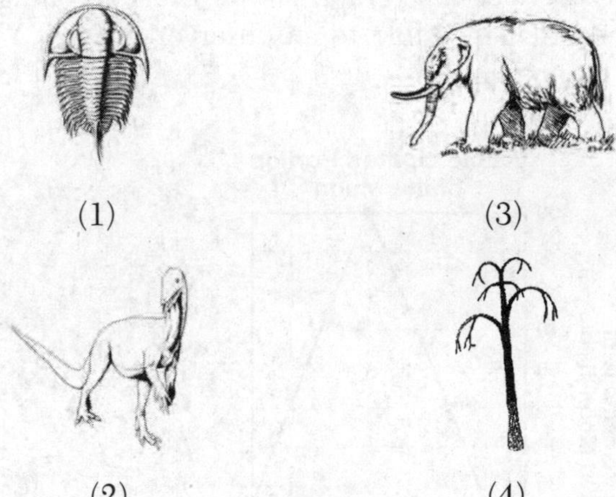

(1) (3)

(2) (4)

50 ____

PART B–2

Answer all questions in this part.

Directions (51–65): Record your answers in the spaces provided. Some questions may require the use of the *2011 Edition Reference Tables for Physical Setting/Earth Science*.

Base your answers to questions 51 through 53 on the graph below and on your knowledge of Earth science. The graph shows the percentage of the lighted portion of the Moon that is visible to an observer in New York State through eight consecutive Moon phases.

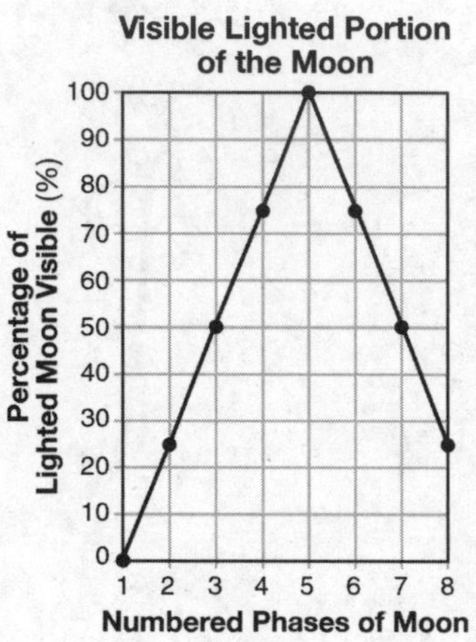

51 The phases of the Moon are said to be waxing when the lighted portion of the Moon gradually increases over time. Identify the numbered phase of the Moon when waxing begins and the numbered phase when waxing ends. [1]

Waxing begins: Phase _______________

Waxing ends: Phase _______________

52 On the diagram *below*, place an **X** on the Moon's orbit to represent the Moon's position at phase 5. [1]

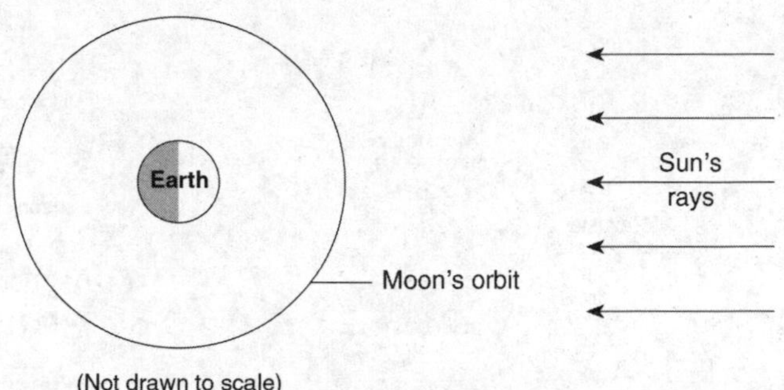

53 The diagram below represents the appearance of the Moon at phase 3 as seen by an observer in New York State.

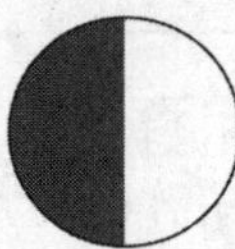

In the circle below, shade the part of the Moon that appears dark to an observer in New York State when the Moon is at phase 7. [1]

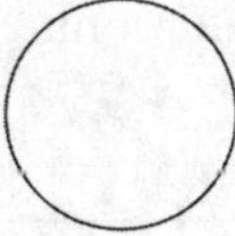

Base your answers to questions 54 through 57 on the flowchart below and on your knowledge of Earth science. The flowchart represents possible pathways in the evolution of stars.

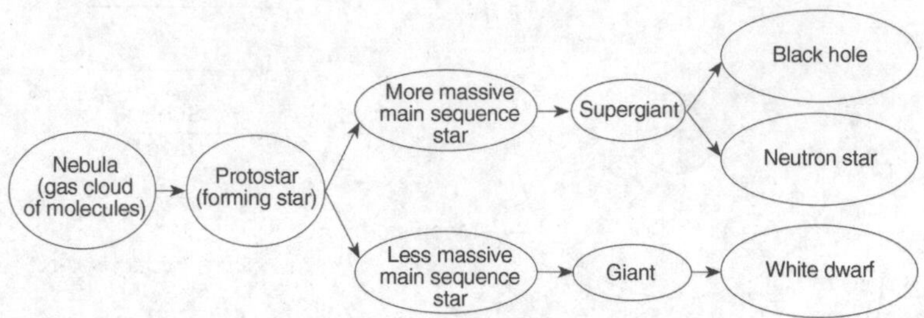

54 Based on this flowchart, identify the characteristic of a main sequence star that determines whether the star becomes a giant or a supergiant. [1]

____

55 State the name of *one* star labeled on the *Characteristics of Stars* graph in the *Earth Science Reference Tables* that may become either a black hole or neutron star. [1]

____

56 Identify the nuclear process that occurs when lighter elements in a star combine to form heavier elements, producing the star's radiant energy. [1]

____

57 Circle the relative surface temperature and relative luminosity of the main sequence star *Sirius* compared with the surface temperature and luminosity of the Sun. [1]

Relative surface temperature of *Sirius* (circle one):

lower **higher** **the same**

Relative luminosity of *Sirius* (circle one):

less **greater** **the same**

Base your answers to questions 58 through 61 on the topographic map below and on your knowledge of Earth science. Points *A*, *B*, *C*, and *D* represent locations on Earth's surface. Elevations are measured in meters.

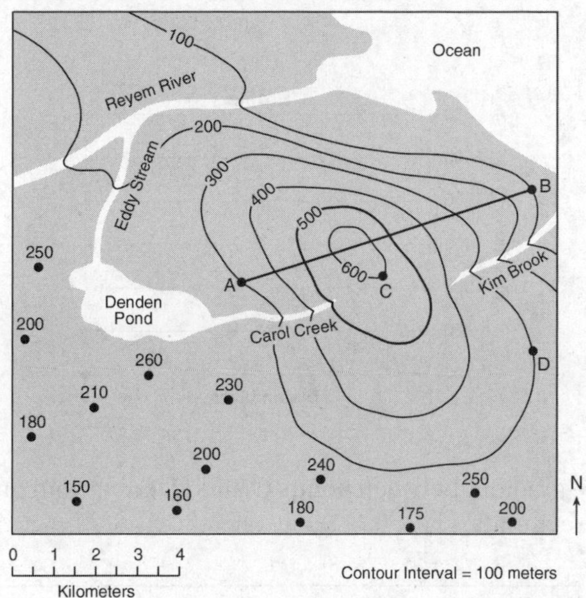

58 On the map *above*, draw the 200-meter contour line in the southern portion of the map. Extend the contour line to the edges of the map. [1]

59 On the grid *below*, construct a topographic profile along line *AB* by plotting the elevation of *each* contour line that crosses line *AB*. The elevations of points *A* and *B* have been plotted on the grid. Connect *all ten* plots with a line from *A* to *B* to complete the profile. [1]

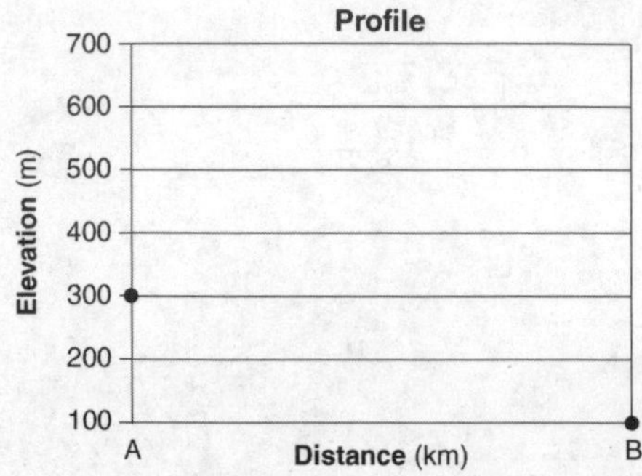

60 Calculate the gradient between points *C* and *D*. Label your answer with the correct units. [1]

____

61 Identify the compass direction toward which Kim Brook flows. Describe the evidence shown on the map that indicates the water flows downhill in that compass direction. [1]

Compass direction: ____________________________

Evidence: __________________________________

____

Base your answers to questions 62 through 65 on the information below and on your knowledge of Earth science.

Adirondack Earthquake

On October 7, 1983, a magnitude 5.3 earthquake occurred in New York State's Adirondack region. The earthquake's epicenter was at Blue Mountain Lake, which is located approximately 32 miles (50 kilometers) southwest of Mt. Marcy.

62 Circle the New York State location (Old Forge *or* New York City) that recorded the greater amount of time between the arrival of the first *P*-wave and the arrival of the first *S*-wave from the Blue Mountain Lake earthquake. Explain why this location had the greater difference between the *P*-wave and *S*-wave arrival times. [1]

Circle one: **Old Forge** **New York City**

Explanation: ____

____

63 Determine how long it took the first *P*-wave to travel from Blue Mountain Lake to a seismic station 1200 kilometers away. [1]

______________ **min** ______________ **s**

64 Identify *one* type of metamorphic surface bedrock where this earthquake epicenter was located. [1]

____

65 Perth, Australia, is located almost directly on the opposite side of Earth from the epicenter of this earthquake. A seismograph in Perth received *P*-waves but *not* *S*-waves from this earthquake. Identify the interior layer of Earth and the characteristic of this layer that prevented the *S*-waves from arriving at Perth. [1]

Earth's interior layer: ______________________________________

Characteristic of this layer: ______________________________

PART C

Answer all questions in this part.

Directions (66–85): Record your answers in the spaces provided. Some questions may require the use of the *2011 Edition Reference Tables for Physical Setting/Earth Science*.

Base your answers to questions 66 through 68 on the diagram below and on your knowledge of Earth science. The diagram represents Earth on the first day of a season. The equator, several lines of longitude, and the North and South Poles have been labeled. Letters *A* through *D* represent locations on Earth's surface.

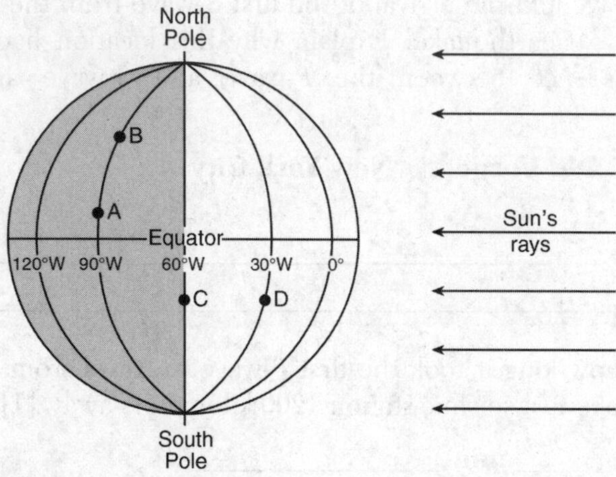

66 Identify *one* possible date that is represented by the position of Earth in this diagram. [1]

____

67 State whether the relative altitude of *Polaris* at location *A* is lower or higher than at location *B*. Explain why this difference is observed. [1]

Relative altitude of *Polaris* at location *A*:

____

Explanation: __________________________

____

68 State the solar time at location *D* if the solar time at location *C* is 6:00 a.m. Indicate a.m. or p.m. in your answer. [1]

____

Base your answers to questions 69 through 72 on the data table below, on the map below, and on your knowledge of Earth science. The data table shows latitude and longitude locations of the center of Hurricane Odile recorded at the same time each day from September 12 to September 18, 2014. The data table also shows the hurricane's barometric pressure in millibars (mb) and wind speed in knots (kt). The location of La Paz, Mexico, is indicated on the map.

Hurricane Odile

Date	**Location**		**Barometric Pressure** (mb)	**Wind Speed** (kt)
	Latitude (° N)	**Longitude** (° W)		
September 12	15	105	993	50
September 13	16	106	983	65
September 14	19	107	918	120
September 15	23	110	941	110
September 16	27	113	987	55
September 17	30	114	995	40
September 18	31	112	1003	25

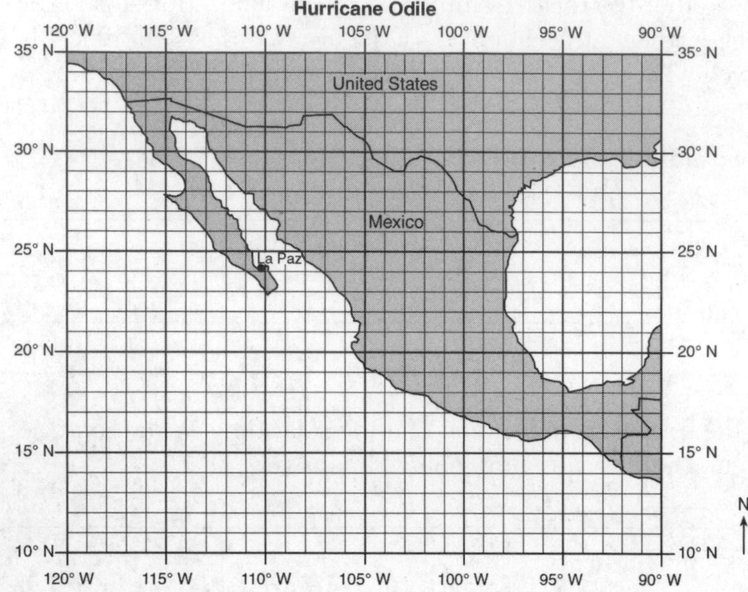

69 On the map *above*, plot the seven locations of Hurricane Odile indicated by the latitudes and longitudes shown in the data table. Connect *all seven* plots with a line. [1]

70 Using the set of axes *below*, draw a line to represent the general relationship between barometric pressure and wind speed associated with Hurricane Odile. [1]

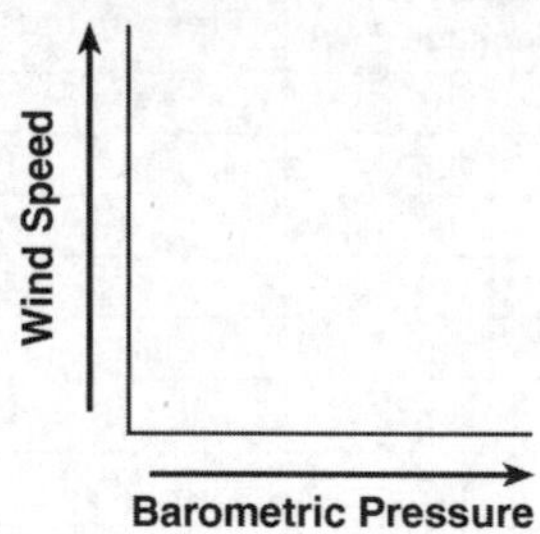

71 Identify *one* weather instrument that was used to measure the wind speed of Hurricane Odile. [1]

____

72 Describe *two* actions that a person living in La Paz, Mexico, could take to prepare for an approaching hurricane. [1]

Action 1: ____________________

Action 2: ____________________

Base your answers to questions 73 through 75 on the cross section below and on your knowledge of Earth science. On the cross section, letters *A* through *F* represent rock units. Line *XX′* indicates an unconformity and line *YY′* indicates a fault. No overturning of rock layers has occurred.

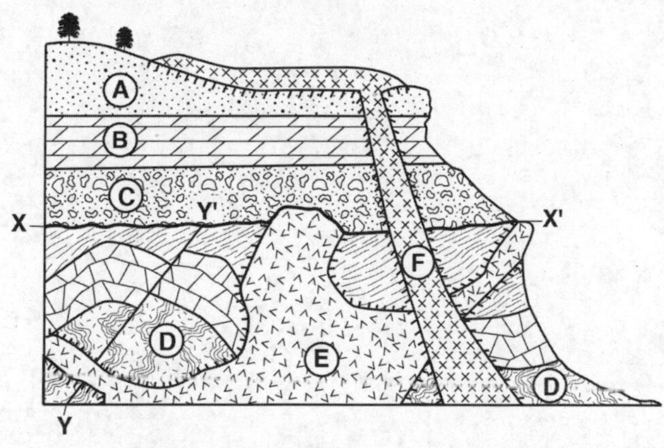

73 Identify the name of the rock formed in the zone of contact metamorphism between rock units *A* and *F*. [1]

____

74 List *two* processes that produced unconformity *XX′*. [1]

Process 1: ____________________

Process 2: ____________________

75 List the letters of rock units *D*, *E*, *F*, and fault *YY′* in the correct sequence from oldest to youngest. [1]

________ → ________ → ________ → ________

Oldest ————————————————→ Youngest

Base your answers to questions 76 through 79 on the passage and cross section below, and on your knowledge of Earth science. The cross section represents the elevation and depth of each Finger Lake in New York State. The gray shading in each lake represents the sediment fill at the bottom of each lake.

Formation of the Finger Lakes

The Finger Lakes originated as a series of south-flowing rivers that existed in what is now central New York State. Around 1.7 to 1.8 million years ago, a continental glacier advanced southward, beginning an ice age that consisted of many advances and retreats of glaciers. The Finger Lakes were carved by several of these advances, which gouged the valleys of the rivers into deep troughs, some of which extend below present-day sea level. As the glaciers advanced, they pushed a great amount of soil and rock ahead of them. During the time when the ice sheets were melting and retreating, glacial moraines were left behind. These deposits dammed the glacial valleys at their southern ends, and the valleys then flooded to form the Finger Lakes.

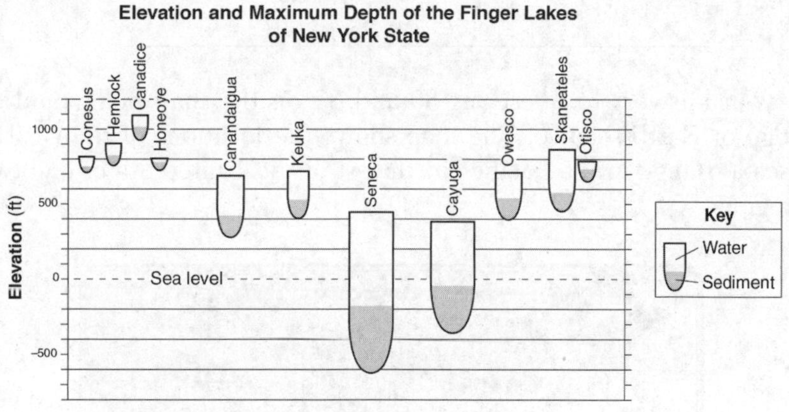

76 Identify *one* Finger Lake that has a maximum water depth of approximately 175 feet. [1]

____________________ **Lake**

77 Identify the geologic time period and epoch when these continental glaciers formed the Finger Lakes of New York State. [1]

________________________________ **Period**

________________________________ **Epoch**

78 Describe *one* way that the arrangement of sediments in the glacial moraines is different from the arrangement of sediments deposited in the earlier river channels. [1]

____

____

79 Describe the cross-sectional shape of the original river valleys before they were gouged by the advancing glaciers. [1]

____

____

Base your answers to questions 80 and 81 on the map below and on your knowledge of Earth science. The map shows the locations of the Coast Range and Cascade Range in the Pacific Northwest of the United States and two cities in this region.

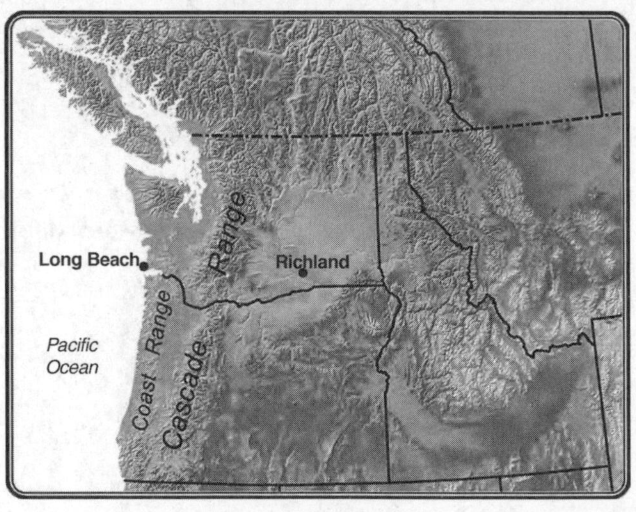

80 Identify the name of the cool surface ocean current that influences the climate of this region. [1]

______________________ **Current**

81 Explain why the difference between average winter and summer temperatures is smaller in Long Beach than in Richland. [1]

___

___

___

Base your answers to questions 82 through 85 on the diagram below and on your knowledge of Earth science. The diagram represents a portion of the scheme for igneous rock identification. Line *AB* represents the percentages of the minerals in igneous rock 1 and line *CD* represents the percentages of the minerals in igneous rock 2.

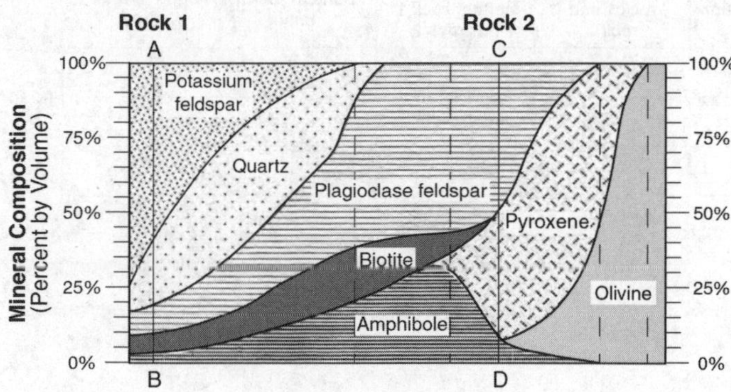

82 Rock 1 is a glassy, vesicular igneous rock. Identify the name of rock 1. [1]

___

83 Compared to the color and density of igneous rock 1, describe how the relative color and density of igneous rock 2 are different. [1]

Relative color of rock 2: ______________________

Relative density of rock 2: ______________________

84 Explain why andesite and diorite can both have the same percentage of mineral composition by volume, and yet be two different igneous rocks. [1]

____

____

85 The table *below* shows the composition of three minerals commonly found in igneous rocks. Complete the table by placing one **X** in each row to indicate if that mineral is found in both rock 1 and rock 2, neither rock 1 nor rock 2, rock 1 only, or rock 2 only. [1]

Mineral Composition	Found in both rock 1 and rock 2	Found in neither rock 1 nor rock 2	Found in rock 1, only	Found in rock 2, only
$(Na,Ca)AlSi_3O_8$				
$KAlSi_3O_8$				
$(Fe,Mg)_2SiO_4$				

Answers August 2017

Physical Setting/Earth Science

Answer Key

PART A

1. 1	**8.** 3	**15.** 4	**22.** 2	**29.** 1
2. 4	**9.** 1	**16.** 1	**23.** 3	**30.** 4
3. 2	**10.** 3	**17.** 3	**24.** 4	**31.** 3
4. 3	**11.** 1	**18.** 1	**25.** 1	**32.** 3
5. 4	**12.** 2	**19.** 2	**26.** 2	**33.** 4
6. 2	**13.** 3	**20.** 4	**27.** 4	**34.** 4
7. 2	**14.** 1	**21.** 1	**28.** 4	**35.** 2

PART B–1

36. 2	**39.** 2	**42.** 3	**45.** 4	**48.** 4
37. 4	**40.** 4	**43.** 1	**46.** 2	**49.** 2
38. 3	**41.** 3	**44.** 2	**47.** 1	**50.** 3

PART B–2 and **PART C**. *See* **Answers Explained**.

Answers Explained

PART A

1. **1** A solar system consists of a single star, and all of the objects (such as planets) that orbit it. All of these objects have mass. Thus, a solar system contains more mass than a planet, such as Saturn. The Milky Way is a galaxy, which is a group of hundreds of billions of stars, many of which have solar systems. Thus, the Milky Way contains more mass than a solar system. The universe is everything that we can perceive and includes all of the stars, gas clouds, galaxies, and other objects that we can detect by the radiation they emit. The universe contains billions of galaxies. Thus, the universe has more mass than a galaxy. Thus, the correct sequence of celestial objects listed in order from smallest mass to largest mass is Saturn, solar system, Milky Way, universe.

2. **4** Find the chart labeled Electromagnetic Spectrum in the *Reference Tables for Physical Setting/Earth Science*. Note that visible light at the red end of the spectrum has a longer wavelength than light at the blue end of the spectrum. If a source of electromagnetic waves is moving *away* from an observer at the same time as it is emitting light of a particular wavelength, fewer wave crests will reach the eye of the observer each second. The eye will interpret this as meaning that the light has a longer wavelength than it actually has. In other words, the light will appear shifted toward the red end of the spectrum. Therefore, the fact that the light from most galaxies shows a red shift is evidence that most galaxies are moving away from Earth. If all galaxies are moving away from Earth, these galaxies are increasing in distance from Earth.

WRONG CHOICES EXPLAINED:

(1) and (2) Galaxies of many different sizes show a red shift because a red shift is the result of the motion of a light source relative to an observer, not the size of the source.

(3) If a galaxy's distance from Earth is decreasing, the galaxy is moving toward Earth. If a galaxy is moving toward an observer on Earth at the same time as it is emitting light of a particular wavelength, more wave crests will reach the eye of the observer each second. The eye will interpret this as meaning that the light has a shorter wavelength than it actually has. In other words, the light will appear shifted toward the blue end of the spectrum. Thus, if distant galaxies are decreasing in distance from Earth, their light would appear blue shifted, not red shifted.

3. **2** In 1851, the French physicist Jean Foucault suspended a heavy iron ball on a long steel wire from the top of the dome of the Pantheon in Paris. As this pendulum swung back and forth, it appeared to change direction slowly, passing over different lines on the floor until eventually coming full circle to its original position after 24 hours. Since Foucault knew that the path of a freely swinging pendulum would not change on its own, he concluded that the apparent shift in the direction of swing of the pendulum was due to the floor (Earth's surface) rotating beneath the pendulum. Thus, the apparent change in direction of swing of a Foucault pendulum provides evidence of Earth's rotation.

Planetary winds tend to blow in a straight line from regions of high pressure toward regions of low pressure. As these winds are blowing, Earth is turning on its axis. This rotation causes the winds to appear to be turning toward the right in the Northern Hemisphere and toward the left in the Southern Hemisphere. This phenomenon is called the Coriolis effect. Since there would be no Coriolis effect if Earth did not rotate, the Coriolis effect is also considered evidence of Earth's rotation.

Thus, the best evidence of Earth's rotation is provided by the Foucault pendulum and Coriolis effect.

WRONG CHOICES EXPLAINED:

(1) The Foucault pendulum does provide evidence of Earth's rotation. However, global warming is an increase in average global temperatures, not evidence that Earth is rotating.

(3) and (4) Moon phases refer to changes in the illuminated portion of the Moon visible to an observer on Earth. Moon phases occur because the Moon revolves into different positions relative to Earth and the Sun. The Moon would revolve around Earth even if Earth did not rotate. Therefore, Moon phases are not evidence of Earth's rotation.

4. **3** Oceans cover more than 70% of Earth's surface. Earth's hydrosphere consists of all the liquid water that rests on the lithosphere. Thus, the sphere of Earth that covers approximately 70% of Earth's surface is the hydrosphere.

WRONG CHOICES EXPLAINED:

(1) Earth's atmosphere is a thin shell of gases, water, dust, and other particles bound to Earth by gravity. Earth's atmosphere is in contact with all of the lithosphere and hydrosphere that comprise Earth's surface. Thus, the atmosphere covers 100% of Earth's surface.

(2) The lithosphere is Earth's solid, outer layer of rock. Earth's crust corresponds to the upper portion of the lithosphere. Thus, the lithosphere comprises 100% of Earth's solid surface.

(4) Find the Inferred Properties of Earth's Interior chart in the *Reference Tables for Physical Setting/Earth Science.* In the cross section, locate the asthenosphere. Note that the asthenosphere is located beneath Earth's surface. Thus, the asthenosphere covers 0% of Earth's surface.

5. **4** As Earth revolves around the Sun, the side of Earth facing the Sun experiences day and the side facing away from the Sun experiences night. Midnight occurs when a location on Earth is directly opposite the Sun in the center of the night side of Earth. At different times of the year, Earth is at different points along its orbit and the night side of Earth faces different portions of the universe. Since stars are visible only at night, only constellations in the portion of the universe facing Earth's night side are visible from Earth. In December, Earth's night side faces one portion of the universe and its constellations. Six months later, in June, Earth has revolved to a position in its orbit directly opposite the Sun from its December position. The night side of Earth faces a different portion of the universe and its constellations. Therefore, the constellations visible to New York State observers at midnight in December are different than those visible at midnight in June because Earth revolves around the Sun.

WRONG CHOICES EXPLAINED:

(1) A constellation is an *imaginary* pattern formed by a group of stars in an area of the sky as seen by an observer on Earth. Therefore, constellations do not have an axis about which they rotate.

(2) Constellations consist of stars. Earth's small mass does not exert enough gravitational force to hold larger, more massive objects such as stars in orbit.

(3) As Earth spins on its axis, the boundary between daylight and darkness does not spin along with it; Earth's night side is always located directly opposite the Sun. Only Earth's motion in relation to the Sun (revolution) affects the direction in which the night side faces and the constellations visible to an observer, not Earth's rotation on its axis.

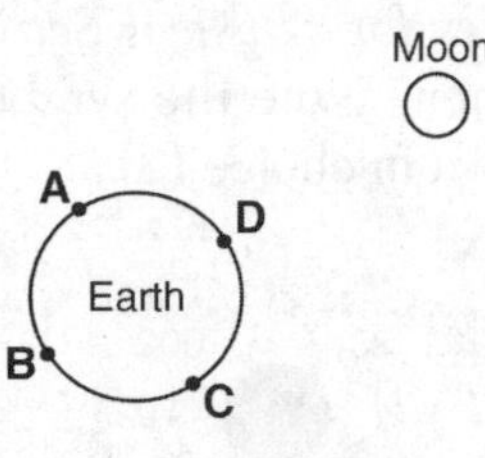

(Not drawn to scale)

6. **2** The gravitational attraction of the Moon creates tidal bulges in Earth's oceans on the side of Earth directly facing the Moon and on the side of Earth directly facing away from the Moon. As Earth rotates on its axis, the positions of the tidal bulges remain fixed in line with the Moon. Locations experience high tide when they align with the bulges and experience low tide when between bulges. According to the diagram, location *D* is on the side of Earth directly facing the Moon and location *B* is on the side of Earth directly facing away from the Moon. Therefore, tidal bulges and their associated high tides will likely be located at *B* and *D*. Locations *A* and *C* are between the bulges and will experience low tides. Thus, high tides will be located at *B* and *D*; low tides at *A* and *C*.

7. **2** Urbanization is the process by which towns and cities are formed and become larger as more and more people begin living and working in central areas. As towns and cities grow, more and more of the land surface is paved over or covered by buildings and other structures. Vegetation is not able to grow on land covered by pavement and buildings. Therefore, urbanization results in a decrease in vegetation. Pavement and the roofs of buildings are impermeable. Therefore, rainwater that falls onto pavement and buildings runs off rather than infiltrates. Thus, urbanization decreases vegetation and increases runoff.

8. **3** A dangerous form of solid precipitation associated with severe thunderstorms is hail. In the upper portion of the troposphere, temperatures below −40°C are the norm year-round. For this reason, thunderstorms whose cloud tops extend into the upper troposphere may produce hail. Hailstones form when water droplets carried into the upper troposphere freeze to form tiny balls of ice (hailstones) and then fall. Again and again, hailstones are hurled up by updrafts and then fall through layers of air that alternate above and below freezing. Each cycle adds a layer to the hailstone. The more violent the updrafts, the larger and heavier the hailstones can become before falling. Large hailstones are dangerous because they can cause great damage when they strike crops, buildings, and vehicles. In some cases, hailstones have resulted in loss of life. Find the Key to Weather Map

Symbols in the *Reference Tables for Physical Setting/Earth Science.* Locate the section labeled "Present Weather." Note the symbol representing hail, Δ, which corresponds to the symbol shown in choice (3).

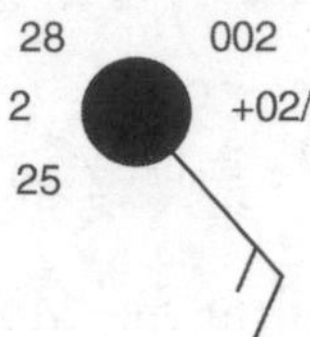

9. **1** Find the Key to Weather Map Symbols in the *Reference Tables for Physical Setting/Earth Science,* and locate the section labeled "Station Model Explanation." Note that barometric trend over the past three hours is shown to the center right of a station model. According to the station model explanation, a "+" indicates an increase in barometric pressure. The two-digit value indicates the number of millibars that the barometric pressure has changed but does not include a decimal point. The decimal should be placed between the first and second digits. Finally, the "Station Model Explanation" indicates that an upward-sloping line after the millibar value means a steady rise. The value representing the barometric trend for the past three hours at the location in New York State shown on the station model in the question is "+02/." Therefore, the barometric trend for the past three hours at this location indicates a steady increase of 0.2 mb.

10. **3** Winter lake-effect snow storms occur when cold winds move across large stretches of warm lake water. Evaporation of the warm lake water adds water vapor to the air. The water vapor is picked up by the cold wind, freezes, and is deposited as snow onto the cold land of the leeward shores of the lake. Find the Generalized Bedrock Geology of New York State map in the *Reference Tables for Physical Setting/Earth Science.* Locate Lake Ontario, and note that it is located at about 43°–44° N latitude. Find the Planetary Wind and Moisture Belts in the Troposphere diagram in the *Reference Tables for Physical Setting/Earth Science.* Note that at the latitudes corresponding to Lake Ontario, the planetary winds generally blow from west to east. Thus, the cold winds are moving from west to east over the lake. So the eastern shores of Lake Ontario are the leeward shores that would experience lake-effect snow storms. On the Generalized Bedrock Geology of New York State map in the *Reference Tables for Physical Setting/Earth Science*, locate the eastern shores of Lake Ontario. Note that of the cities listed as answer choices, only Oswego is located on the eastern shores of Lake Ontario. Thus, the New York State location most often affected by lake-effect snow storms caused by winds blowing over Lake Ontario is Oswego.

11. **1** The characteristics of an air mass are the result of the geographical region over which it formed, or its *source region*. Air resting on or moving very slowly over a region tends to take on the characteristics of that region. In general, air masses that form near the poles are cold and are called polar air masses; air masses that form near the equator are warm and are called tropical air masses. Air masses that form over water are moist and are called maritime air masses; air masses that form over land are dry and are called continental air masses.

The type of air mass that would most likely form over the Pacific Ocean is a maritime air mass. Find the Tectonic Plates map in the *Reference Tables for Physical Setting/Earth Science.* Locate the Aleutian Trench. Trace horizontally right to the latitude scale along the vertical edge of the map. Note that the Aleutian Trench is located at about 60° N latitude (in the Arctic Circle). Thus, the type of air mass that would form north of the Aleutian Trench is a polar air mass. Therefore, an air mass that formed over the Pacific Ocean north of the Aleutian Trench would most likely be a maritime polar air mass. Find the Key to Weather Map Symbols in the *Reference Tables for Physical Setting/Earth Science.* In the section labeled "Air Masses," note that the symbol for a maritime polar air mass is mP, as shown in choice (1).

12. **2** Temperature is a major climate factor. Mountains are, by definition, regions of high elevation. Find the Selected Properties of Earth's Atmosphere chart in the *Reference Tables for Physical Setting/Earth Science.* Locate the graph labeled "Temperature (°C)." Note that temperature decreases rapidly with increasing altitude (elevation) in the troposphere. Elevation is a climate factor because there is a decrease in temperature with an increase in elevation. The fact that temperature decreases about 1°C for every 100-meter rise in elevation explains why high mountains may have tropical vegetation at their bases but permanent ice and snow at their peaks. Thus, the climate factor that best explains the presence of permanent snow on the peak of Mount Kilimanjaro is elevation (altitude above sea level).

WRONG CHOICES EXPLAINED:

(1) It is given that the latitude of Mount Kilimanjaro is 3° S, which is near the equator. Near the equator, insolation strikes Earth's surface almost vertically. Therefore, throughout the year, the insolation reaching the tropics is more concentrated (i.e., more intense) than that reaching midlatitude or polar regions. The more intense the insolation, the more Earth's surface is warmed by that insolation. More warming would likely melt snow, not allow the snow to remain permanently. Therefore, it is unlikely that the latitude of Mount Kilimanjaro is the climate factor that best explains the permanent snow on the mountain's peak.

(3) Find the Planetary Winds and Moisture Belts in the Troposphere diagram in the *Reference Tables for Physical Setting/Earth Science*. Note that 3° S latitude lies in the planetary wind belt labeled "S.E. Winds." Find the Surface Ocean Currents map in the *Reference Tables for Physical Setting/Earth Science*. In the key, note the symbols for warm currents and cool currents. Note that the surface ocean currents off the coast of east Africa are all warm ocean currents. Thus, the prevailing winds that reach Mount Kilimanjaro would come from over warm water. Air moving over warm water becomes warm and moist. Warm, moist winds would not explain the cold temperatures necessary to support the permanent snow on the peak of Mount Kilimanjaro.

(4) Find the Surface Ocean Currents map in the *Reference Tables for Physical Setting/Earth Science*. In the key, note the symbols for warm currents and cool currents. Note that the surface ocean currents off the coast of east Africa are all warm ocean currents. Warm ocean currents would not explain the cold temperatures necessary to support permanent snow on the peak of Mount Kilimanjaro.

13. **3** The wavelength of radiation emitted by a source depends on the temperature of the source. The cooler the source, the longer the wavelength of the radiation it emits. At Earth's average surface temperature, the radiation emitted would be in the range of long-wavelength infrared radiation. Thus, the maximum intensity of Earth's outgoing radiation is in the infrared portion of the electromagnetic spectrum.

14. **1** A major component of insolation is visible light. Color is a fairly good indicator of whether the substance absorbs more light than it reflects. Light-colored substances tend to reflect light. Dark-colored substances tend to absorb light. A smooth surface is more likely to reflect light than is a rough surface. The irregularities on a rough surface can cause some of the light to hit the surface several times before leaving. Each time the light strikes the irregular surface, a little more of the light's energy is absorbed. Therefore, a dark-colored and rough surface would most likely absorb the greatest amount of visible light. Thus, a dark-colored and rough tubing exterior will best absorb insolation.

15. **4** Find the Specific Heats of Common Materials chart in the *Reference Tables for Physical Setting/Earth Science*. Note that specific heat is measured as J/g · °C. This means that the numerical value of each material's specific heat is the number of joules of heat energy that must be added to 1 gram of the substance to increase that material's temperature by 1°C. For example, if 0.79 J/g of heat energy are added to 1 gram of granite, the temperature of the granite will increase

by 1°C. However, if 0.79 J of heat energy is added to 1 gram of lead, the temperature of the lead will increase by about 6°C (0.79 J/g ÷ 0.13 J/g · °C = 6.07°C). Thus, the smaller the specific heat value, the smaller the amount of heat energy needed to raise the temperature of the material by 1°C and the faster the material will warm up. It is given that equal masses of the materials were placed in sunlight. It is reasonable to infer that all received the same amount of energy from the sunlight per unit of time. Therefore, the material that will warm up the fastest is the one with the smallest specific heat value. Of the choices given, the material with the smallest specific heat is lead.

16. **1** Volcanic ash consists of tiny, solid particles of igneous rock. These solid particles block sunlight, thereby decreasing the amount of insolation that reaches Earth's surface. Thus, the most likely effect of this ash cloud is a decrease in insolation reaching Earth's surface.

WRONG CHOICES EXPLAINED:

(2) and (4) Find the Selected Properties of Earth's Atmosphere chart in the *Reference Tables for Physical Setting/Earth Science*. Note that Earth's atmospheric layers correspond to temperature zones within the atmosphere. In each zone, the temperature profile is different—with temperature either increasing or decreasing with altitude. The tiny particles of igneous rock in a volcanic ash cloud cool rapidly. By the time they reach the stratosphere, they would have cooled to the surrounding temperature and spread out. Therefore, a volcanic ash cloud would not change the temperature profile upon which the thickness of the stratosphere is based. Thus, the ash cloud would not change the thickness of the stratosphere.

(3) Solid particles of volcanic ash block sunlight. The more sunlight that is blocked by particles, the less that reaches Earth's surface. Thus, the most likely effect of this ash cloud is to decrease, not increase the insolation reaching Earth's surface.

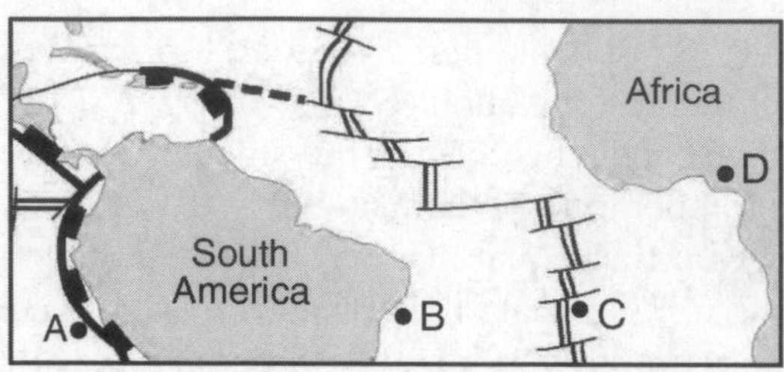

17. **3** Find the Tectonic Plates map in the *Reference Tables for Physical Setting/Earth Science.* Locate the region corresponding to the map in the question. Note that letter *A* is located on the Pacific Ocean floor of the Nazca Plate. Locations *B* and *C* are located on ocean floor on either side of the Mid-Atlantic Ridge. Location *D* is on the continent of Africa. The youngest crustal bedrock is formed when molten rock emerges from volcanoes and in the rift zones of the mid-ocean ridges and then hardens into bedrock. As the tectonic plates diverge along a mid-ocean ridge, the bedrock fractures and new molten rock emerges, pushing the bedrock aside. As this process is repeated, new bedrock is formed, and the bedrock on either side of the mid-ocean ridges is pushed sideways. Thus, the youngest, most recently formed bedrock is found nearest a mid-ocean ridge, and the oldest bedrock is found on either side farthest from the mid-ocean ridge. Therefore, the location that most likely has the youngest bedrock will be nearest a mid-ocean ridge. Location *C* is nearest a mid-ocean ridge and will most likely have the youngest bedrock.

WRONG CHOICES EXPLAINED:

(1) Find the Tectonic Plates map in the *Reference Tables for Physical Setting/Earth Science.* Note that the tectonic plate boundary along the west coast of South America near location *A* corresponds to a convergent plate boundary (subduction zone). Note that location *A* is near the edge of the subducting plate, which is being forced beneath the overriding South American Plate. Therefore, the bedrock near location *A* is already existing bedrock, not newly formed, young bedrock.

(2) Recall that the youngest, most recently formed bedrock is found nearest a mid-ocean ridge and the oldest bedrock is found on either side farthest from the ridge. Note that *B* is farther from the Mid-Atlantic Ridge than *C*. Thus, the bedrock at *B* is older than the bedrock at *C* and, therefore, is not the youngest bedrock.

(4) Find the Geologic History of New York State chart in the *Reference Tables for Physical Setting/Earth Science*. Locate the column labeled "Inferred Positions of Earth's Landmasses." Note that the African landmass existed for

hundreds of millions of years prior to the initial opening of the Atlantic Ocean. Thus, it is unlikely that the bedrock at *D* on the African landmass is composed of the youngest crustal bedrock.

18. **1** Find the Inferred Properties of Earth's Interior chart in the *Reference Tables for Physical Setting/Earth Science.* Locate the "Density (g/cm^3)" scale along the right side of the cross section. Note that granitic continental crust has a density of 2.7 and that basaltic oceanic crust has a density of 3.0. Thus, continental crust is granitic while oceanic crust is basaltic. Note, too, that continental crust is less dense than oceanic crust. Find the Scheme for Igneous Rock Identification in the *Reference Tables for Physical Setting/Earth Science.* Locate granite and basalt in the upper portion of the chart. Trace the columns corresponding to granite and basalt downward to the arrow labeled "Composition" in the center portion of the chart labeled "Characteristics." Note that granite is near the end of the arrow labeled "Felsic (rich in Si, Al)" and that basalt is near the end of the arrow labeled "Mafic (rich in Fe, Mg)." Thus, granitic continental crust is more felsic than basaltic ocean crust. Therefore, compared to the average density and composition of oceanic crust, continental crust is less dense and more felsic.

19. **2** Find the Tectonic Plates map in the *Reference Tables for Physical Setting/Earth Science.* Note the symbol labeled "Hawaii Hot Spot" over the islands of Hawaii. Locate this symbol in the key along the bottom of the Tectonic Plates map, and note that it corresponds to a mantle hot spot. A hot spot is a long-lasting zone of rising hot magma beneath moving plates. Large plumes of magma rise from the mantle at these hot spots and work their way upward through the plate above. The magma rises because it is more buoyant than the surrounding rock, wedges apart cracks in the plate, and melts through to erupt, forming a volcano. When the volcano rises above the level of the ocean, it forms an island. As the plate continues to move, the volcano that formed over the hot spot is carried away from the hot spot in the direction in which the plate is moving. At the same time, the plate motion carries a new section of the plate over the hot spot and a new volcano forms. As this process continues, a series of volcanic islands forms with the youngest located nearest the hot spot and the oldest farthest from the hot spot. Thus, the Hawaiian Islands were formed as a result of an oceanic plate moving over a mantle hot spot.

WRONG CHOICES EXPLAINED:

(1) A plate boundary where two tectonic plates move apart is called a divergent plate boundary. Divergent boundaries occur along plate edges. Find the Tectonic Plates map in the *Reference Tables for Physical Setting/Earth Science.*

Note that the Hawaiian Islands are located in the center of the Pacific Plate far from any divergent boundaries along the edges of the plate. Thus, it is unlikely that the Hawaiian Islands formed as a result of lava flowing over Earth's surface where two tectonic plates move apart.

(3) A collision between two oceanic plates would take place at the edges of the two plates. Find the Tectonic Plates map in the *Reference Tables for Physical Setting/Earth Science.* Note that the Hawaiian Islands are located in the center of the Pacific Plate far from the edges of the plate. Therefore, it is unlikely that the Hawaiian Islands formed as a result of two oceanic plates colliding to form an island arc.

(4) Tectonic plates slide past each other along their edges. Find the Tectonic Plates map in the *Reference Tables for Physical Setting/Earth Science.* Note that the Hawaiian Islands are located in the center of the Pacific Plate far from the edges of the plate. Therefore, it is unlikely that the Hawaiian Islands formed as a result of tectonic plates sliding past each other.

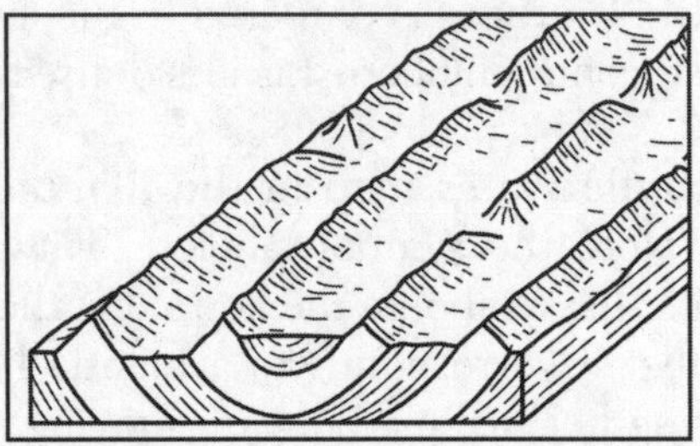

20. **4** Water flows from higher elevations to lower elevations. Therefore, water flows downhill in channels along either side of the tops of the long, parallel ridges shown in the block diagram into the valleys between the ridges, forming larger streams flowing down the centers of the valleys parallel to the ridges. Thus, the most likely stream drainage pattern on the surface of this landscape is shown in diagram (4).

21. **1** Mountain landscape regions consist of mountains. By definition, mountains are parts of Earth's crust that project at least 300 meters above the surrounding land and have great relief (steep slopes), a restricted summit, and considerable bare-rock surface. Most mountains are formed by crustal motions that fold or fault rock and have an underlying bedrock structure composed of folded and faulted rock layers. Thus, the characteristics that identify mountain landscape regions are steep slopes with deformed bedrock.

22. **2** Sandblasting refers to using a stream of sand projected by compressed air or steam to clean, polish, or decorate the surface of something by abrasion. In nature, windblown sand directed against bedrock can have a similar effect. Smooth, polished surfaces on bedrock in dry, sandy areas are most likely the result of abrasion by wind-borne sand. Thus, the agent of erosion that causes sandblasting of bedrock is wind.

23. **3** On steep slopes, the dominant type of erosion that occurs is mass wasting. In erosion by mass wasting, fragments of rock broken loose by weathering move downslope under the direct influence of gravity. As you can see in the photograph, the road is at the bottom of the steep slope. Any rocks on the steep slope that are broken loose by weathering would be carried downhill to the roadway. If rocks fell on or in the path of cars traveling on the roadway, they could cause serious damage to the cars and possibly loss of life. Accidents resulting from cars swerving to avoid the rocks could also cause loss of property or life. Thus, the wire netting was installed to prevent rocks that were broken loose by weathering from falling onto the roadway. Therefore, the wire netting has been installed to prevent loss of property or life resulting from weathering and erosion.

WRONG CHOICES EXPLAINED:

(1) Crosscutting refers to a geologic feature that cuts across another geologic feature. It does not involve rocks on a steep slope that would be held back from falling by wire netting. Downwarping refers to a bending of Earth's crust that results in a bowl-like depression, not rocks on a steep slope that would be held back from falling by wire netting.

(2) Folding and faulting result from enormous forces acting on Earth's crust. These forces are too enormous for wire netting to constrain the crust from folding or faulting.

(4) The wire netting would not prevent high winds or flooding because wind and water would be able to easily pass through the large openings in the wire netting.

24. **4** A naturally occurring solid material from which a metal or valuable mineral can be profitably extracted is called an ore. Find the Properties of Common Minerals table in the *Reference Tables for Physical Setting/Earth Science.* In the column labeled "Use(s)," locate the phrase "ore of iron." Note that "ore of iron" occurs in two rows. Trace each of these rows to the right to the column labeled "Mineral Name." Note that two minerals that are ores of iron are magnetite and hematite. Thus, two minerals that are commercial sources of iron are hematite and magnetite.

25. **1** Find the Scheme for Metamorphic Rock Identification in the *Reference Tables for Physical Setting/Earth Science.* In the column labeled "Rock Name," locate phyllite. From phyllite, trace left horizontally to the column labeled "Composition." Note that phyllite is composed of quartz, feldspar, amphibole, garnet, and pyroxene. Now find the Scheme for Sedimentary Rock Identification in the *Reference Tables for Physical Setting/Earth Science.* In the column labeled "Rock Name," locate sandstone. From sandstone, trace left horizontally to the column labeled "Composition." Note that sandstone is "mostly quartz, feldspar, and clay minerals." Finally, find the Scheme for Igneous Rock Identification in the *Reference Tables for Physical Setting/Earth Science*. Locate granite in the upper part of the chart. Trace the column containing granite vertically down to the section labeled "Mineral Composition (relative by volume)." Note that granite is composed of the minerals potassium feldspar, quartz, plagioclase feldspar, biotite, and amphibole. Thus of the choices, the mineral that can be found in the rocks phyllite, sandstone, and granite is quartz.

26. **2** Earth's axis of rotation is tilted at an angle of 23.5° from a perpendicular to Earth's orbital plane. This position relative to Earth's orbital plane is best represented in diagram (2).

WRONG CHOICES EXPLAINED:

(1) In this diagram, Earth's axis is perpendicular to Earth's orbital plane instead of being tilted 23.5° from a perpendicular to the orbital plane.

(3) In this diagram, Earth's axis is tilted 23.5° from the orbital plane itself, not 23.5° from a perpendicular to the orbital plane.

(4) In this diagram, Earth's axis of rotation is aligned with Earth's orbital plane instead of being tilted 23.5° from a perpendicular to the orbital plane.

27. **4** On a hot, sunny, summer afternoon, sunlight causes land surfaces to increase in temperature more than water surfaces because water has a higher specific heat than land. The cooler air over the water exerts more pressure than the warmer air over the land. Therefore, air pressure over the water will be high, and air pressure over the land will be low. As a result, the air will move from the region of higher air pressure over the water toward the region of lower air pressure over the land, as shown in cross section (4).

Type of Precipitation	**Average Diameter** (millimeters)	**Falling Velocity** (meters/ second)
drizzle	0.96	4.1
light rain	1.24	4.8
moderate rain		
heavy rain	2.05	6.7
excessive rain	2.40	7.3

28. **4** By interpolation, it can be inferred that the average diameter and falling velocity of moderate rain will be between the average diameters and falling velocities of light rain and heavy rain. According to the table, the average diameter of light rain is 1.24 millimeters and the average diameter of heavy rain is 2.05 millimeters. Therefore, the average diameter of moderate rain will be between 1.24 and 2.05 millimeters. According to the table, the falling velocity of light rain is 4.8 meters/second and the falling rate of heavy rain is 6.7 meters/second. Therefore, the falling velocity of moderate rain will be between 4.8 and 6.7 meters/second. Only choice (4) has values that fall into these ranges: average diameter = 1.60 mm, falling velocity = 5.7 m/s.

29. **1** Find the Geologic History of New York State chart in the *Reference Tables for Physical Setting/Earth Science*. Among the diagrams of index fossils along the bottom of the chart, locate the four index fossils shown in the answer choices. Note the circled letter associated with each of the four index fossils. Note the index fossil corresponding to each choice: 1—Ⓣ, 2—Ⓟ, 3—Ⓘ, and 4—Ⓧ. In the section of the chart labeled "Time Distribution of Fossils (including important fossils of New York)," locate the vertical gray bar labeled "Corals." Note that

the circled letters representing index fossils on this bar are Ⓣ, Ⓤ, and Ⓥ. Thus, the index fossils at the bottom of the chart corresponding to the circled letters Ⓣ, Ⓤ, and Ⓥ are corals. Therefore, of the choices given, the New York State index fossil that is classified as a coral is Ⓣ, which corresponds to the index fossil shown in choice (1).

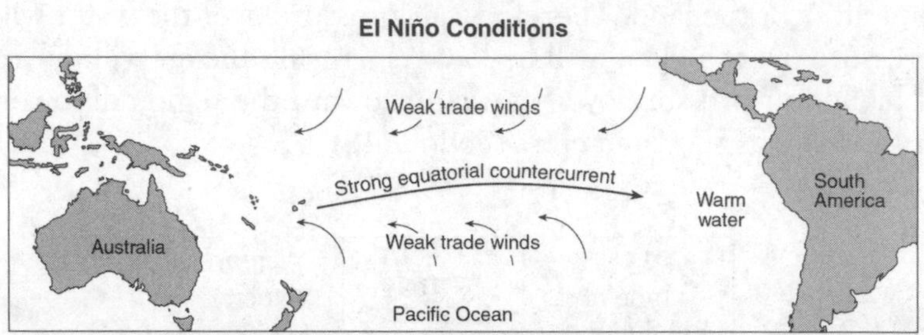

30. **4** Note the region in the map labeled "Warm water" adjacent to the west coast of South America. The warm water along the west coast of South America warms the air that comes into contact with it. Therefore, atmospheric temperature along the west coast of South America increases. The higher the temperature, the higher the evaporation rate. So the moisture level of the air over the warm water increases. The higher temperature and moisture levels result in air with a low density. As this low-density air rises and cooler, denser air rushes in to replace it, numerous convection cells and thunderstorms form. Therefore, there is an increase in precipitation. Thus, the most likely changes to atmospheric temperature and precipitation along the west coast of South America during El Niño conditions are higher temperatures and higher amounts of precipitation.

WRONG CHOICES EXPLAINED:

(1) and (2) The presence of warm water along the west coast of South America during El Niño conditions would raise temperatures, not lower temperatures.

(3) The presence of warm water along the west coast of South America warms the air and results in higher temperatures. Higher temperatures cause an increase in evaporation rates, which adds more moisture to the air. The more moisture the air contains, the greater the chance of precipitation. Therefore, during El Niño conditions, the amounts of precipitation would be higher, not lower.

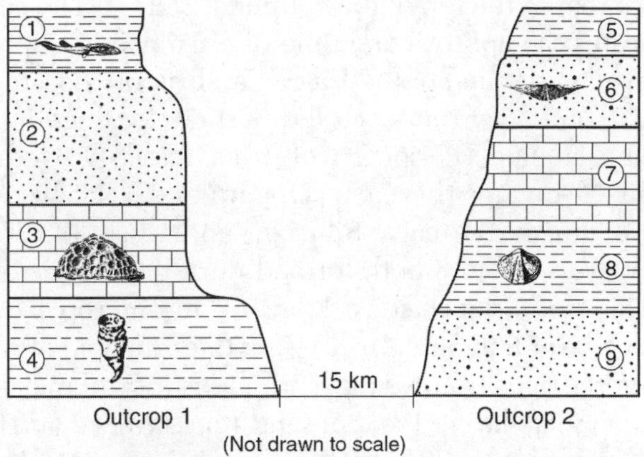

(Not drawn to scale)

31. **3** Note that in the outcrops, all of the rock units are flat and horizontal. The presence of fossils indicates that these are sedimentary rocks. The principle of superposition states that the bottom layer of a sedimentary series is the oldest, unless it was overturned or had older rock thrust over it, because the bottom layer was deposited first. Similarly, in a sequence of rock layers, a rock layer is older than those above it and is younger than those below it. Therefore, in outcrop 1, the layers from oldest to youngest are 4, 3, 2, and 1. In outcrop 2, the layers from oldest to youngest are 9, 8, 7, 6, and 5.

Rock layers can sometimes be correlated on the basis of distinct similarities in physical characteristics such as composition, color, thickness, and fossil remains. Thus, it is reasonable to correlate the rock layers in these two outcrops by their composition and fossil content.

Find the Scheme for Sedimentary Rock Identification in the *Reference Tables for Physical Setting/Earth Science.* In the column labeled "Map Symbol," locate the map symbols representing the rock layers in the outcrops. From each map symbol, trace left to the column labeled "Rock Name." Note the name of the rock corresponding to each map symbol present in the outcrops: rock layers 1, 4, 5, and 8—shale; rock layers 2, 6, and 9—sandstone; rock layers 3 and 7—limestone.

Now find the Geologic History of New York State chart in the *Reference Tables for Physical Setting/Earth Science*. Among the diagrams of index fossils along the bottom of the chart, locate the five index fossils shown in rock layers 1, 3, 4, 6, and 8. Note the circled letter associated with each of the five index fossils: layer 1—Ⓡ, layer 3—Ⓥ, layer 4—Ⓤ, layer 6—Ⓩ, and layer 8—Ⓨ. Locate the column labeled "Time Distribution of Fossils (including important fossils of

New York)." Note that at the top of the column it states: "The center of each lettered circle indicates the approximate time of existence of a specific index fossil (e.g., Fossil Ⓐ lived at the end of the Early Cambrian)." Locate the lettered circles Ⓡ, Ⓥ, Ⓤ, Ⓨ, and Ⓩ. From each lettered circle, trace horizontally to the left to the column labeled "Period." Note that fossil Ⓡ lived during the late Devonian, fossil Ⓥ during the early Devonian, fossil Ⓤ during the early Silurian, fossil Ⓨ during the early Silurian, and fossil Ⓩ during the middle Devonian. Since layers 4 and 8 both formed during the early Silurian, they are the same age. Therefore, the shale rock layer 4 in outcrop 1 can be correlated with the shale rock layer 8 in outcrop 2. Using the principle of superposition, we can now correlate the ages from youngest to oldest of the rock layers of the two outcrops as follows: shale layers 1 and 5; sandstone layers 2 and 6; limestone layers 3 and 7; shale layers 4 and 8; sandstone layer 9. Thus, of the choices given, the two layers that most likely formed during the same geologic time period are 3 and 7.

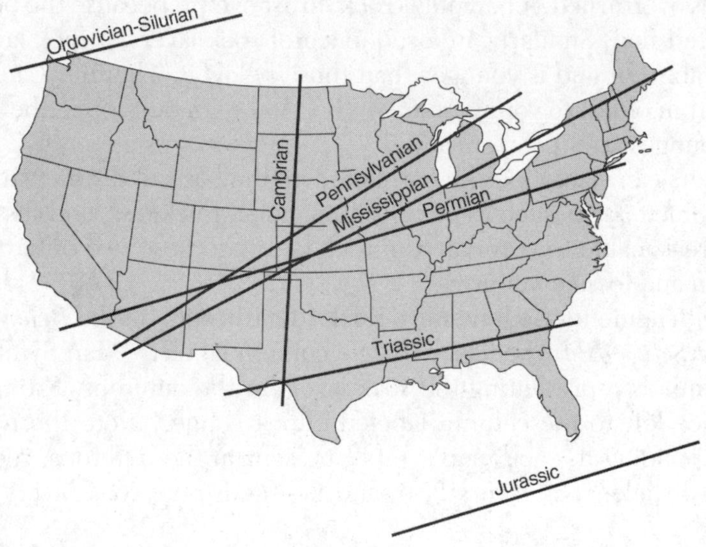

32. **3** Find the Generalized Bedrock Geology of New York State map in the *Reference Tables for Physical Setting/Earth Science.* Locate Watertown, New York, and note its relative position in New York State. On the map in the question, locate the position that corresponds to Watertown, New York. Note that the line showing the inferred position of Earth's equator during the middle of several geologic periods closest to this position is labeled "Mississippian." Find the Geologic History of New York State chart in the *Reference Tables for Physical Setting/*

Earth Science. In the column labeled "Period," locate "Mississippian." From Mississippian, trace right to the time scale labeled "Million years ago," and note that the Mississippian extended from 318 to 359 million years ago. Of the choices, only 340 mya falls within this range. Thus, the region around current-day Watertown, New York, was located nearest to the equator 340 million years ago.

33. **4** Find the Scheme for Sedimentary Rock Identification in the *Reference Tables for Physical Setting/Earth Science.* Find the column labeled "Rock Name," and locate the names of the sedimentary rocks listed in the tables. From each rock name, trace left to the column labeled "Texture," and note the texture of that rock. For example, in choice (1), sandstone has a clastic texture, breccia has a clastic texture, and rock gypsum has a crystalline texture. Thus, the table in choice (1) does not correctly match rock textures with a sedimentary rock that exhibits each texture. In this way, check the names and textures of the rocks in each of the tables. Note that only the table in choice (4) correctly matches rock textures with a sedimentary rock that exhibits each texture.

Temperatures Measured with a Psychrometer

Day	1	2	3	4
Dry-bulb temperature (°C)	0	5	10	15
Wet-bulb temperature (°C)	–5	0	5	10

34. **4** According to the data table, the difference between the wet-bulb and the dry-bulb temperatures on each of the four days was 5°C. Find the Relative Humidity (%) chart in the *Reference Tables for Physical Setting/Earth Science.* Note the vertical scale labeled "Dry-Bulb Temperature (°C)" along the left side of the chart. Now locate the column labeled "5" along the horizontal scale at the top of the chart labeled "Difference Between Wet-Bulb and Dry-Bulb Temperatures (°C)." Trace this column downward. Note that as the dry-bulb temperature increases, the values in this column also increase. Thus, the higher the dry-bulb temperature when the difference between the wet-bulb and dry-bulb temperature is 5°C, the higher the relative humidity. Thus, the day that had the highest relative humidity was day 4.

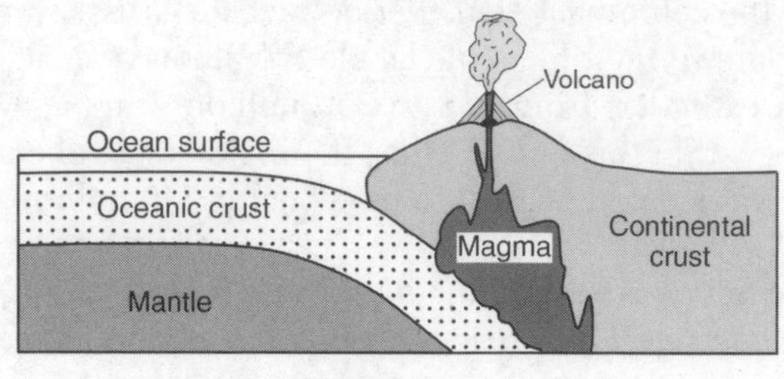

(Not drawn to scale)

35. **2** According to the cross section, the oceanic crust is plunging beneath the continental crust. This is consistent with a subduction zone. Recall that in a subduction zone, denser ocean crust collides with and is pushed beneath less dense continental crust. During this collision, the edge of the continental crust crumples and is uplifted as the denser ocean crust plunges beneath it. As the plunging plate is forced into the mantle and melts, molten plate material rises beneath the overlying crust, causing further uplift. If the molten material reaches the surface through cracks, volcanoes form. Find the Tectonic Plates map in the *Reference Tables for Physical Setting/Earth Science.* In the key along the bottom of the map, locate "Convergent plate boundary (subduction zone)." At a convergent plate boundary, the plates are moving toward one another and colliding. Therefore, the arrows showing the relative directions of plate movement should be pointed toward one another as in diagram (2).

PART B–1

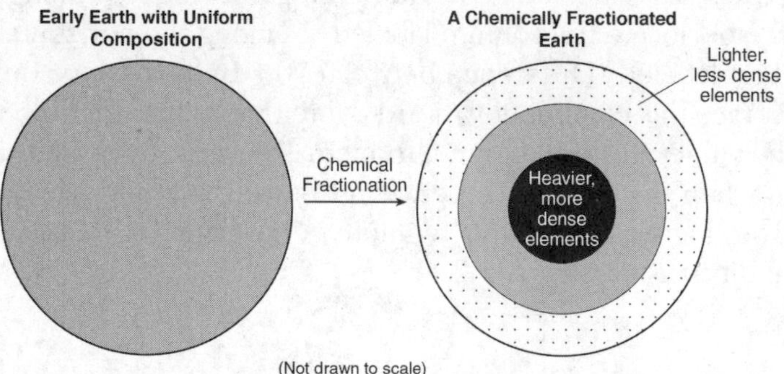

(Not drawn to scale)

36. **2** The force of gravitational attraction between any two objects is directly proportional to their masses and indirectly proportional to the square of the distance between them. Density is the ratio of mass to volume. Denser elements

have more mass per unit volume than less dense elements. Chemical fractionation occurred while Earth was molten, that is, while Earth was a vast globe of hot liquid. Now imagine that in this vast, hot liquid, there is a volume of denser elements and an equal volume of less dense elements that are the same distance from Earth's center. The volume of denser elements will have more mass. Therefore, there will be a greater force of gravitational attraction between Earth and the volume of denser elements than between Earth and the volume of less dense elements. The denser elements will be drawn more forcefully toward Earth's center and sink in the liquid Earth, displacing the less dense elements upward. Thus, chemical fractionation is most likely caused by gravity.

WRONG CHOICES EXPLAINED:

(1) Solidification is a change from the liquid phase to the solid phase. Chemical fractionation occurred while Earth was in the liquid (molten) phase, before solidification occurred. Therefore, chemical fractionation was not caused by solidification.

(3) A magnetic force would act on only magnetic elements. Most elements are not magnetic. Additionally, the shape of Earth's magnetic field is very different from the layering shown by the chemically fractionated Earth. Therefore, it is unlikely that chemical fractionation was caused by magnetic force.

(4) Chemical weathering is the breaking down of rock by changing its chemical composition. Rocks are solids. Chemical fractionation occurred while Earth was in the liquid (molten) phase. Therefore, chemical fractionation was not caused by chemical weathering.

37. **4** The question states that chemical fractionation occurred early in Earth's formation. Find the Geologic History of New York State chart in the *Reference Tables for Physical Setting/Earth Science*. In the column labeled "Era," locate the statement "Estimated time of origin of Earth and the solar system." Trace left to the time scale labeled "Million years ago," and note that this occurred 4600 million years ago. Thus, Earth and the other planets in our solar system began the process of chemical fractionation approximately 4600 million years ago.

38. **3** According to the diagram labeled "A Chemically Fractionated Earth," the heavier, denser elements are located nearest Earth's center. Find the Inferred Properties of Earth's Interior chart in the *Reference Tables for Physical Setting/Earth Science*. In the cross section at the top of the chart, note that the layer nearest Earth's center is labeled "Inner core (iron & nickel)." Thus, the pair of elements that sank to Earth's center during chemical fractionation was iron and nickel.

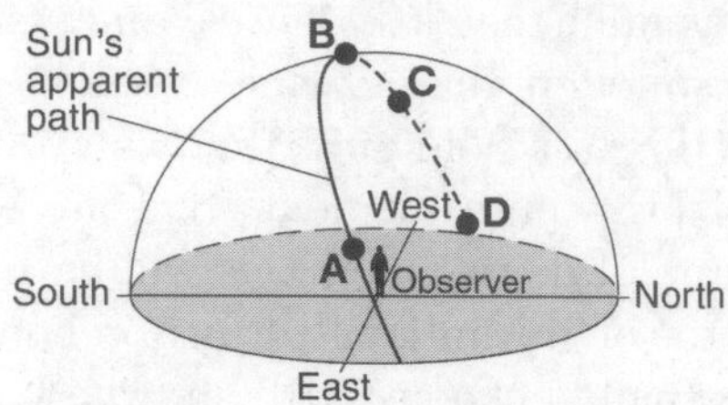

39. **2** The Sun's apparent path through the sky from sunrise to sunset is an arc, or part of a circle, like all other celestial objects. The Sun appears to move across the sky from east to west at a constant rate of 15° per hour, or one complete circle every day (24 hours/day × 15°/hour = 360°/day) due to Earth's rotation on its axis.

WRONG CHOICES EXPLAINED:

(1) Earth revolves around the Sun once every 365.26 days, or about 1° per day. If the apparent path of the Sun was caused by Earth's revolution around the Sun, the Sun would appear to move at a rate of 1° per day. However, the Sun appears to move across the sky from east to west at a constant rate of 15° per hour. Therefore, the apparent path of the Sun is not caused by Earth's revolution around the Sun.

(3) The Sun does not revolve around Earth. Therefore, this motion could not cause the apparent rising and setting of the Sun as viewed from Earth.

(4) In order for the Sun to change position in the sky, it must change position relative to an observer on Earth. The Sun's rotation on its axis does not change its position relative to an observer on Earth. The Sun could remain stationary to an observer on Earth and still rotate. Therefore, the apparent path of the Sun through the sky cannot be caused by the Sun's rotation on its axis.

40. **4** As the angle at which the Sun's rays strike Earth's surface decreases, the length of the shadow cast by an object increases. Thus, an observer's shadow is longest when the altitude of the Sun is lowest, that is, when the Sun is closest to the horizon. According to the diagram, the Sun is closest to the horizon at position *D*. Thus, the observer has the longest shadow when the Sun is at position *D*.

41. **3** The Sun appears to move across the sky from east to west at a constant rate of 15° per hour. Thus, sunrise occurs along the eastern horizon, and sunset occurs along the western horizon. Therefore, the Sun moves across the sky from position *A*, to *B*, to *C*, and then to *D*. Solar noon occurs when the Sun reaches its highest point in the sky for the day. At this moment, the Sun crosses an observer's meridian, a semicircle connecting the North and South Poles that passes through the observer's zenith. Therefore, position *B* represents solar noon. Since the Sun

arrives at position *C* after position *B*, position *C* is in the afternoon. Thus, the approximate time of day when the Sun is at position *C* is 3 p.m.

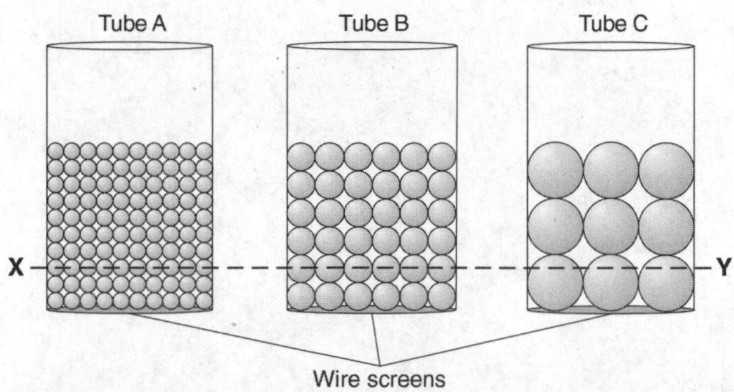

42. **3** The infiltration rate of a substance depends on two factors: the size of its pores (the openings between the particles) and the degree to which the pores are interconnected. Small pores constrict the flow of water, slowing the rate of infiltration. If water cannot get from one pore to another, it cannot flow through a substance. Since all four tubes contain uniform-sized, spherical beads, all four have interconnecting pores. Thus, the size of the pores determines the infiltration rate. In general, the larger the sediment particles, the larger the pore spaces are between them. Therefore, as particle diameter increases, the rate of infiltration increases. This relationship is best shown by graph (3).

43. **1** Capillarity is the tendency of water to rise in narrow openings in solid particles. The water rises because its attraction to the surrounding particles pulls the water upward and surface tension drags the water surface between the materials upward along with it. The narrower the opening is, the greater the force of attraction is between the water and the surrounding materials compared to the weight of the water column. So the narrower the opening between particles, the higher the water rises. Smaller particles fit closer together than larger ones. Therefore, the smaller the particles, the narrower the openings between particles and the higher water rises due to capillarity. Tube *A* contains the smallest beads. Thus, after one hour, the height of the water above line *XY* will be highest in tube *A*.

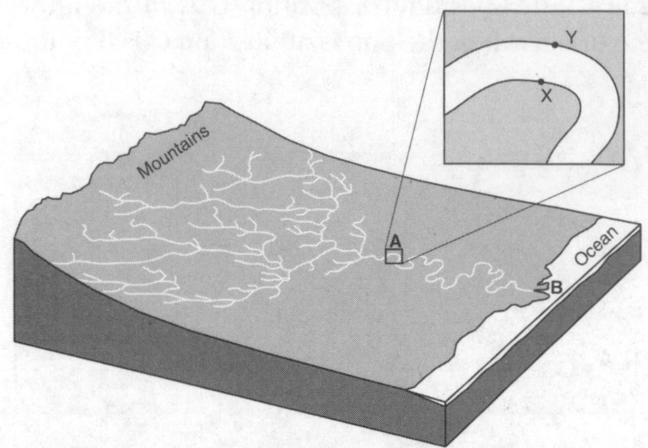

44. **2** By definition, a watershed is the area drained by a stream and its tributaries. A river is a stream. The question states that the block diagram represents a river drainage system. Therefore, the area of land drained by this river and its tributaries is best described as the river's watershed.

WRONG CHOICES EXPLAINED:

(1) Topography refers to the shape of Earth's surface, not the area drained by a river.

(3) The water table is the upper surface of underground pore spaces that are filled with groundwater, not the area drained by a river.

(4) A floodplain is a low plain adjacent to a stream that is likely to be flooded if a stream overflows, not the area drained by a river.

45. **4** Stream erosion is greatest where the velocity of the water is greatest. Along a straight section of a stream's channel, velocity is greatest in the center of the stream farthest from the friction of the banks or bed. When the stream curves, or meanders, velocity is greatest along the outside of the curve and lowest along the inside of the curve. In the diagram, point *Y* is located at the outside of a curve in the stream where the velocity of the water and erosion are the greatest. Point *X* is located at the inside of a curve in the stream where the velocity of the water and erosion are the least. Therefore, the streambed will be more deeply eroded near point *Y* than it is near point *X*—a shape best represented by the cross section in choice (4).

46. **2** As the water flowing in the river enters the standing water of the ocean, the velocity of the river decreases. Find the Relationship of Transported Particle Size to Water Velocity graph in the *Reference Tables for Physical Setting/Earth*

Science. Note that as stream velocity decreases, the particle size that can be transported also decreases. Therefore, as distance from where the river enters the ocean increases, stream velocity decreases and the sediment size deposited decreases. Thus, the arrangement of the sediments deposited where the river enters the ocean is largest particles deposited in the shallow waters near the mouth of the river and smallest particles deposited farther out in the ocean in deeper water. This pattern of sediments deposited where the river enters the ocean at location *B* is best represented by the cross section in choice (2).

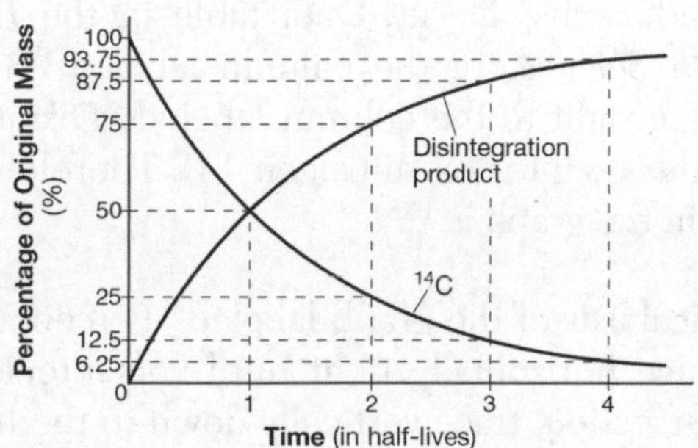

47. **1** Find the Radioactive Decay Data table in the *Reference Tables for Physical Setting/Earth Science.* Note that carbon-14 is a radioactive isotope. Radioactive isotopes break apart, or decay, at a steady, constant rate that is not affected by outside factors such as changes in temperature, pressure, or chemical state. Therefore, the decay process may be used as a clock to determine the actual age of geologic samples. Radioactive isotopes start to decay when the geologic sample forms. The decay of a radioactive isotope occurs at a statistically predictable rate known as a half-life. The half-life is the time required for one-half of the unstable radioisotope to change into a stable decay product. By comparing the relative amounts of radioactive isotope and stable decay product in a rock, the number of half-lives that elapsed may be determined and may be used as a clock to determine the actual ages of these geologic samples. Therefore, radioactive carbon-14 is often useful in determining the absolute age of geologic samples because radioactive isotopes decay at a regular rate.

WRONG CHOICES EXPLAINED:

(2) When a radioactive isotope decays, its atoms break apart and form more stable atoms of a different element. Therefore, radioactive isotopes become more stable during decay, not less stable.

(3) The most common isotopes of the elements that comprise rocks are stable, meaning their atoms do not change. However, some isotopes are unstable.

In a process called radioactive decay, these atoms break apart. Such isotopes are called radioactive isotopes. Thus, radioactive isotopes are, by definition, unstable and do not remain unchanged over time.

(4) Radioactive isotopes continue to decay at a constant rate until all of the radioactive atoms have decayed. According to the graph, after four half-lives, 6.25% of the original mass of radioactive carbon-14 still remains. Therefore, radioactive carbon-14 does not stabilize after four half-lives because some still remains undecayed.

48. **4** Find the Radioactive Decay Data table in the *Reference Tables for Physical Setting/Earth Science.* In the column labeled "Radioactive Isotope," locate carbon-14. Trace right to the column labeled "Disintegration," and note that ^{14}C (carbon-14) decays into ^{14}N (nitrogen-14). Therefore, the disintegration product represented in the graph is ^{14}N.

49. **2** On the vertical axis of the graph labeled "Percentage of Original Mass (%)," locate "25." Trace horizontally right until you intersect the line labeled "14C." From this intersection, trace vertically down to the horizontal axis of the graph labeled "Time (in half-lives)." Note the value "2." Thus, if a sample contains 25% of its original carbon-14, 2 half-lives have passed.

50. **3** Find the Radioactive Decay Data table in the *Reference Tables for Physical Setting/Earth Science.* In the column labeled "Radioactive Isotope," locate carbon-14. Trace right to the column labeled "Half-life (years)," and note that carbon-14 has a half-life of 5.7×10^3 (5700) years. After about 10 half-lives (~50,000 years), there is too little carbon-14 left in a fossil to accurately date it. Therefore, carbon-14 can be used to date only relatively recent organic remains. Now find the Geologic History of New York State chart in the *Reference Tables for Physical Setting/Earth Science*. Among the diagrams of index fossils along the bottom of the chart, locate the four index fossils shown as choices. Note the circled letter associated with each of the four index fossils: 1—Ⓐ, 2—Ⓛ, 3—Ⓞ, and 4—Ⓠ. Locate the column labeled "Time Distribution of Fossils (including important fossils of New York)." Note that at the top of this column it states: "The center of each lettered circle indicates the approximate time of existence of a specific index fossil (e.g., Fossil Ⓐ lived at the end of the Early Cambrian)." Locate the lettered circles Ⓐ, Ⓛ, Ⓞ, and Ⓠ. From each lettered circle, trace horizontally to the left to the time scale labeled "Million years ago." Note how many million years ago each index fossil existed: Ⓐ—488–542; Ⓛ—200–251; Ⓞ—0–1.8; and Ⓠ—359–416. Of the choices, index fossil Ⓞ existed most recently. The drawing of index fossil Ⓞ is labeled "Mastodont." Mastodonts existed until

the end of the last ice age about 10,000 years ago, well within the 50,000-year range during which carbon-14 can be used to date a fossil. Thus, the only index fossil whose age could be determined by using carbon-14 is the mastodont shown in choice (3).

PART B–2

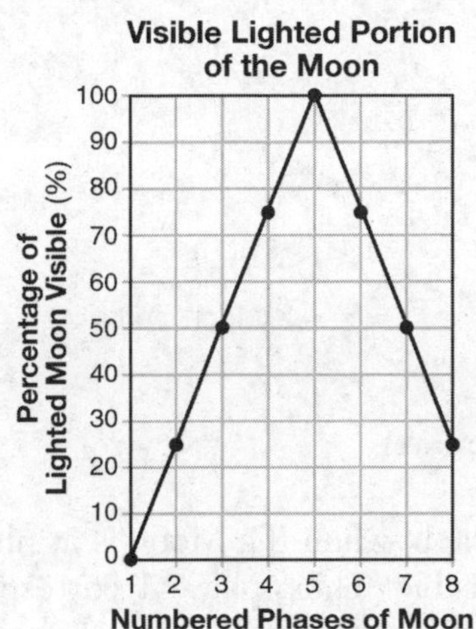

51. Locate the vertical axis of the graph labeled "Percentage of Lighted Moon Visible (%)." Note that the percentage increases from the bottom of the graph to the top. Therefore, the phases of the Moon begin waxing when the line on the graph begins to ascend. The waxing ends when the line has reached the maximum percentage of lighted Moon visible and begins to descend. Locate the point on the graph when the line begins to ascend and note that it corresponds to "1" on the horizontal axis labeled "Numbered Phases of Moon." Locate the point at which the line reaches its maximum and begins to descend. Trace vertically downward to the horizontal axis labeled "Numbered Phases of Moon," and note that it corresponds to the numbered phase "5." Thus, the numbered phase when waxing begins is 1, and the numbered phase when waxing ends is 5. The phase of the Moon at which 0% of the lighted Moon is visible is called the New Moon phase. The phase of the Moon at which 100% of the lighted portion of the Moon is visible is called the Full Moon Phase.

One credit is allowed for **waxing begins at Phase 1 *or* New Moon and waxing ends at Phase 5 *or* Full Moon.**

52. The full Moon phase occurs when the Moon is on the opposite side of Earth from the Sun. Thus, the **X** should be placed on the orbit, as shown below.

One credit is allowed if **the center of the X is within or touches the clear banded region shown below**.

Note: Allow credit if a symbol other than **X** is used.

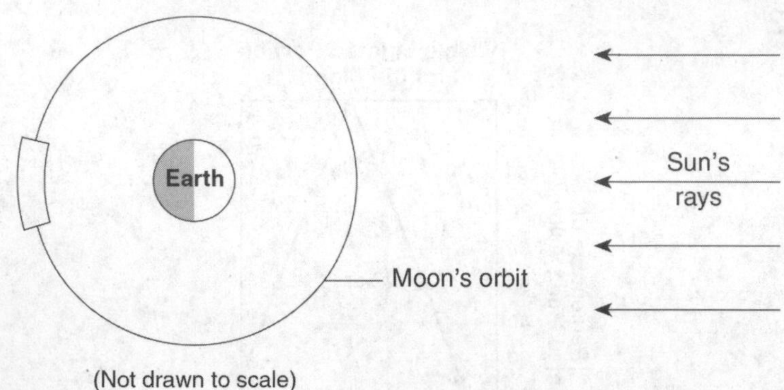

53. According to the graph, when the Moon is at phase 1, 0% of the lighted portion of the Moon is visible. Thus, phase 1 corresponds to the New Moon phase. When the Moon is at phase 5, 100% of the lighted portion of the Moon is visible. Thus, phase 5 corresponds to the Full Moon phase. Therefore, phase 3 occurs midway between the New Moon and Full Moon phases and corresponds to the First Quarter phase. Similarly, phase 7 occurs midway between the Full Moon and New Moon phases and corresponds to the Last Quarter phase. Note that at the First Quarter phase, the 50% of the lighted portion of the Moon visible to an observer in New York State is to the observer's right and the left side of the Moon is shaded. When the Moon is at the Last Quarter phase, it is on the side of Earth directly opposite where it was at the First Quarter phase. Thus, an observer would see the lighted portion of the Moon on the opposite side of the Moon, or to the observer's left. Therefore, the right side of the circle should be shaded.

One credit is allowed for **shading half of the circle on the right side. The edge of the shading should be within or touching the clear rectangle shown below**.

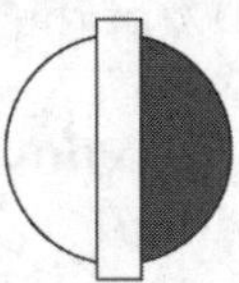

Examples of 1-credit responses:

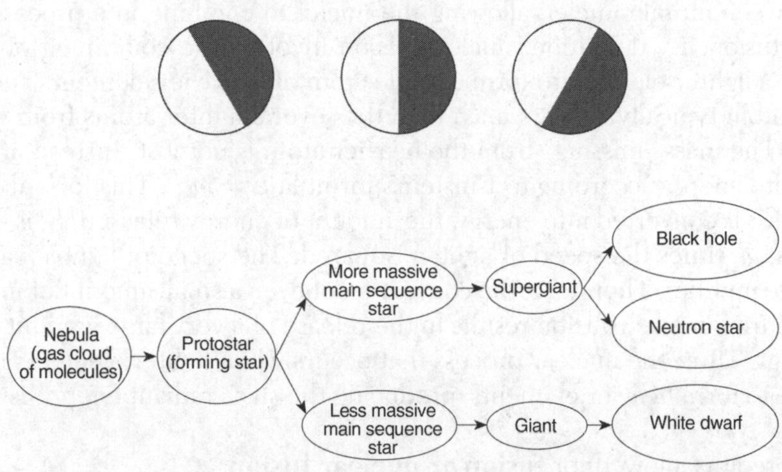

54. According to the flowchart, supergiants form from more massive main sequence stars and giants form from less massive main sequence stars. Thus, the characteristic of a main sequence star that determines whether the star becomes a giant or a supergiant is mass. Mass is a measure of the amount of matter in a star. Typically, the more mass a main sequence star contains, the larger the star.

One credit is allowed. Acceptable responses include but are not limited to:

- **mass**
- **amount of matter in a star**
- **how massive the star is**
- **size**

Note: Credit is *not* allowed for "luminosity" because even though luminosity is related to the mass of a star, this is not how scientists classify the life cycle of stars based on this flowchart.

55. According to the flowchart, black holes and neutron stars form from supergiants. Find the Characteristics of Stars chart in the *Reference Tables for Physical Setting/Earth Science.* Locate the row labeled "Supergiants (Intermediate stage)" at the top of the chart. Note the names of stars in this row: *Rigel, Deneb, Betelgeuse, Spica,* and *Polaris.*

One credit is allowed for ***Rigel*, *Deneb*, *Betelgeuse*, *Spica*, or *Polaris*.**

56. Most of the radiant energy given off by stars is the result of nuclear fusion. Within stars, the force of gravity is strong enough to overcome the force of repulsion between atomic nuclei, allowing the nuclei to combine in a process called nuclear fusion. By definition, nuclear fusion involves the combining of several atoms of a lighter element to form a single atom of a heavier element. The single heavier atom typically has less mass than the several lighter atoms from which it formed. The mass "missing" from the heavier atom is not lost. Instead, it is converted into energy according to Einstein's formula $E = mc^2$. This formula means that if mass is converted into energy, the amount of energy released, E, is equal to the mass, m, times the speed of light, c, squared. The speed of light squared is a very large number. Therefore, the conversion of even a small amount of mass into energy during nuclear fusion results in the release of a very large amount of radiant energy. Thus, the nuclear process that occurs when lighter elements in a star combine to form heavier elements, producing the star's radiant energy, is nuclear fusion.

One credit is allowed for **fusion *or* nuclear fusion**.

57. Find the Characteristics of Stars chart in the *Reference Tables for Physical Setting/Earth Science*. Locate "*Sun.*" Trace horizontally left to the vertical scale along the left-hand side of the chart labeled "Luminosity (Rate at which a star emits energy relative to the Sun)." Note that the Sun has a luminosity of 1. From the Sun, trace vertically downward to the horizontal scale along the bottom of the chart labeled "Surface Temperature (K)." Note that the Sun has a surface temperature of about 5,700K. Repeat this procedure for *Sirius*. Note that *Sirius* has a luminosity of about 25 and a surface temperature of nearly 10,000K. Thus, *Sirius* has a higher relative surface temperature and a greater relative luminosity than the Sun.

One credit is allowed if ***both*** **higher relative surface temperature and greater relative luminosity are circled**.

58. Contour lines are isolines that connect points of equal elevation on a topographic map. To draw the 200-meter isoline, connect all of the points labeled "200" with a smooth curve. Not every point that has an elevation of 200 meters has been labeled. However, between any locations, the elevation changes steadily from one location to the other. Begin with the point labeled "200" near the lower right corner of the map. Make sure you extend this line to the right edge of the map, as shown below. Note the two nearby points labeled "175" and "250." Somewhere between these two points, the elevation is 200 meters. Therefore extend the 200-meter contour line westward between these two points. Farther west are two points labeled "240" and "180." Extend the 200-meter contour line

between these two points to the point labeled “200.” From this point, extend the 200-meter contour line westward and then north between the two points labeled “210” and “180” to the point labeled “200” near the west edge of the map and extend the line to the edge of the map, as shown below.

One credit is allowed for **correctly drawing the 200-meter contour line extended to the edges of the map**.

Note: If additional contour lines are drawn, all must be drawn correctly to receive credit.

Example of a 1-credit response:

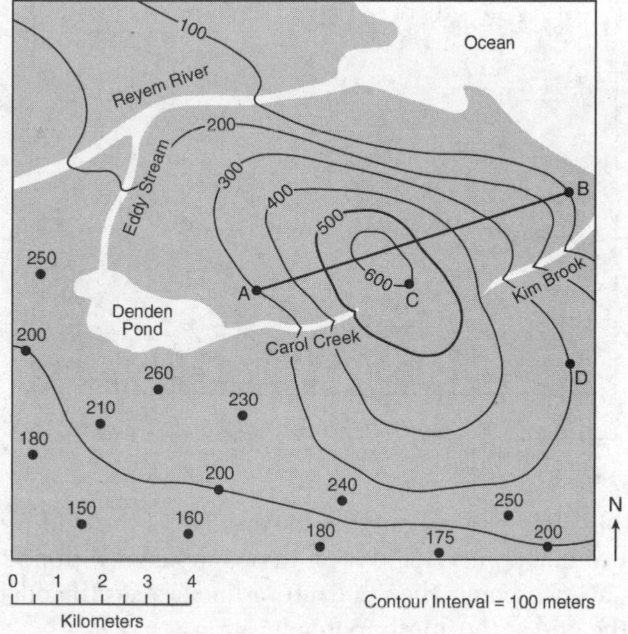

59. To construct a topographic profile along line *AB*, proceed as follows. Place the straightedge of a piece of scrap paper along the solid line connecting point *A* to point *B*. Mark the edge of the paper at points *A* and *B* and wherever the paper intersects a contour line. Wherever the paper intersects a contour line, label the mark with the elevation of that contour line, as shown below.

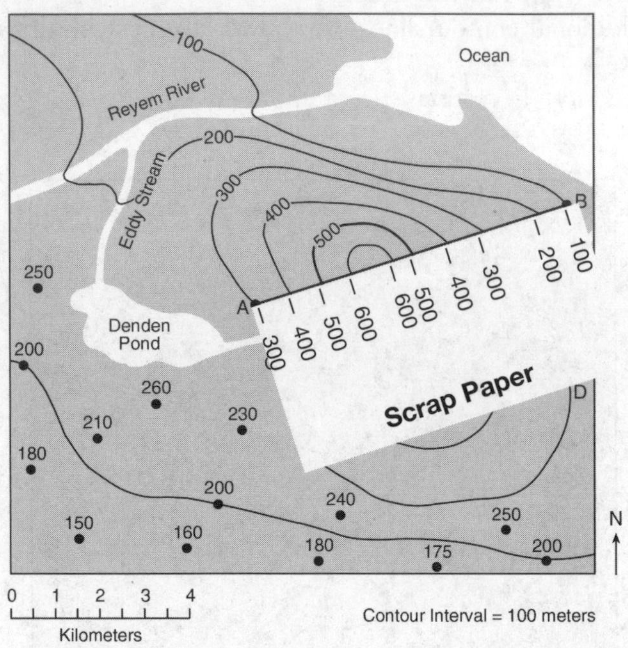

Now place this scrap paper along the lower edge of the grid provided in your answer booklet so that points *A* and *B* on the paper align with points *A* and *B* on the grid. Then, at each point where a contour line crosses the edge of the paper, draw a dot on the grid at the appropriate elevation.

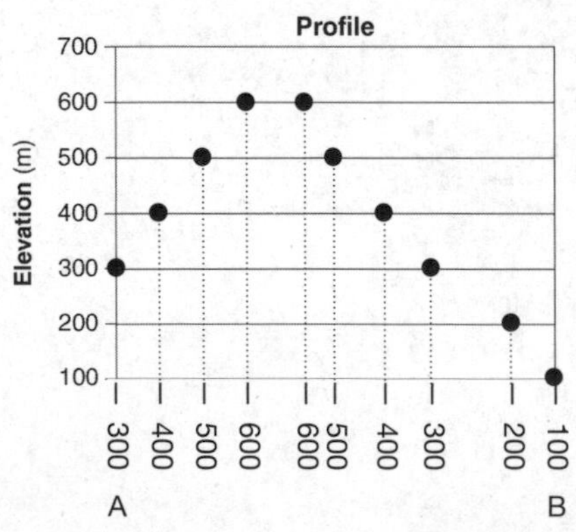

Finally, connect all of the dots in a smooth curve to form the finished profile, as shown below. Since the contour interval is 100 meters and no additional contour line exists within the 600-meter contour, the elevation in that area is greater than 600 meters but less than 700 meters. Therefore, the high point of the profile must be greater than 600 meters but less than 700 meters.

One credit is allowed if **the centers of *all eight* plots are within or touch the clear rectangles shown below and *all ten* plots are correctly connected with a line that passes within or touches the rectangles. The line must show the highest elevation above 600 m but below 700 m.**

Note: Credit is allowed if the line does not pass through the plots but is still within or touches the rectangles.

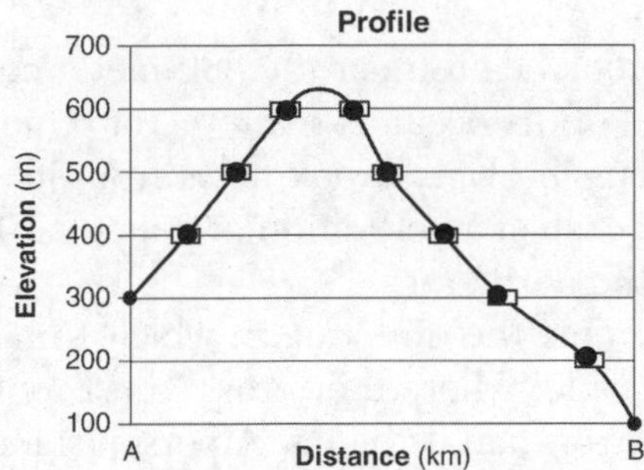

60. Find the Equations section of the *Reference Tables for Physical Setting/ Earth Science*, and note the equation for gradient:

$$\text{Gradient} = \frac{\text{change in field value}}{\text{change in distance}}$$

The map shown is a topographic map. The field value on a topographic map is elevation. Note that point *C* is located directly on the 600-meter contour line and that point *D* is located directly on the 300-meter contour line. Thus, the change in field value (elevation) from *C* to *D* is 200 meters.

On a piece of scrap paper, mark off the distance along the straight line between points *C* and *D*. Compare the distance marked off on the scrap paper to the scale printed beneath the map to determine the distance between points *C* and *D*—about 4 kilometers. Substitute these values in the equation, as shown below:

$$\text{Gradient} = \frac{600 \text{ meters} - 300 \text{ meters}}{4 \text{ kilometers}}$$

Solve the equation, as shown below:

$$\text{Gradient} = \frac{300 \text{ m}}{4 \text{ km}} = 75 \frac{\text{m}}{\text{km}}$$

Thus, the approximate gradient along the straight line between points *C* and *D* is 75 meters/kilometer.

One credit is allowed for **any value from 73 to 77 *or* –73 to –77 with the correct units**.

Acceptable responses include but are not limited to:

- **m/km**
- **meters/kilometers**

61. Note that Kim Brook begins near the 300-meter contour line and ends in the ocean. The elevation of the ocean is sea level, or 0 meters. Water flows from higher to lower elevations. So Kim Brook flows from its source at an elevation near 300 meters to the ocean at an elevation of 0 meters. Thus, Kim Brook flows toward the ocean or to the northeast.

Another way to determine the direction in which Kim Brook flows is to note the shape of the contour lines where they cross Kim Brook. The bed of a stream slopes downhill and is lower than its banks. A person standing on a streambed is at a lower elevation than a person standing on the stream's bank. A person standing on the streambed has to walk uphill along the streambed, that is, upstream, to reach the same elevation as a person on the bank. On a topographic map, contour lines connect points of equal elevation. Therefore, the contour line has to bend upstream of the stream's bank to intercept the same elevation at the center of the streambed and then downstream on the opposite side to reach the same elevation on the opposite bank. Therefore, contour lines bend upstream when they cross a streambed, forming a distinctive V-shaped curve with the apex pointing upstream and the sides opening toward the downstream direction. Note that where the contour lines cross Kim Brook, they bend with the apex of each bend pointing southwest and the sides opening toward the northeast. Therefore, Kim Brook is flowing toward the northeast.

One credit is allowed if *both* responses are correct. Acceptable responses include but are not limited to:

Compass direction:

- **NE**
- **east northeast**
- **NNE**
- **from SW to NE**

Evidence:

- **Contour lines bend upstream when they cross Kim Brook.**
- **Contour lines make a V shape at Kim Brook and the V points uphill.**
- **The brook flows out of the Vs.**
- **Kim Brook flows from a higher to a lower elevation.**
- **Elevations decrease toward the northeast.**
- **Kim Brook flows to sea level/the ocean.**

Note: Do *not* accept the response "from the SW" only, because this does *not* indicate the direction *toward* which Kim Brook flows. Do *not* accept the response "water flows downhill" because this is given in the question.

62. When an earthquake occurs, both types of seismic waves, *P*-waves and *S*-waves, start moving outward from the focus at the same time. However, since they travel at different speeds, these two different types of seismic waves do not arrive at a seismic station at the same time. The faster *P*-waves arrive first, followed by the slower *S*-waves some time later. The farther a seismic station is located from the epicenter, the greater the difference between the arrival times of the *P*-waves and the *S*-waves recorded on a seismogram. Find the Generalized Bedrock Geology of New York State map in the *Reference Tables for Physical Setting/Earth Science*. Locate the map scale in the lower right-hand corner of the map. On the straightedge of a piece of scrap paper, mark off a distance equivalent to 50 kilometers. Now locate Mt. Marcy. Using the scrap paper, mark a point 50 kilometers to the southwest of Mt. Marcy. This point represents the location of the epicenter of the Blue Mountain Lake earthquake. Next locate Old Forge and New York City. Note that Old Forge is much closer to the epicenter than New York City. Therefore, the difference between the arrival time of the first *P*-wave and the arrival of the first *S*-wave is greater in New York City than in Old Forge because New York City is farther from the epicenter than Old Forge.

One credit is allowed for ***both*** **circling New York City and providing an acceptable explanation.**

Acceptable responses include but are not limited to:

- **New York City is farther from the epicenter, so there is a greater difference between the arrival of the first *P*-wave and the first *S*-wave.**
- **As distance to the epicenter increases, the difference in arrival times increases.**
- **Old Forge is closer to the epicenter near Blue Mountain Lake (*or* Mt. Marcy), so the arrival times are closer together.**

Note: Credit is allowed if New York City is not circled but is correctly used in the explanation. All responses must correctly refer to the earthquake epicenter or earthquake origin to receive credit.

63. It is given that the seismic station is 1200 kilometers from the epicenter. Find the Earthquake P-Wave and S-Wave Travel Time graph in the *Reference Tables for Physical Setting/Earth Science.* Locate the point corresponding to 1200 kilometers along the horizontal axis at the bottom of the graph labeled "Epicenter Distance ($\times$ 10^3 km)." From 1200 km, trace vertically until you intersect the bold line labeled "*P*." From this intersection, trace horizontally to the left to the vertical axis labeled "Travel Time (min)." Read the *P*-wave travel time—about 2 minutes 35 seconds. Thus, the time needed for a *P*-wave to travel the 1200 km from the epicenter to the seismic station is 2 min 35 s.

One credit is allowed for **any value from 2 min 30 s to 2 min 40 s**.

64. Find the Generalized Bedrock Geology of New York State map in the *Reference Tables for Physical Setting/Earth Science.* Then find the point representing the Blue Mountain Lake earthquake epicenter you located in question 62. Note the map symbol for the surface bedrock at that point. Depending on how precisely you determined the direction to the southwest of Mt. Marcy, you might have located your point on one of two different map symbols. Now locate the two map symbols closest to the epicenter in the key labeled "Geologic Periods and Eras in New York." Note that one is labeled "MIDDLE PROTEROZOIC gneisses, quartzites, and marbles" and the other is labeled "MIDDLE PROTEROZOIC anorthositic rocks." Thus, the type of metamorphic surface bedrock where this earthquake was located is gneiss, quartzite, marble, or anorthositic rocks.

One credit is allowed for an acceptable response. Acceptable responses include but are not limited to:

- **gneiss**
- **quartzite**
- **marble**
- **anorthositic rocks**

65. In order to reach Perth, Australia, on the opposite side of Earth, the seismic waves from this earthquake would have to pass through Earth's center and all of Earth's interior layers. Find the Inferred Properties of Earth's Interior chart in the *Reference Tables for Physical Setting/Earth Science*. In the upper section of the diagram, find the layer labeled "Outer Core (Iron & Nickel)." Next trace the dashed lines marking the upper and lower boundaries of the outer core downward to the graph of temperature versus depth at the bottom of the chart. Note that in the sec-

tion of the graph corresponding to the outer core, the interior temperature (solid line) is higher than the melting temperature (dotted line), indicating that the outer core is liquid. *S*-waves are transverse waves that twist rock back and forth, deforming the rock's shape in a direction perpendicular to that of wave travel. *S*-waves can be transmitted only through solids. *S*-waves cannot travel through liquids and gases because when liquids and gases are deformed, they do not return to their original shape. Therefore, *S*-waves from this earthquake that travel toward Earth's center are absorbed by the liquid outer core rather than being transmitted.

One credit is allowed for ***both*** **outer core and an acceptable characteristic**.

Acceptable responses of characteristics of this layer include but are not limited to:

- **liquid/fluid**
- **molten rock**
- **melted iron and nickel**
- **temperature greater than melting point**

Note: Do *not* accept the response "core" alone because it does not indicate the liquid part of the core.

PART C

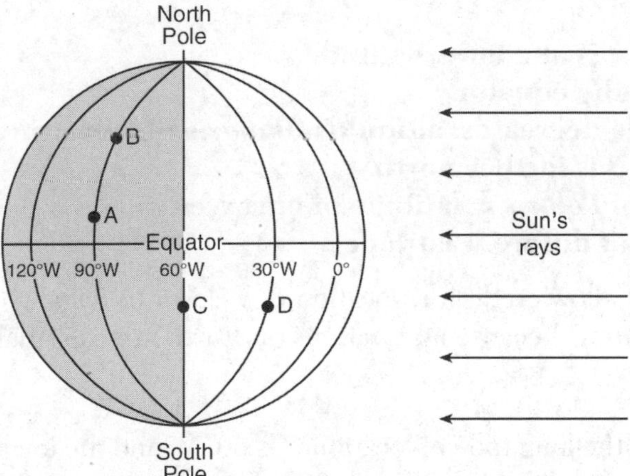

66. When sunlight strikes the spherical Earth, exactly one-half of Earth's sphere is illuminated and experiences daylight and exactly one-half is in darkness. The boundary between daylight and darkness forms a great circle that bisects Earth's sphere. On an equinox, the Sun's direct rays are at the equator. On that

day, the boundary between daylight and darkness passes through both the North and South Poles as it bisects Earth's spherical surface. Thus, the date represented by the position of Earth in this diagram is an equinox.

One credit is allowed. Acceptable responses include but are not limited to:

- **March 19, 20, 21, or 22**
- **September 21, 22, 23, or 24**
- **Spring/vernal equinox**
- **Autumn/fall equinox**
- **Equinox**
- **3/20**

67. Recall that the latitude of the North Pole is 90° and the latitude of the equator is 0°. Thus, location *A*, which is nearer the equator, has a lower latitude than location *B*, which is nearer the North Pole. Simple geometry shows that to an observer in the Northern Hemisphere, the altitude of *Polaris* is the same as the observer's latitude. (Refer to Figure 4.5 in *Let's Review: Earth Science, The Physical Setting* for a fuller explanation.) Therefore, the relative altitude of *Polaris* at location *A* is lower than at location *B* because the latitude of location *A* is lower than the latitude of location *B*.

Allow 1 credit for ***both*** **lower and an acceptable explanation**.

Acceptable responses include, but are not limited to:

- **Location *A* is farther from the North Pole, where *Polaris* is directly overhead.**
- **Location *A* is at a lower latitude.**
- **Closer to the equator**
- **As latitude decreases, altitude of *Polaris* decreases.**
- **Location *B* is farther north.**
- **Altitude of *Polaris* = latitude of observer**
- **They are at different latitudes.**

Note: Do *not* allow credit for "location *B* is closer to *Polaris*" or "location *A* is farther from *Polaris*" because all locations on Earth are essentially the same distance from *Polaris*.

68. Note that the longitude of location *C* is 60° W and the longitude of location *D* is 30° W. Thus, locations *C* and *D* are 30° of longitude apart. Since Earth makes one 360° rotation every 24 hours, in one hour Earth rotates 360°/24, or 15°. Thus, every 15° difference in longitude corresponds to a 1-hour difference in solar time. Since points *C* and *D* are 30° of longitude apart, the solar time difference between points *C* and *D* is 30/15, or 2 hours. Note that location *D* is to the east of location *C*.

Since Earth rotates from west to east, for every 15° longitude you travel east of a location, the solar time is 1 hour later. Therefore, when the solar time at location *C* is 6:00 a.m., the solar time at location *D* is two hours later or 8:00 a.m.

One credit is allowed for **a response that indicates a time value of 8 a.m**. Acceptable responses include but are not limited to:

- **8:00 a.m.**
- **8 a.m.**
- **8 o'clock in the morning**
- **0800**

Hurricane Odile

Date	Location		Barometric Pressure (mb)	Wind Speed (kt)
	Latitude (° N)	Longitude (° W)		
September 12	15	105	993	50
September 13	16	106	983	65
September 14	19	107	918	120
September 15	23	110	941	110
September 16	27	113	987	55
September 17	30	114	995	40
September 18	31	112	1003	25

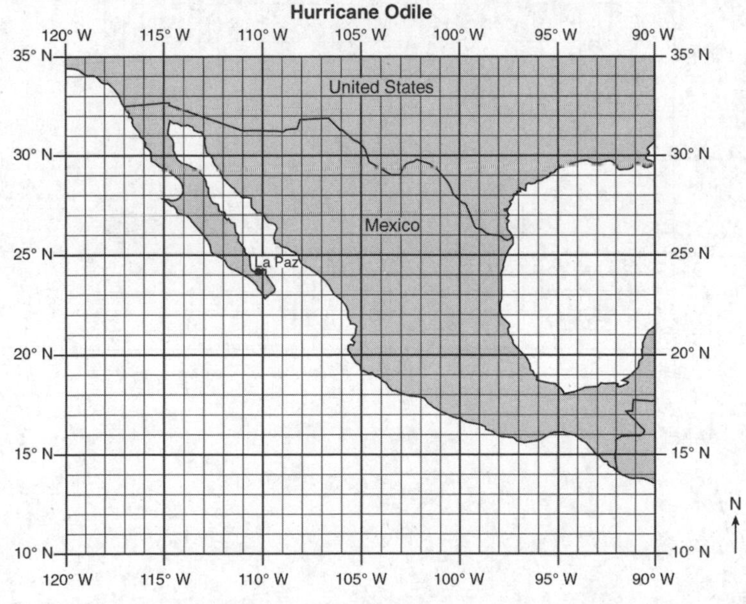

69. To plot the seven locations of Hurricane Odile indicated by the latitudes and longitudes shown in the data table, proceed as follows. From September 12, trace right to the column labeled "Latitude (°N)." Note the value "15." Continue tracing right to the column labeled "Longitude (°W)." Note the value "105." Thus on September 12, the coordinates of Hurricane Odile were 15° N, 105° W. On the map provided, locate 15° N on the vertical axis along the left side of the map. Now locate 105° W along the horizontal axis at the bottom of the map. Trace horizontally right from 15° N and vertically upward from 105° W until the lines intersect. Plot this point. Repeat this procedure for each of the dates listed in the data table. When completed, connect the plots with a solid line, as shown below.

One credit is allowed if **the centers of *all seven* plots are within or touch the circles shown and are correctly connected with a line that passes within or touches each circle**.

Note: Credit is allowed if the line does *not* pass through the student's plots but is still within or touches the circles.

Example of a 1-credit response:

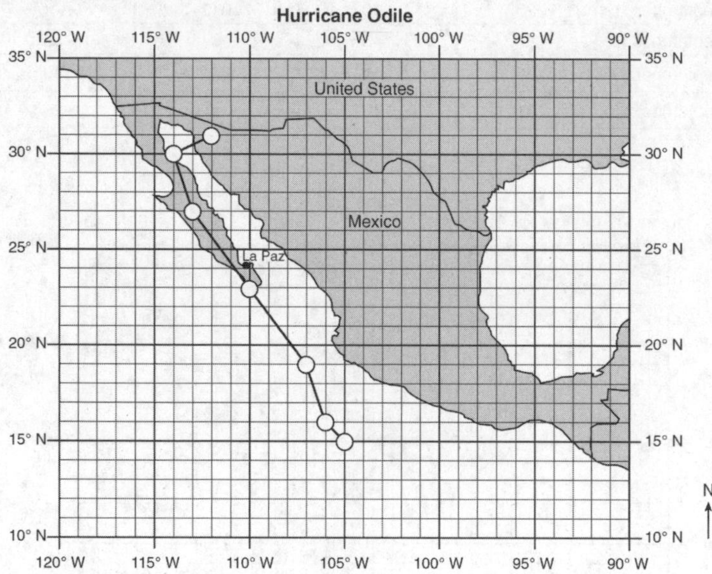

70. Note that the data table is ordered according to date. This makes it easy to track the motion of the hurricane over time. However, in order to see the relationship between barometric pressure and wind speed more easily, these data should be ordered by barometric pressure, as shown below:

Barometric Pressure (mb)	Wind Speed (kt)
918	120
941	110
983	65
987	55
993	50
995	40
1003	25

In this format, the relationship is readily apparent. As barometric pressure increases, wind speed decreases—an inverse relationship. Thus, the line on the graph should be drawn with a negative slope, that is, sloping downward from top left to bottom right.

One credit is allowed for **a line that shows a negative slope**.

Examples of 1-credit responses:

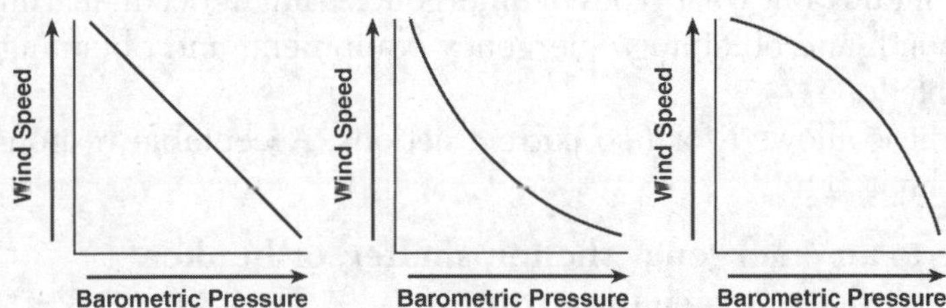

71. One of the most common instruments used to measure wind speed is an anemometer. A simple anemometer has cups mounted on a shaft that cause the shaft to spin when wind blows against them. The faster the wind blows, the faster the cups spin the shaft and the higher the number that registers on the anemometer's scale. Small handheld anemometers, called wind gauges, contain small fan blades that spin in the wind. There are also a variety of other instruments that can be used to measure wind speed. Some handheld digital wind speed meters contain a metal wire that is heated by an electrical current. The wire is cooled by air flow, and the meter instantly calculates wind speed. Wind speed can also be

measured using a pitot tube. A pitot tube is a narrow tube containing a fluid that is closed at one end. The open end of the tube faces toward the wind, which exerts a pressure on the fluid that compresses the fluid. The higher the wind speed, the more the liquid is compressed. Doppler radar can also be used to determine wind speed. For Doppler radar, atmospheric objects moving inbound (toward the radar) produce a positive shift in frequency of the radar signal. Objects moving away from the radar (outbound) produce a negative shift in frequency. This change in frequency can then be analyzed to determine wind speed.

One credit is allowed. Acceptable responses include but are not limited to:

- **anemometer**
- **wind speed meter**
- **wind gauge**
- **Doppler radar/radar**
- **pitot tube**

72. According to the National Weather Service, the major hazards associated with hurricanes are storm surge and storm tide, heavy rains, inland flooding, high winds, rip currents, and tornadoes. Actions that people who live in hurricane-prone areas should take in order to prepare for future hurricanes include finding out if you live in an evacuation area and learning your evacuation routes/shelter locations; assessing your risks and knowing your home's vulnerabilities to wind, storm surge, or flooding; having on hand the materials needed to secure your home; finding out what types of emergencies might occur, learning how you should respond, and obtaining emergency equipment; and obtaining water and nonperishable food.

One credit is allowed for *two* correct actions. Acceptable responses include but are not limited to:

- **Move to an emergency shelter, shelter, or bunker.**
- **Evacuate/learn evacuation routes.**
- **Stock up on supplies of food, water, and medicine.**
- **Assemble an emergency kit.**
- **Cover windows/board up windows.**
- **Secure outdoor furniture.**
- **Get a generator/make sure generator is working.**
- **Ensure vehicle has full tank of gas.**
- **Charge or stock up on batteries for electronic equipment.**
- **Stack sandbags where needed.**

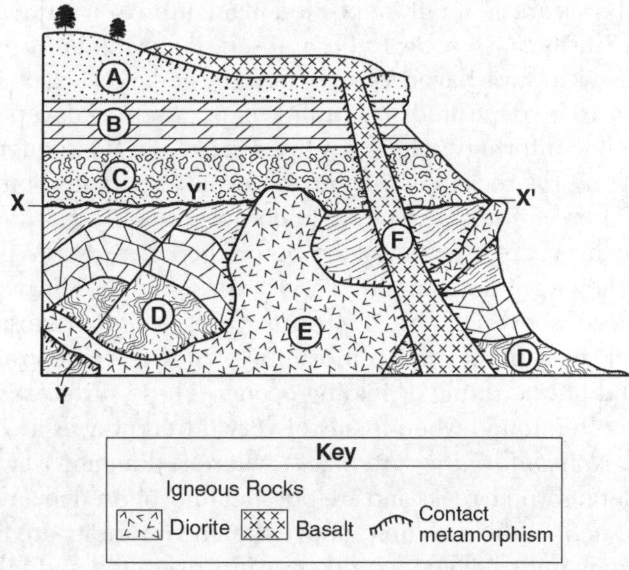

73. Note the map symbol for rock unit *F*. According to the key, the map symbol for rock unit *F* corresponds to basalt, an igneous rock. Note the map symbol for rock unit *A*. Find the Scheme for Sedimentary Rock Identification in the *Reference Tables for Physical Setting/Earth Science.* Locate the map symbol for rock unit *A*, trace left to the column labeled "Rock Name," and note that rock unit *A* is sandstone. From sandstone, trace left to the column labeled "Composition." Note that sandstone is composed mostly of quartz. Thus, the rock formed in the zone of contact metamorphism between rock units *A* and *F* is contact metamorphosed quartz sandstone. Find the Scheme for Metamorphic Rock Identification in the *Reference Tables for Physical Setting/Earth Science.* In the column labeled "Comments," locate the statement "Metamorphism of quartz sandstone." Trace this row left to the column labeled "Type of Metamorphism," and note the statement "Regional or contact." Now trace this row right to the column labeled "Rock Name." Note that the rock formed by the contact metamorphism of quartz sandstone is quartzite. Additionally, in the column labeled "Comments," locate the statement "Various rocks changed by heat from nearby magma/lava." Trace this row left to the column labeled "Type of Metamorphism," and note the statement "Contact (heat)." Now trace this row right to the column labeled "Rock Name." Note that the rock formed by the contact metamorphism of various rocks by heat from nearby magma or lava is hornfels.

One credit is allowed for **quartzite *or* hornfels**.

74. Layers of rock are generally deposited in an unbroken sequence. However, if forces within Earth cause rocks to be uplifted, deposition ceases. Weathering and erosion may wear away layers of rock before the land surface is low enough for another layer to be deposited. The uplift may cause rock layers to tilt or fold. The result is an unconformity, which is a break or gap in the sequence of a series of rock layers. Thus, the rocks above an unconformity are quite a bit younger than those below it. There are several types of unconformities. *Angular unconformities* form when rock layers are tilted or folded before being eroded. When new layers are deposited, they form horizontally. The layers below the unconformity are at an angle to those above it. *Disconformities* are irregular erosional surfaces between parallel layers of rock. Disconformities occur when deposition stops and layers are eroded but no tilting or folding occurs. These surfaces are not easy to discern and are often found when fossils of very different ages are discovered in adjacent layers. *Nonconformities* are places where sedimentary layers lie on top of igneous or metamorphic rocks and are not metamorphosed in any way. According to the cross section, the rock units above the unconformity are flat, horizontal layers and the rock units below the unconformity are tilted and folded. There is also a fault along line *YY′* and the layers to the right of the fault moved upward relative to the layers to the left of the fault. Thus, unconformity *XX′* is most likely an angular unconformity that resulted when tilted and folded rock layers were uplifted and then worn down by weathering and erosion. This was followed by subsidence and submersion during which time rock layers *A*, *B*, and *C* were deposited. Then the land was again uplifted, exposing these rocks at the surface.

One credit is allowed for *two* correct responses. Acceptable responses include but are not limited to:

- **uplift/emergence**
- **weathering**
- **erosion**
- **subsidence/submergence**
- **deposition**
- **burial**

75. Note that fault *YY′* cuts across rock unit *D* and that rock unit *D* has been displaced along the fault. In order to be displaced, a rock must already have existed when the fault occurred. Therefore, fault *YY′* is younger than rock unit *D*. Note, however, that rock unit *E* cuts across fault *YY′* and has not been displaced. Therefore, rock layer *E* had not yet formed when the fault occurred. So rock layer *E* is younger than fault *YY′*. Note the map symbol for rock unit *E*. Locate this map symbol in the key. Note that it corresponds to diorite, which is an igneous rock. Note, too, the symbol for contact metamorphism along the boundary between

rock units D and E. Thus, rock unit E is an igneous intrusion that cuts across and contact metamorphosed rock unit D. An igneous intrusion is younger than any rock layer or structure that it cuts across. In order to be contact metamorphosed, a rock layer must have already existed when the igneous intrusion occurred. Therefore, rock unit E is younger than the rock unit D. Now note the map symbol for rock unit F. According to the key, the map symbol for rock unit F corresponds to basalt, which is an igneous rock. Note that rock unit F cuts across rock unit E and that rock unit E has been contact metamorphosed. Therefore, rock unit F is younger than rock unit E. Thus, from oldest to youngest, the rock units and fault YY' are D, YY', E, and F.

One credit is allowed for $\boldsymbol{D \rightarrow}$ ***fault YY'*** $\boldsymbol{\rightarrow E \rightarrow F}$.

Note: Credit is allowed for fault, *or* YY' alone, *or* Y alone in place of fault YY'. Credit is allowed if the correct rock names are substituted for D (gneiss), E (diorite), and F (basalt).

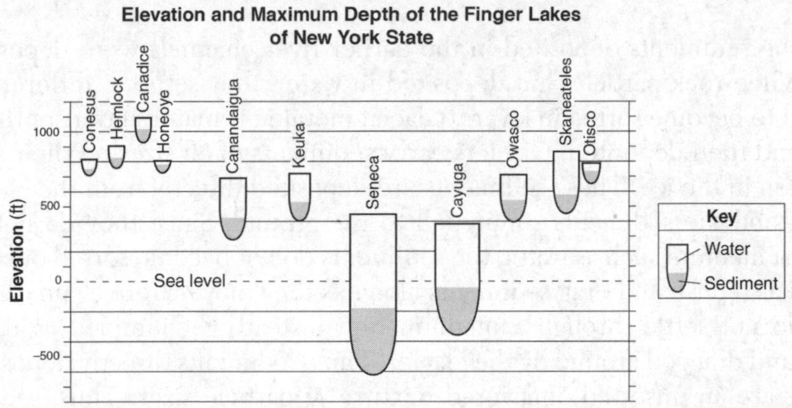

76. In the key along the right side of the cross section, note how water and sediment in the Finger Lakes are indicated. The maximum water depth of a Finger Lake is the vertical distance between the top of the sediment at the bottom of a lake and the surface of that lake in the cross section (the region corresponding to "Water" in the key). Find the vertical scale labeled "Elevation (ft)" along the left side of the cross section. Note that the scale is marked off in subdivisions of 100. Using the straightedge of a piece of scrap paper, mark off a distance equivalent to 175 ft, or about 1¾ subdivisions. Holding the piece of scrap paper vertically, compare the marked off distance of 175 ft on the scrap paper with the distance between the top of the sediment at the bottom of each lake and the water surface of the lake. Note that Keuka Lake and Owasco Lake have a maximum water depth of about 175 ft.

One credit is allowed for **Owasco Lake *or* Keuka Lake**.

77. According to the reading passage, "Around 1.7 to 1.8 million years ago, a continental glacier advanced southward, beginning an ice age that consisted of many advances and retreats of glaciers. The Finger Lakes were carved by some of these advances." Thus, the continental glaciers formed the Finger Lakes around 1.7 to 1.8 million years ago. Find the Geologic History of New York State chart in the *Reference Tables for Physical Setting/Earth Science*. On the vertical time scale labeled "Million years ago" to the right of the column labeled "Epoch," locate the row corresponding to between 0.01 and 1.8 million years ago. This range includes 1.7 to 1.8 million years ago. Trace this row left to the column labeled "Epoch," and note that this range corresponds to the Pleistocene epoch. Now trace this row left to the column labeled "Period," and note that this range corresponds to the Quaternary period. Thus, these continental glaciers formed the Finger Lakes of New York State during the Pleistocene epoch of the Quaternary Period.

One credit is allowed for ***both* the Quaternary period and Pleistocene epoch**.

78. The sediments deposited in the earlier river channels were deposited in water. When rock particles are deposited in water, they settle at different rates and tend to become sorted in layers. Glacial moraine is material transported by a glacier and then deposited. Glaciers carry sediments of all sizes on their surface and frozen in the ice. These sediments are deposited directly from the glacier as it melts, and the sediments simply fall to the ground. Since they do not settle through a medium such as water, the sediments do not become sorted or layered. Glaciers also push and drag sediments along as they move. Here again, the sediments do not settle through a medium but, instead, get jumbled as they get pushed and dragged around by the glacier. Thus, sediments directly deposited by a glacier are an unsorted, unlayered mixture of particle sizes. Thus, sediments deposited in glacial moraines are unsorted and unlayered, while sediments deposited in earlier river channels are sorted and layered.

One credit is allowed. Acceptable responses include but are not limited to:

- **Glacial moraine deposits are unsorted/mixed.**
- **Glacial deposits in a moraine are unlayered.**
- **Unsorted**
- **River deposits are sorted.**

79. In most streams, water continues to flow even when there is no rainfall or runoff from the surrounding slopes. Therefore, the stream erodes downward into the underlying bedrock even when the sides of the valley are not being eroded by runoff. As a result, the stream valley becomes deeper faster than it becomes wider, forming a narrow valley bottom with sloping sides, or a V shape. Thus, the

cross-sectional shape of the original river valleys before they were gouged by the advancing glacier was most likely a V shape, with a narrow valley bottom with sloping sides.

One credit is allowed. Acceptable responses include but are not limited to:

- **V-shaped valleys**
- **narrow valley bottom with gently sloped sides**

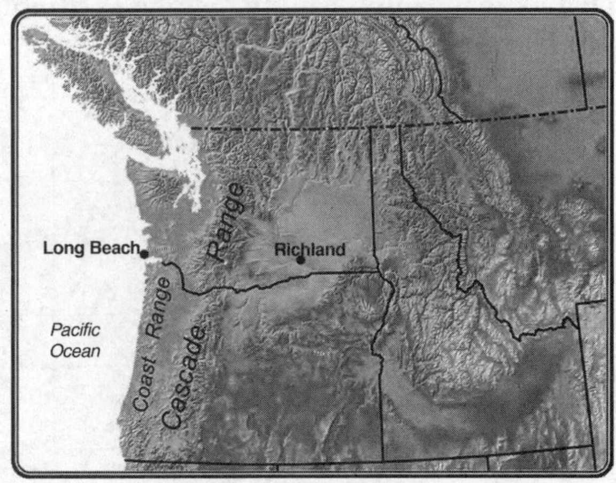

80. Find the Surface Ocean Currents map in the *Reference Tables for Physical Setting/Earth Science*. Locate the region corresponding to the Pacific Northwest Coast of the United States. Note that the California Current flows southward along the coast in this region. According to the key, the California Current is a cool current. Therefore, the name of the cool surface ocean current that influences the climate of this region is the California Current.

One credit is allowed for **California Current**.

81. According to the map, Long Beach and Richland are located at about the same latitude. Therefore, both cities receive the same intensity of insolation. However, Long Beach is located on the coast next to the Pacific Ocean while Richland is an inland location. Due to the high specific heat of water, land surfaces increase in temperature more than water surfaces when insolation strikes. Land surfaces also cool more rapidly than water surfaces. Furthermore, moist air has a higher specific heat than dry air, so dry air heats and cools more rapidly than moist air. Therefore, large bodies of water tend to moderate the temperatures of coastal regions of nearby landmasses by warming them in winter and cooling them in summer. Thus Long Beach, which is a coastal city, experiences a smaller range of temperatures throughout the year because the nearby Pacific Ocean moderates its temperatures.

One credit is allowed for an acceptable response. Acceptable responses include but are not limited to:

- **The Pacific Ocean moderates the climate of Long Beach.**
- **Water has a higher specific heat, and temperatures change more slowly.**
- **Long Beach is near a large body of water.**
- **Richland is farther inland.**
- **Long Beach has a marine climate, whereas Richland's is more continental.**
- **It is located closer to a large body of water.**

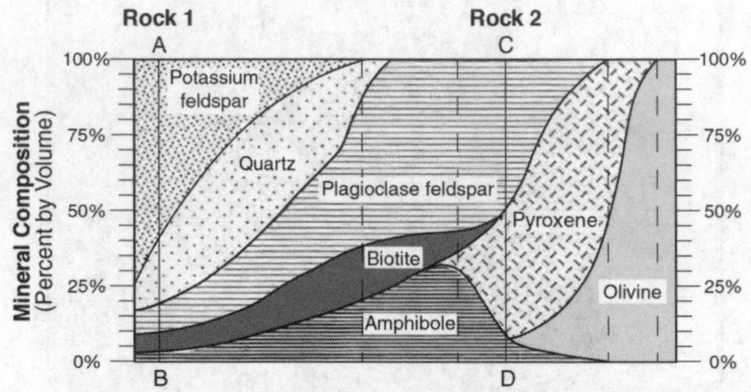

82. The question states that rock 1 is a glassy, vesicular igneous rock. Find the Scheme for Igneous Rock Identification in the *Reference Tables for Physical Setting/Earth Science*. On the graph labeled "Mineral Composition (relative by volume)," locate the position corresponding to the line representing rock 1 on the diagram in the question. Trace vertically upward from this position to the section of the Scheme for Igneous Rock Identification labeled "Igneous Rocks." In the column labeled "Texture," locate the row corresponding to both a "Glassy" and "Vesicular (gas pockets)" texture. Trace this row to the left to where it intersects with the position corresponding to rock 1. Note the name "Pumice." Thus, the name of rock 1 is pumice.

One credit is allowed for **pumice**.

83. Find the Scheme for Igneous Rock Identification in the *Reference Tables for Physical Setting/Earth Science*. On the graph labeled "Mineral Composition (relative by volume)," locate the position corresponding to the line representing rock 1 on the diagram in the question. Trace upward to the section of the Scheme for Igneous Rock Identification labeled "Characteristics." Note that an igneous rock with the composition of rock 1 intersects the arrow labeled "Color" near the end of the range labeled "Lighter" and intersects the arrow labeled "Density" near the end of the range labeled "Lower."

On the graph labeled "Mineral Composition (relative by volume)," locate the position corresponding to the line representing rock 2 on the diagram in the question. Trace upward to the section of the Scheme for Igneous Rock Identification labeled "Characteristics." Note that an igneous rock with the composition of rock 2 intersects the arrow labeled "Color" near the end of the range labeled "Darker" and intersects the arrow labeled "Density" near the end of the range labeled "Higher." Thus, compared to the color and density of igneous rock 1, igneous rock 2 is darker in color and has a higher density.

Another way to compare color is to compare the colors of the major minerals that comprise rocks 1 and 2. According to the diagram, rock 1 is composed mainly of potassium feldspar, quartz, and plagioclase feldspar. Rock 2 is composed mainly of plagioclase feldspar, pyroxene, and amphibole. Find the Properties of Common Minerals table in the *Reference Tables for Physical Setting/Earth Science.* In the column labeled "Mineral Name," locate potassium feldspar, quartz, plagioclase feldspar, pyroxene, and amphibole. From each of these minerals, trace left to the column labeled "Common Colors." Note the common colors of these minerals: potassium feldspar—white to pink, quartz—colorless or variable, plagioclase feldspar—white to gray, pyroxene—black to dark green, and amphibole—black to dark green. Therefore, rock 1 is composed mainly of light-colored minerals, and rock 2 is composed mainly of dark-colored minerals, particularly black to dark green pyroxene.

One credit is allowed if *both* responses are acceptable. Acceptable responses include but are not limited to:

Relative color of rock 2:

- **Igneous rock 2 is darker/dark.**
- **Blacker/black**
- **Greener/green**
- **Rock 1 is lighter than rock 2.**

Relative density of rock 2:

- **Igneous rock 2 has a higher density.**
- **Greater**
- **More dense/denser**
- **Rock 1 is less dense than rock 2.**

Note: The response "more mafic" is *not* accepted for either relative color or relative density of rock 2 as this refers to composition.

84. Find the Scheme for Igneous Rock Identification in the *Reference Tables for Physical Setting/Earth Science*. In the upper section of the scheme labeled "Igneous Rocks," locate andesite and diorite. Trace these two rows left to the column labeled "Environment of Formation." Note that the row corresponding to andesite is in the range labeled "Extrusive" and that the row corresponding to diorite is in the range labeled "Intrusive." Thus, andesite is an extrusive (volcanic) rock formed by the cooling of lava, and diorite is an intrusive (plutonic) rock formed by the cooling of magma. Now trace these two rows right to the column labeled "Texture." Note that the row corresponding to andesite is in the range labeled "Fine" and that the row corresponding to diorite is in the range labeled "Coarse." Therefore, andesite and diorite can both have the same percentage of mineral composition by volume and yet be two different igneous rocks because they formed in different environments and have different textures.

One credit is allowed. Acceptable responses include but are not limited to:

- **The rocks may have different environments of formation.**
- **Diorite is plutonic, and andesite is volcanic.**
- **One forms from cooling magma, and one forms from cooling lava.**
- **They cooled at different rates.**
- **They have different grain sizes or textures.**
- **One is extrusive, and one is intrusive.**
- **They formed differently.**

85. Find the Properties of Common Minerals table in the *Reference Tables for Physical Setting/Earth Science*. In the column labeled "Composition," locate "$(Na,Ca)AlSi_3O_8$." From this chemical formula, trace right to the column labeled "Mineral Name." Note that this composition corresponds to plagioclase feldspar. Repeat this procedure for the other two chemical formulas listed in the table. Note that "$KAlSi_3O_8$" corresponds to potassium feldspar and that "$(Fe,Mg)_2SiO_4$" corresponds to olivine.

According to the diagram in the question, rock 1 is composed of potassium feldspar, quartz, plagioclase feldspar, biotite, and amphibole. Rock 2 is composed of plagioclase feldspar, pyroxene, and amphibole. Therefore, plagioclase feldspar [$(Na,Ca)AlSi_3O_8$] is found in both rock 1 and rock 2, potassium feldspar [$KAlSi_3O_8$] is found in rock 1 only, and olivine [$(Fe,Mg)_2SiO_4$] is found in neither rock 1 nor rock 2. This information should be recorded on the answer table, as shown below:

One credit is allowed if ***all three*** **Xs are only placed in the correct boxes, as shown below**.

Note: Allow credit if a symbol other than an **X** is used.

Mineral Composition	Found in both rock 1 and rock 2	Found in neither rock 1 nor rock 2	Found in rock 1, only	Found in rock 2, only
$(Na,Ca)AlSi_3O_8$	X			
$KAlSi_3O_8$			X	
$(Fe,Mg)_2SiO_4$		X		

Topic	Question Numbers (Total)	Wrong Answers (x)	Grade
Standards 1, 2, 6, and 7: Skills and Application			
Skills Standard 1 Analysis, Inquiry, and Design	2, 6, 8–10, 12, 13, 15, 17–19, 24–26, 28, 29, 32–40, 47–51, 55–57, 60, 62–71, 73, 74, 76, 77, 79–85		$\frac{100(55-x)}{55} = \%$
Standard 2 Information Systems	72		$\frac{100(1-x)}{1} = \%$
Standard 6 Interconnectedness, Common Themes	1, 6, 7, 9, 16, 17, 20, 23, 26, 27, 30–32, 34–36, 39–46, 52–55, 57–59, 61, 62, 64, 66, 67, 69, 73, 75, 76, 78, 79, 81		$\frac{100(43-x)}{43} = \%$
Standard 7 Interdisciplinary Problem Solving	72		$\frac{100(1-x)}{1} = \%$
Standard 4: The Physical Setting/Earth Science			
Astronomy The Solar System (MU 1.1a, b; 1.2d)	6, 40, 51–53, 66		$\frac{100(6-x)}{6} = \%$
Earth Motions and Their Effects (MU 1.1c, d, e, f, g, h, i)	3–5, 26, 39, 41, 67–69		$\frac{100(9-x)}{9} = \%$
Stellar Astronomy (MU 1.2b)	1, 54–57		$\frac{100(5-x)}{5} = \%$
Origin of Earth's Atmosphere, Hydrosphere, and Lithosphere (MU 1.2e, f, h)			
Theories of the Origin of the Universe and Solar System (MU 1.2a, c)	2, 36, 37		$\frac{100(3-x)}{3} = \%$

Topic	Question Numbers (Total)	Wrong Answers (x)	Grade
Meteorology			
Energy Sources for Earth Systems (MU 2.1a, b)			
Weather (MU 2.1c, d, e, f, g, h)	8–11, 27, 28, 34, 70–72		$\frac{100(10-x)}{10} = \%$
Insolation and Seasonal Changes (MU 2.1i; 2.2a, b)	13–16		$\frac{100(4-x)}{4} = \%$
The Water Cycle and Climates (MU 1.2g; 2.2c, d)	7, 12, 30, 42, 43, 80, 81		$\frac{100(7-x)}{7} = \%$
Geology			
Minerals and Rocks (MU 3.1a, b, c)	24, 25, 33, 64, 73, 82–85		$\frac{100(9-x)}{9} = \%$
Weathering, Erosion, and Deposition (MU 2.1s, t, u, v, w)	22, 23, 44–46, 76, 78, 79		$\frac{100(8-x)}{8} = \%$
Plate Tectonics and Earth's Interior (MU 2.1j, k, l, m, n, o)	17–19, 32, 35, 38, 62, 63, 65		$\frac{100(9-x)}{9} = \%$
Geologic History (MU 1.2i, j)	29, 31, 47–50, 74, 75, 77		$\frac{100(9-x)}{9} = \%$
Topographic Maps and Landscapes (MU 2.1p, q, r)	20, 21, 58–61		$\frac{100(6-x)}{6} = \%$
ESRT			
2011 Edition Reference Tables for Physical Setting/ Earth Science	8–11, 13, 15, 18, 19, 24, 25, 29, 32–35, 37, 38, 48, 50, 55, 57, 60, 62–65, 73, 77, 80, 82–85		$\frac{100(33-x)}{33} = \%$

To further pinpoint your weak areas, use the Topic Outline in the front of the book.
MU = Major Understanding (see Topic Outline)

Examination June 2018

Physical Setting/Earth Science

PART A

Answer all questions in this part.

Directions (1–35): For *each* statement or question, choose the word or expression that, of those given, best completes the statement or answers the question. Some questions may require the use of the *2011 Edition Reference Tables for Physical Setting/Earth Science*. Record your answers in the space provided.

1 The photographs below show two types of solar eclipses. Letters *A* and *B* represent two celestial objects.

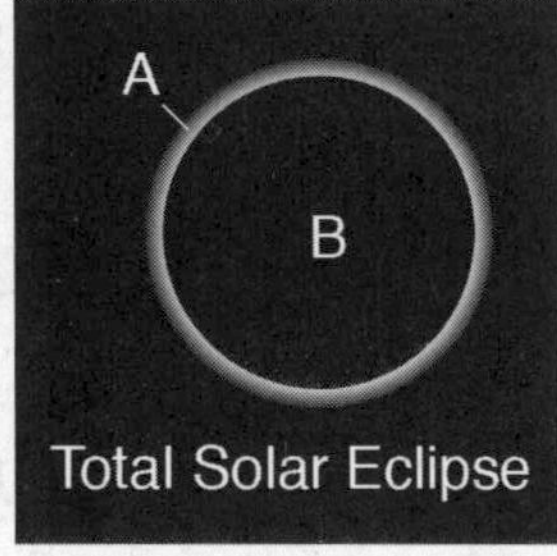

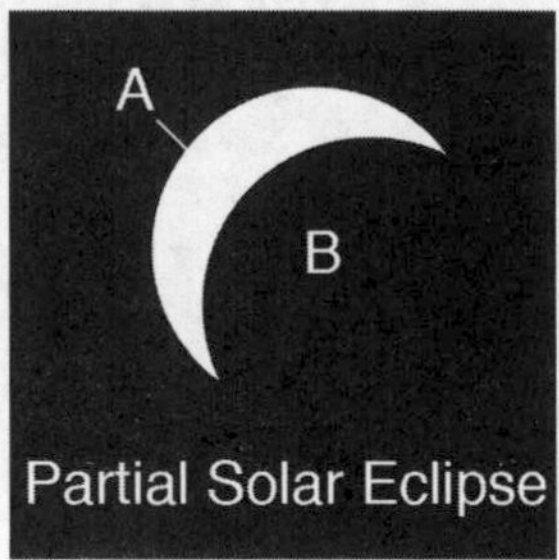

Which two celestial objects are represented by letters *A* and *B*?

(1) *A*-Moon; *B*-Sun
(2) *A*-Moon; *B*-Earth
(3) *A*-Sun; *B*-Moon
(4) *A*-Sun; *B*-Earth

1 _____

2 Compared to the terrestrial planets, the Jovian planets

(1) are less massive
(2) are more dense
(3) have greater orbital velocities
(4) have shorter periods of rotation 2____

3 Which event occurred more than 10 billion years ago?

(1) Big Bang
(2) origin of life on Earth
(3) Pangaea begins to break up
(4) origin of Earth and its Moon 3____

4 In 1851, French physicist Léon Foucault used a swinging pendulum to demonstrate that Earth

(1) is rotating
(2) is revolving
(3) has a curved surface
(4) has a gravitational pull 4____

5 Approximately how many degrees does Earth travel in its orbit in one month?

(1) 1° (3) 30°
(2) 15° (4) 360° 5____

6 What is the relative humidity when the drybulb temperature is 16°C and the wet-bulb temperature is 10°C?

(1) 6% (3) 33%
(2) 14% (4) 45% 6____

7 Boarding up windows would be one emergency action most likely taken to prepare for which natural disaster?

(1) earthquake
(2) hurricane
(3) flood
(4) tsunami

7 _____

8 Which diagram best represents the general position and direction of flow of the polar front jet stream in the Northern Hemisphere during the winter months?

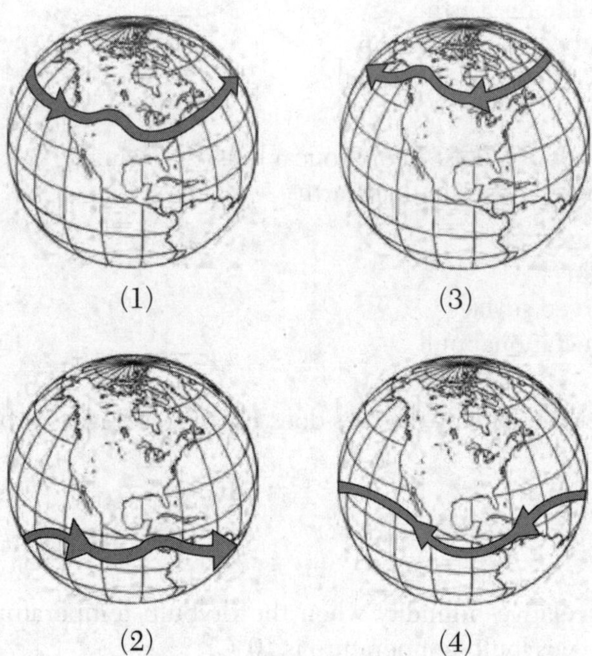

8 _____

9 The diagram below represents four positions of the Moon, labeled *A* through *D*, as it orbits Earth.

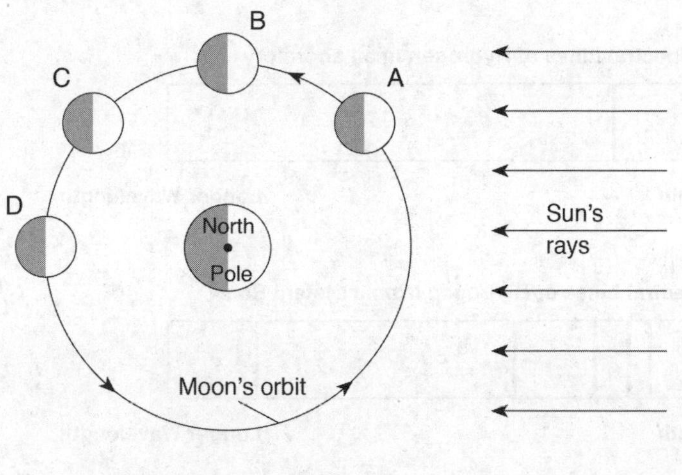

Which diagram best represents the sequence of Moon phases, as seen by an observer in New York State, when the Moon travels from position *A* to position *D* in its orbit around Earth?

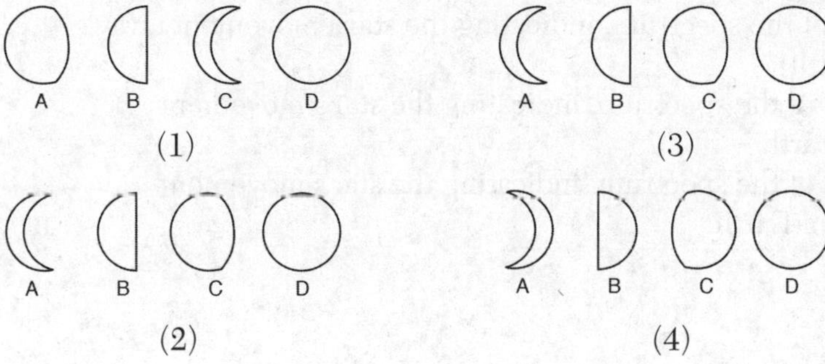

9 ____

10 The diagrams below represent spectral lines of hydrogen gas observed in a laboratory and the spectral lines of hydrogen gas observed in the light from a distant star.

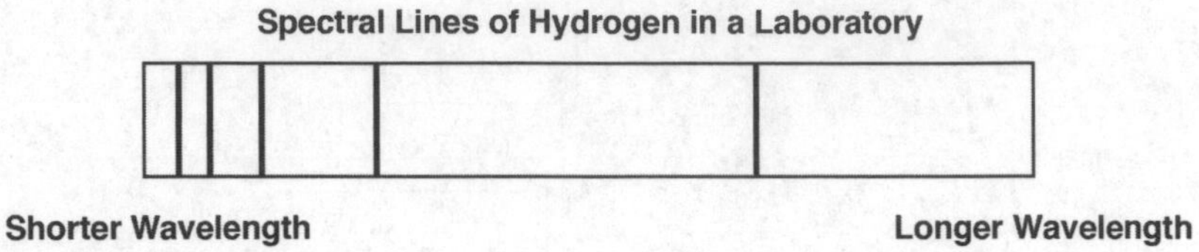

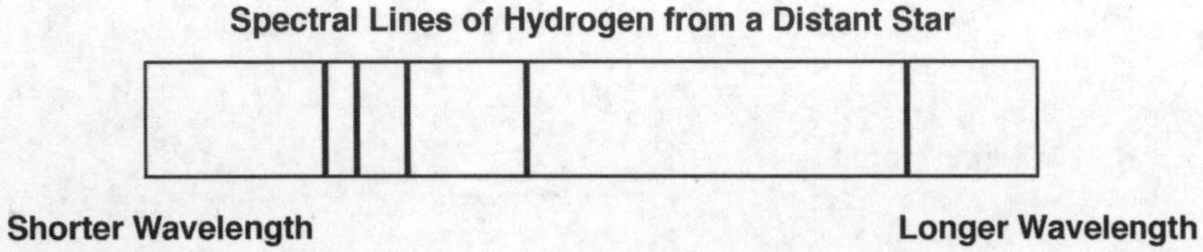

Spectral Lines of Hydrogen in a Laboratory

Shorter Wavelength **Longer Wavelength**

Spectral Lines of Hydrogen from a Distant Star

Shorter Wavelength **Longer Wavelength**

Compared to the spectral lines observed in the laboratory, the spectral lines observed in the light from the distant star have shifted toward the

(1) red end of the spectrum, indicating the star's movement toward Earth
(2) red end of the spectrum, indicating the star's movement away from Earth
(3) blue end of the spectrum, indicating the star's movement toward Earth
(4) blue end of the spectrum, indicating the star's movement away from Earth

10 ____

11 The diagram below represents a cross-sectional view of the plane of Earth's orbit around the Sun. A line drawn perpendicular to the plane of Earth's orbit is shown on the diagram.

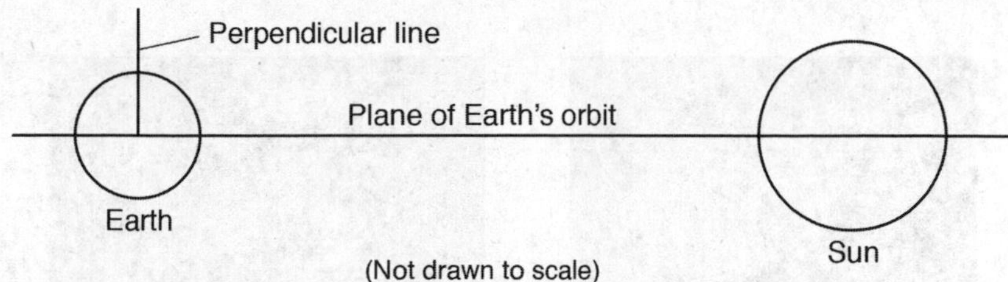

How many degrees is Earth's rotational axis tilted with respect to the perpendicular line shown in the diagram?

(1) 15° (3) 90°
(2) 23.5° (4) 180° 11 ____

12 The larger white dots in the diagrams below represent stars in the constellations Scorpius and Orion. Information indicating when these constellations are visible from New York State is provided below the diagrams.

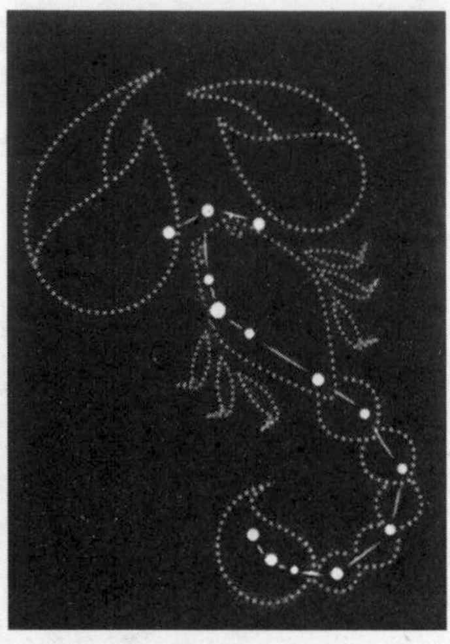

Scorpius

Visible in the New York State nighttime sky during July; not visible at all in January

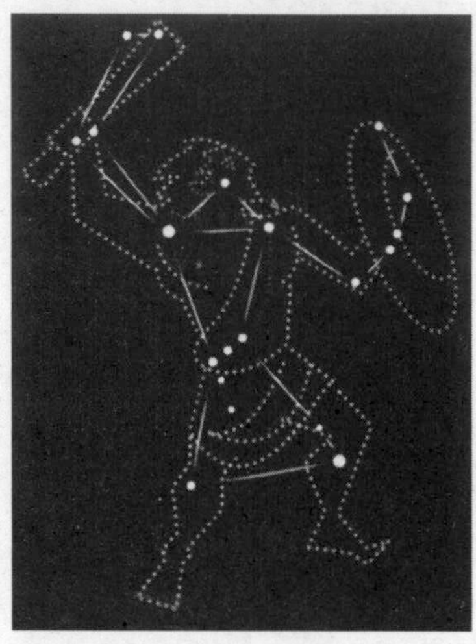

Orion

Visible in the New York State nighttime sky during January; not visible at all in July

Which statement best explains why these two constellations are visible in the night sky in the months identified?

(1) Earth spins on its axis at a constant rate during a 24-hour period.
(2) Earth spins on its axis at a variable rate during the year.
(3) The nighttime side of Earth is facing different parts of our galaxy as Earth orbits the Sun.
(4) The nighttime side of Earth is facing different parts of our galaxy as the stars orbit Earth.

12 _____

13 Which table correctly shows the interior temperature, melting point, and state (phase) of matter of the materials located 4000 kilometers below Earth's surface?

Interior Temperature (°C)	Melting Point (°C)	State of Matter
5700	5400	solid

(1)

Interior Temperature (°C)	Melting Point (°C)	State of Matter
5700	5400	liquid

(2)

Interior Temperature (°C)	Melting Point (°C)	State of Matter
5400	5700	solid

(3)

Interior Temperature (°C)	Melting Point (°C)	State of Matter
5400	5700	liquid

(4)

13 ____

14 Which gas is a greenhouse gas that has increased in Earth's atmosphere partly as a result of deforestation over the last 100 years?

(1) ozone
(2) oxygen
(3) nitrogen
(4) carbon dioxide

14 ____

15 Which ocean current brings warm water to the southeastern tip of Africa?

(1) Brazil Current
(2) Agulhas Current
(3) Guinea Current
(4) Benguela Current

15 ____

16 Which pie graph is shaded to best represent the approximate percentage of time that humans have existed during Earth's entire history?

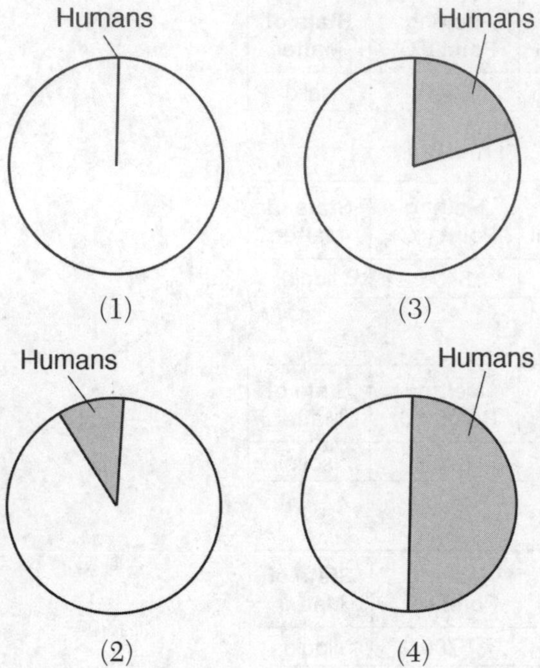

16 ____

17 Volcanic ash can be used as a time marker to correlate rock layers because the ash

(1) is deposited rapidly over a large area
(2) represents a buried erosional surface
(3) forms intrusive igneous rock
(4) cuts across rock layers

17 ____

18 The cross section below represents a mountain range. Points *A* and *B* represent locations on Earth's surface.

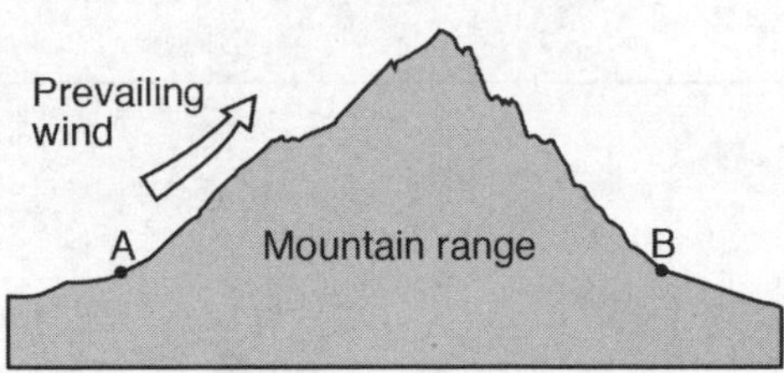

Compared to the climate of location *A*, the climate of location *B* is most likely

(1) cooler and wetter
(2) cooler and drier
(3) warmer and wetter
(4) warmer and drier

18 _____

19 The photograph below shows conglomerate composed of pebbles cemented together with calcite.

Compared to the ages of the calcite cement and the conglomerate, the relative age of the pebbles is

(1) younger than both the calcite cement and the conglomerate
(2) younger than the calcite cement, but the same age as the conglomerate
(3) older than both the calcite cement and the conglomerate
(4) older than the calcite cement, but the same age as the conglomerate

19 _____

20 The cross section below represents some parts of Earth's water cycle. Letters *A*, *B*, *C*, and *D* represent processes that occur during the cycle.

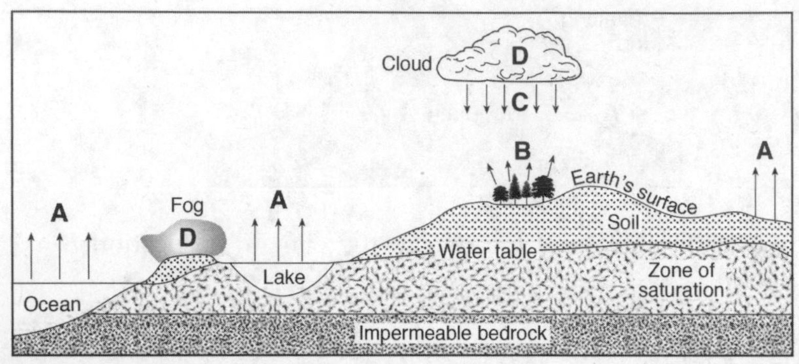

Which table correctly matches each letter with the process it represents?

(1)

Letter	Process
A	Condensation
B	Transpiration
C	Precipitation
D	Evaporation

(2)

Letter	Process
A	Evaporation
B	Precipitation
C	Transpiration
D	Condensation

(3)

Letter	Process
A	Condensation
B	Evaporation
C	Precipitation
D	Transpiration

(4)

Letter	Process
A	Evaporation
B	Transpiration
C	Precipitation
D	Condensation

20 ______

21 Which table best shows the relationship between latitude and general climate conditions on Earth?

Latitude	Climate Conditions
90°N	Arid
60°N	Arid
30°N	Humid
0°	Humid
30°S	Humid
60°S	Arid
90°S	Arid

(1)

Latitude	Climate Conditions
90°N	Humid
60°N	Arid
30°N	Humid
0°	Humid
30°S	Humid
60°S	Arid
90°S	Humid

(3)

Latitude	Climate Conditions
90°N	Arid
60°N	Humid
30°N	Arid
0°	Humid
30°S	Arid
60°S	Humid
90°S	Arid

(2)

Latitude	Climate Conditions
90°N	Humid
60°N	Arid
30°N	Humid
0°	Arid
30°S	Humid
60°S	Arid
90°S	Humid

(4)

21 ____

22 The photograph below shows different-sized rounded sediment.

Which table shows the most likely process and agent of erosion responsible for this rounded sediment?

Process	Agent of Erosion
sandblasting	running water

(1)

Process	Agent of Erosion
abrasion	wave action

(2)

Process	Agent of Erosion
land slide	mass movement

(3)

Process	Agent of Erosion
deposition	wind

(4)

22 ____

23 The photograph below shows an outcrop with two basaltic intrusions, labeled *A* and *B*, in a rock unit, labeled *C*.

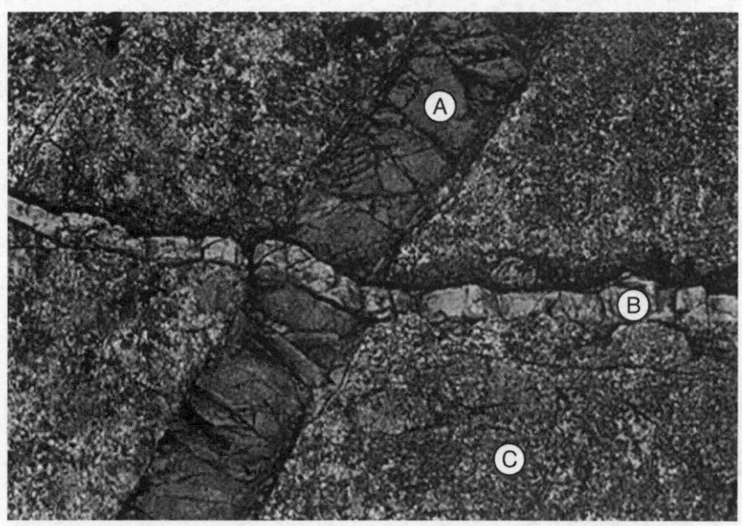

What is the relative age of these three rock units from oldest to youngest?

(1) $B \rightarrow A \rightarrow C$
(2) $B \rightarrow C \rightarrow A$
(3) $C \rightarrow A \rightarrow B$
(4) $C \rightarrow B \rightarrow A$

23 ____

24 The world map below shows Earth's major tectonic plate boundaries. Letters *A* through *D* represent four surface locations.

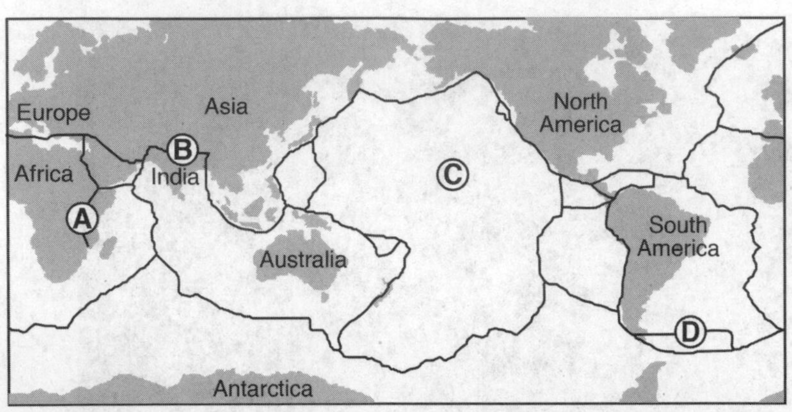

Which location is on a major rift valley?

(1) *A* (3) *C*
(2) *B* (4) *D* 24 ____

25 The first *P*-wave of an earthquake took 11 minutes to travel to a seismic station from the epicenter of the earthquake. What is the seismic station's distance to the epicenter of the earthquake and how long did it take for the first *S*-wave to travel that distance?

(1) Distance to epicenter: 3350 km
S-wave travel time: 4 min 50 sec
(2) Distance to epicenter: 3350 km
S-wave travel time: 6 min 10 sec
(3) Distance to epicenter: 7600 km
S-wave travel time: 9 min
(4) Distance to epicenter: 7600 km
S-wave travel time: 20 min 25 ____

26 The Catskills are commonly called mountains, but are actually part of the Allegheny Plateau. The Catskills are classified as a plateau because of their

(1) low elevation
(2) bedrock structure
(3) bedrock age
(4) high degree of metamorphism 26 ____

27 The minimum stream velocity necessary to transport a sediment particle that is 0.1 centimeter in diameter is closest to

(1) 0.1 cm/s (3) 5.5 cm/s
(2) 0.002 cm/s (4) 10.0 cm/s 27 ____

28 Which rock is classified as an evaporite?

(1) clastic shale
(2) foliated phyllite
(3) nonfoliated marble
(4) crystalline rock salt 28 ____

29 Which pair of elements makes up most of Earth's crust by volume?

(1) nitrogen and potassium
(2) oxygen and silicon
(3) hydrogen and oxygen
(4) potassium and oxygen 29 ____

30 The cross section below represents zones of soil labeled *A*, *B*, and *C*. Letter *D* represents underlying bedrock.

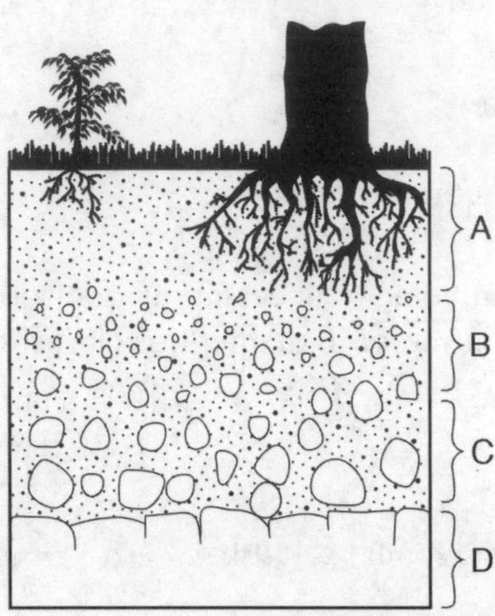

Which letter identifies the zone having the most organic and weathered material?

(1) *A*
(2) *B*
(3) *C*
(4) *D* 30 ____

31 Which type of surface bedrock is most commonly found in the Utica, New York area?

(1) sedimentary, with limestone, shale, sandstone, and doloston
(2) sedimentary, with limestone, shale, sandstone, and conglomerate
(3) metamorphic, with quartzite, dolostone, marble, and schist
(4) metamorphic, with gneiss, quartzite, marble, and slate 31 ____

32 The diagram below represents a geologic landscape.

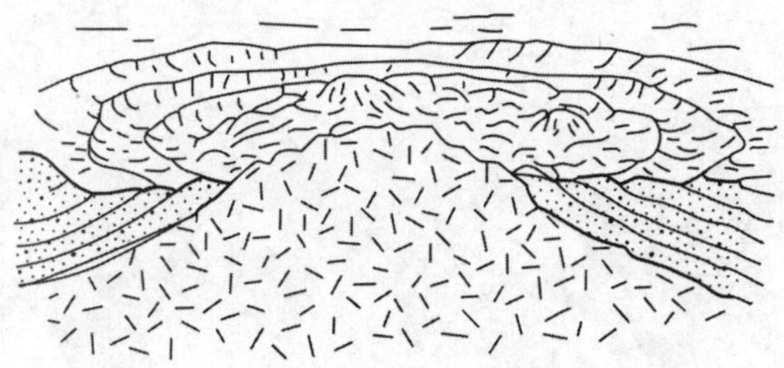

Which type of stream drainage pattern formed on this landscape?

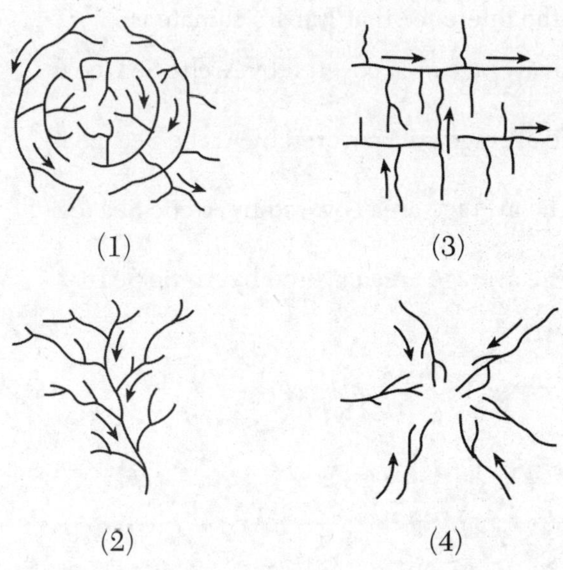

32 ______

33 The north polar view maps below show the average area covered by Arctic Sea ice in September of 1980, 2000, and 2011.

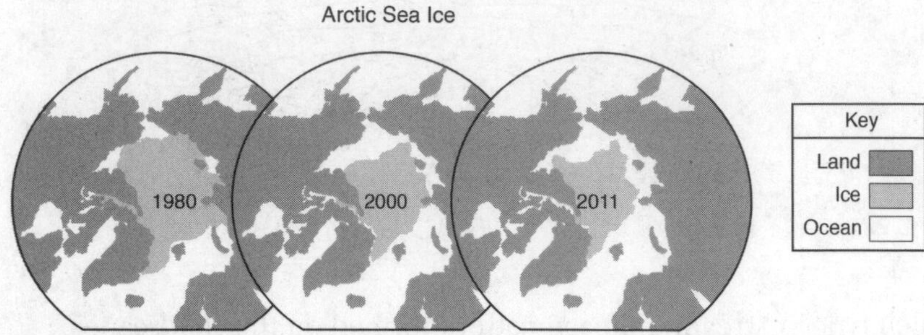

The maps best support the inference that Earth's climate is

(1) cooling, because the average area covered by Arctic Sea ice is decreasing
(2) cooling, because the average area covered by Arctic Sea ice is increasing
(3) warming, because the average area covered by Arctic Sea ice is decreasing
(4) warming, because the average area covered by Arctic Sea ice is increasing

33 _____

34 The photographs below show two depositional features labeled *A* and *B*.

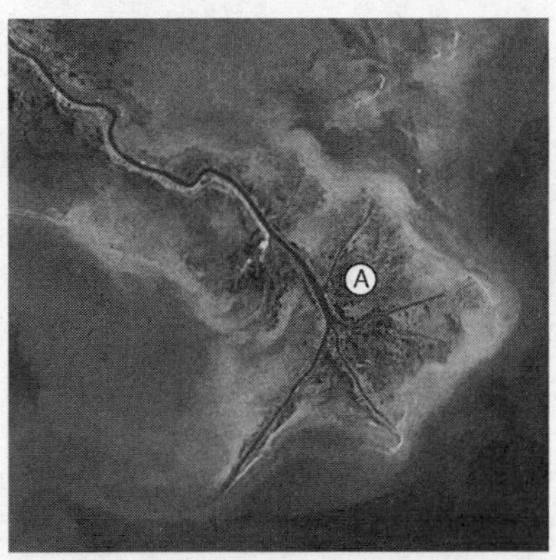

Which terms correctly identify depositional features *A* and *B*?

(1) *A*-delta; *B*-barrier island
(2) *A*-sand bar; *B*-island arc
(3) *A*-barrier island; *B*-delta
(4) *A*-island arc; *B*-sand bar 34 ____

35 Diagrams *A* and *B* represent magnified views of the arrangement of mineral crystals in a rock before and after being subjected to geologic processes.

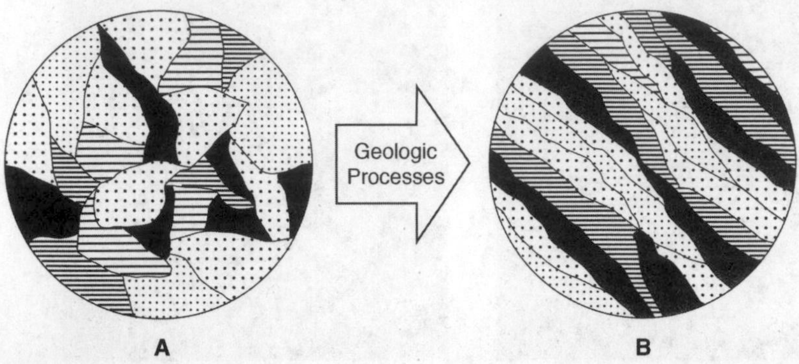

Which geologic processes are most likely responsible for the banding and alignment of mineral crystals represented in diagram *B*?

(1) melting and solidification
(2) heating and increasing pressure
(3) compaction and cementation
(4) weathering and erosion 35 ____

PART B–1

Answer all questions in this part.

Directions (36–50): For *each* statement or question, choose the word or expression that, of those given, best completes the statement or answers the question. Some questions may require the use of the *2011 Edition Reference Tables for Physical Setting/Earth Science*. Record your answers in the space provided.

Base your answers to questions 36 through 39 on the graph below and on your knowledge of Earth science. The graph shows the observed water levels, in feet (ft), for a tide gauge located at Montauk, New York, on the easternmost end of Long Island, from January 24, 2008 to noon on January 25, 2008.

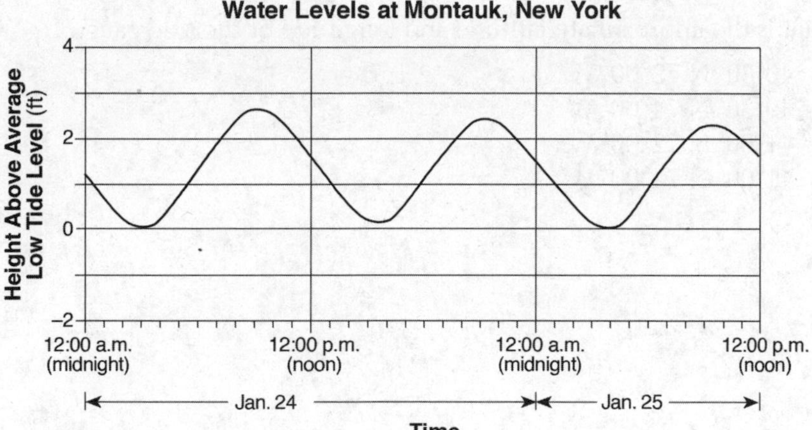

36 What was the height of the water above average low tide level at noon on January 24?

(1) 1.2 ft
(2) 1.6 ft
(3) 2.2 ft
(4) 2.6 ft

36 ____

37 These changing water levels at Montauk can best be described as

(1) cyclic and predictable
(2) cyclic and not predictablee
(3) noncyclic and predictable
(4) noncyclic and not predictable 37 ____

38 What causes the water-level variation pattern shown by the graph?

(1) changes in wind velocity produced by coastal storms
(2) changes in magnetic orientation of the North American Plate
(3) Earth's revolution and the distance from the equator
(4) Earth's rotation and the gravitational pull of the Moon 38 ____

39 What is the approximate latitude and longitude of the tide gauge?

(1) 40°30′ N 72°00′ W
(2) 40°30′ N 74°00′ W
(3) 41°00′ N 72°00′ W
(4) 41°00′ N 74°00′ W 39 ____

Base your answers to questions 40 through 42 on the diagram below and on your knowledge of Earth science. The diagram represents some of the inferred stages in the life cycle of stars according to their original mass.

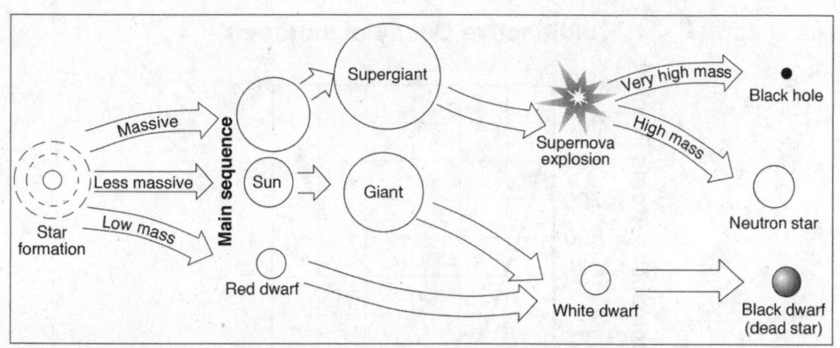

(Not drawn to scale)

40 The final stage in the life cycle of the most massive stars is a

(1) black hole
(2) black dwarf
(3) supergiant
(4) white dwarf

40 _____

41 Which star may once have been similar to our Sun in mass and luminosity?

(1) *Deneb*
(2) *Spica*
(3) *Procyon B*
(4) *Proxima Centauri*

41 _____

42 Energy is produced in the cores of main sequence stars when

(1) lighter elements undergo fusion into heavier elements
(2) heavier elements undergo fusion into lighter elements
(3) cosmic background radiation is absorbed
(4) cosmic background radiation is released

42 _____

Base your answers to questions 43 and 44 on the graph below and on your knowledge of Earth science. The graph shows the number of radioactive Isotope *X* atoms present as a sample of the isotope undergoes radioactive decay.

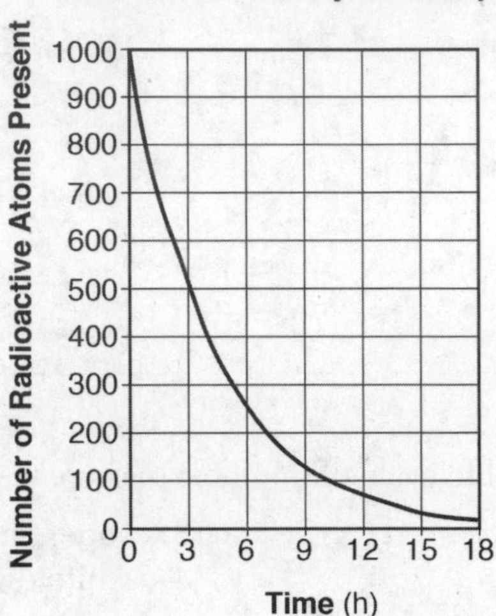

43 Based on the graph, the half-life of this radioactive isotope is

(1) 6 h
(2) 9 h
(3) 3 h
(4) 12 h

43 ____

44 Based on the graph, what is the approximate number of radioactive atoms of Isotope *X* that are present when 8 hours of decay has occured?

(1) 90
(2) 115
(3) 155
(4) 200

44 ____

Base your answers to questions 45 through 47 on the diagram below and on your knowledge of Earth science. The arrows in the diagram show air movement in a thunderstorm cloud. Point *A* represents a location in the atmosphere.

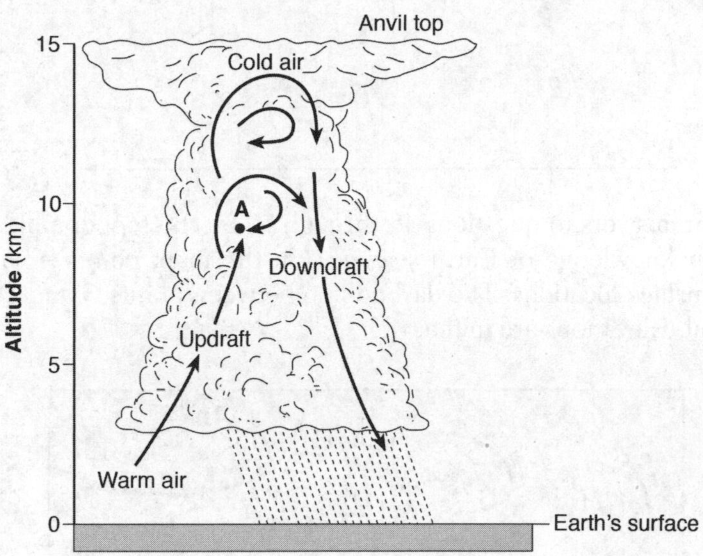

45 In which temperature zone of the atmosphere is point *A* located?

(1) thermosphere
(2) mesosphere
(3) stratosphere
(4) troposphere

45 _____

46 The updrafts and downdrafts represented within this cloud are primarily caused by differences in

(1) altitude above sea level
(2) air density
(3) relative humidity
(4) specific heat

46 _____

47 Which weather symbol would be placed on a station model to represent this weather event?

(1) (2) (3) (4) 47 ____

Base your answers to questions 48 through 50 on the topographic map below and on your knowledge of Earth science. On the map, points *A*, *B*, *C*, and *D* represent surface locations. The dashed line between points *C* and *D* represents a hiking trail. Elevations are in feet (ft).

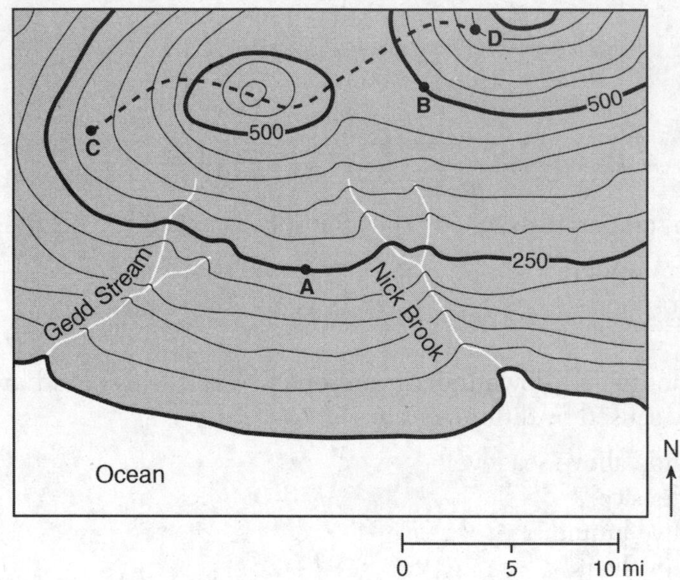

48 What is the contour interval on this map?

(1) 25 ft
(2) 50 ft
(3) 150 ft
(4) 250 ft

48 ____

49 The gradient between location *A* and location *B* is approximately

(1) 0.04 ft/mi
(2) 25 ft/mi
(3) 40 ft/mi
(4) 50 ft/mi

49 ____

50 A person walks along the trail from location *C* to location *D*. The person will be walking

(1) downhill then uphill, only
(2) downhill, then uphill, then downhill again
(3) uphill then downhill, only
(4) uphill, then downhill, then uphill again

50 ____

PART B–2

Answer all questions in this part.

Directions (51–65): Record your answers in the spaces provided. Some questions may require the use of the *2011 Edition Reference Tables for Physical Setting/ Earth Science.*

Base your answers to questions 51 through 54 on the passage below and on your knowledge of Earth science.

The Mica Family

The familiar term "mica" is not the name of a specific mineral, but rather the name for a family of more than 30 minerals that share the same properties. All members of the mica family have high melting points and are similar in density, luster, hardness, streak, type of breakage, and crystal shape. As a result, telling the micas apart can be difficult. However, some common members of the family can be identified by color. For example, biotite is black to dark brown while muscovite can be light shades of several colors, or even colorless. When less common members of the mica family have any of these colors, or have similar colors, chemical tests are needed to tell them apart.

51 Identify the *two* chemical elements present in biotite mica that are *not* present in muscovite mica. [1]

________________ and ________________

52 Identify the luster, hardness, and dominant form of breakage for members of the mica family. [1]

Luster: ________________

Hardness: ________________

Dominant form of breakage: ________________

53 State the name of the igneous rock in which crystals of biotite mica are larger than 10 millimeters in diameter. [1]

____

54 Large crystals of mica, sometimes weighing several hundred tons, have been found in igneous rock in Canada. Identify the environment of formation and the relative rate of cooling of the magma that formed the igneous rock containing these large crystals. [1]

Environment of formation: ____________________

Relative rate of cooling: ____________________

Base your answers to questions 55 through 58 on the diagram below and on your knowledge of Earth science. The diagram represents the Sun's apparent daily path for the first day of three seasons at 43° North latitude. The solid lines represent daytime paths as seen by an observer at this latitude. The dashed lines represent the nighttime paths that can *not* be seen by the observer.

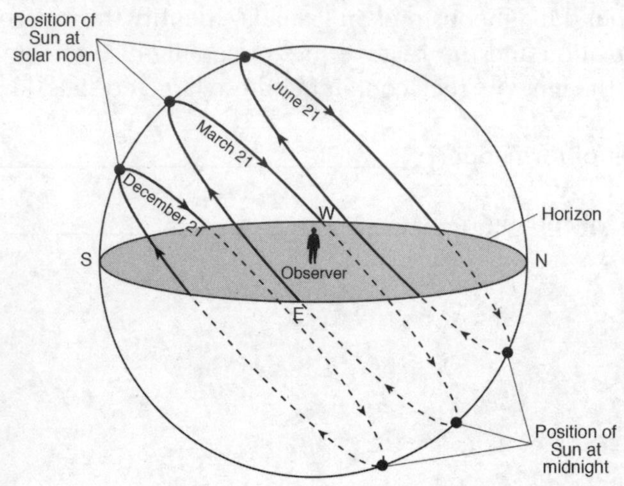

55 On the diagram *above*, draw an **X** to represent the solar noon position of the Sun as seen by the observer on April 21. [1]

56 Identify the rate of the Sun's apparent movement, in degrees per hour, along its path on December 21. [1]

______________________ °/h

57 Identify the compass direction toward which the observer's shadow would point at solar noon on March 21. [1]

____

58 List the three dates shown on the diagram from the least number of nighttime hours to the greatest number of nighttime hours. [1]

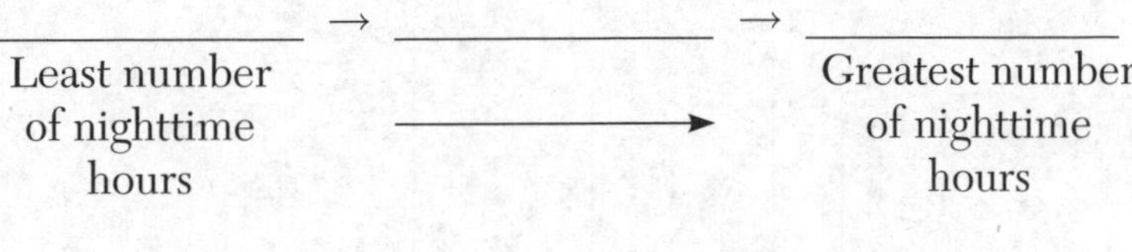

Base your answers to questions 59 through 62 on the information below, on the map below, and on your knowledge of Earth science. The map shows a portion of the tectonic plates map from the *2011 Edition Reference Tables for Physical Setting/Earth Science*. Letters *A* and *B* represent locations on the ocean floor.

The area between North America and South America is a tectonically active region of Earth. This region contains all of the types of tectonic plate boundaries, and it has frequent earthquake and volcanic activity. The tectonic plates on either side of the East Pacific Ridge move at an average rate of 7.5 cm/year.

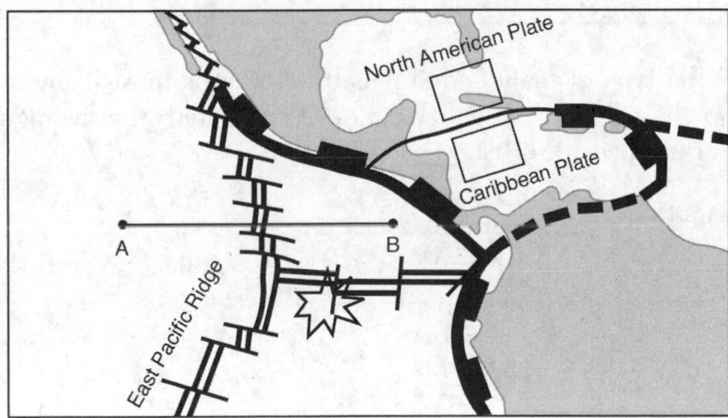

59 On the map *above*, draw *one* arrow in each of the two boxes to show the relative motion of the Caribbean Plate and the North American Plate. [1]

60 On the set of axes *below*, draw a line to represent the relative age of the ocean floor bedrock from location *A* to location *B*. [1]

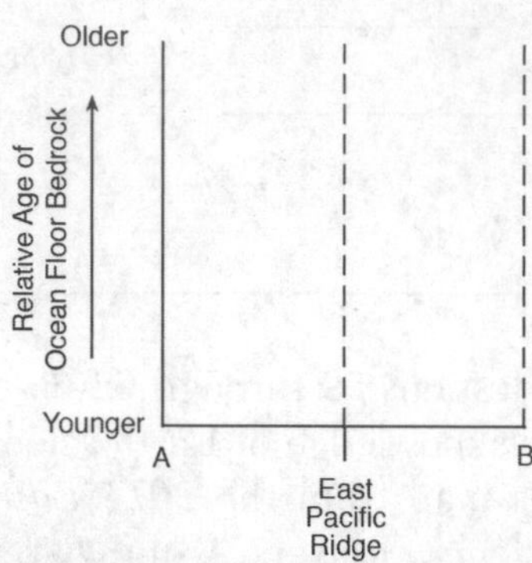

61 Identify the name of the hot spot shown on the map, and identify the name of the tectonic plate under which the center of this hot spot is located. [1]

________________________________ **Hot Spot**

________________________________ **Plate**

62 Identify the type of mafic igneous bedrock that is most likely to make up the oceanic crust at location *A*, and state the average density of this oceanic crust. [1]

Type of bedrock: ____________________

Density: ____________________ **g/cm³**

Base your answers to questions 63 through 65 on the diagram and data table below and on your knowledge of Earth science. The diagram represents laboratory materials used for an investigation of the effects of particle diameter on permeability and porosity (percentage of pore space). Four separate plastic tubes were filled to the same level with different particles.

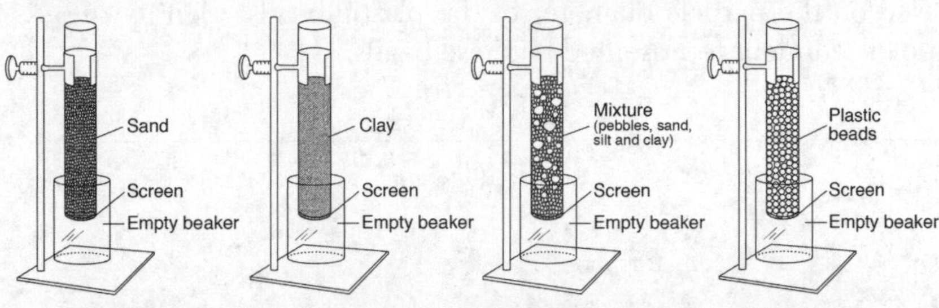

(Not drawn to scale)

Particle Type	Particle Diameter (cm)	Time for Water to Infiltrate (s)	Porosity (%)
Sand	0.1	7	42.0
Clay	0.0003	322	40.0
Mixture	from 0.0003 to 0.8	15	34.0
Plastic beads	0.4	4	44.0

63 Explain why the particle sizes fit together more closely in the mixture, resulting in the lowest porosity of all these particle types. [1]

____

____

64 The height of the column of sand is 28 centimeters. Calculate the rate of infiltration, in centimeters per second, for the water that flowed through the column of sand. [1]

____________________ **cm/s**

65 Based on the particle diameter of the plastic beads, identify the type of sediment represented by these beads. [1]

____

PART C

Answer all questions in this part.

Directions (66–85): Record your answers in the spaces provided. Some questions may require the use of the *2011 Edition Reference Tables for Physical Setting/Earth Science*.

Base your answers to questions 66 through 68 on the weather map below and on your knowledge of Earth science. The map shows the location of a wintertime low-pressure system over Lake Ontario with two fronts extending into New York State. Isobar values are recorded in millibars. Partial weather station data are shown for several locations.

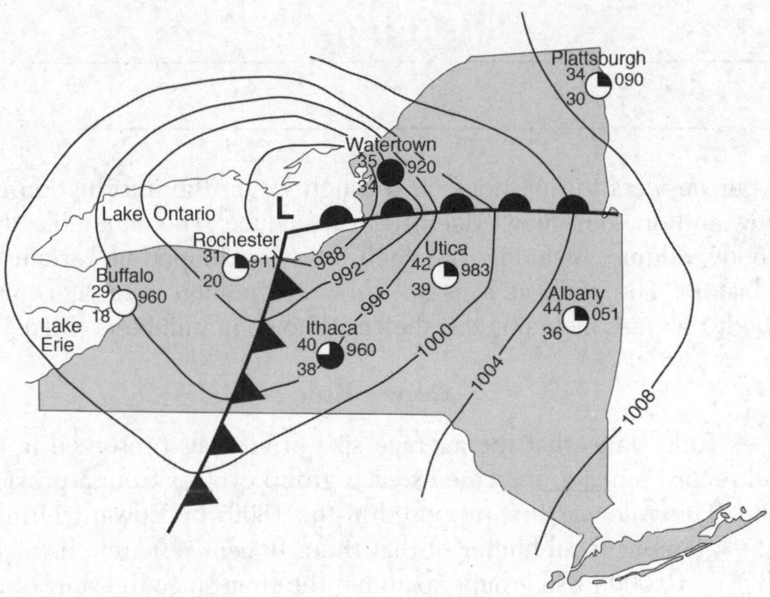

66 Describe the evidence shown on the map that indicates that the highest wind speeds occurred near Watertown, New York. [1]

____

____

67 Complete the table *below* by recording the weather data shown on the station model for Albany, New York. [1]

Albany, New York

Weather Variable	Weather Data
Dewpoint	°F
Cloud cover	%
Actual barometric pressure	mb

68 State the compass direction toward which the center of this low-pressure system moved over the next two days if the low followed a normal storm track. [1]

Base your answers to questions 69 through 72 on the information and data table below and on your knowledge of Earth science. The data table shows the average body volume, including the shell, of a brachiopod at certain times in geologic history. The geologic ages are shown in million years ago (mya). The average body volumes including the shell are shown in milliliters (mL).

Cope's Rule

Cope's Rule states that the average size of animals preserved in the fossil record tends to increase as each group evolves from a previous group. This rule was first proposed in the 1800s by Edward Drinker Cope, a famous fossil hunter of that time. Recent research, involving well over 10,000 fossil groups spanning the time since the start of the Cambrian Period until today, has shown that Cope's Rule is accurate for most animal groups. Brachiopod data support Cope's Rule.

Brachiopod Data Table

Geologic Age (mya)	**Average Body Volume Including the Shell** (mL)
480	0.1
460	0.2
430	0.6
410	1.0
380	1.1

69 On the grid *below*, plot the average brachiopod body volume for each of the geologic ages listed in the data table. Connect *all five* plots with a line. [1]

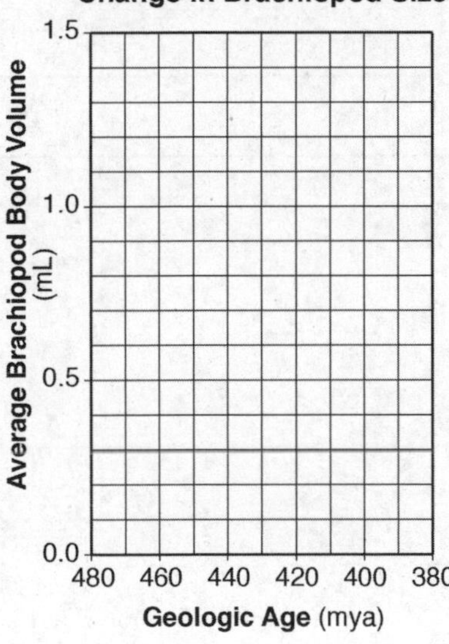

70 Identify, by name, *two* geologic periods when the brachiopods represented in the data table were living. [1]

_______________ **Period**

and _______________ **Period**

71 State the names of the *two* brachiopod index fossils found in New York State bedrock. [1]

_______________ and _______________

72 The earliest horses appeared in the Eocene epoch and were about the size of a large dog of today. Explain how the evolution of horses supports Cope's Rule. [1]

____

____

____

Base your answers to questions 73 through 75 on the snowfall map below and on your knowledge of Earth science. The snowfall map shows some average yearly snowfall values, measured in inches, recorded for a portion of New York State. Some average yearly snowfall isolines have been drawn. Line *XY* is a reference line on the map. The cities of Watertown and Oswego are shown on the map.

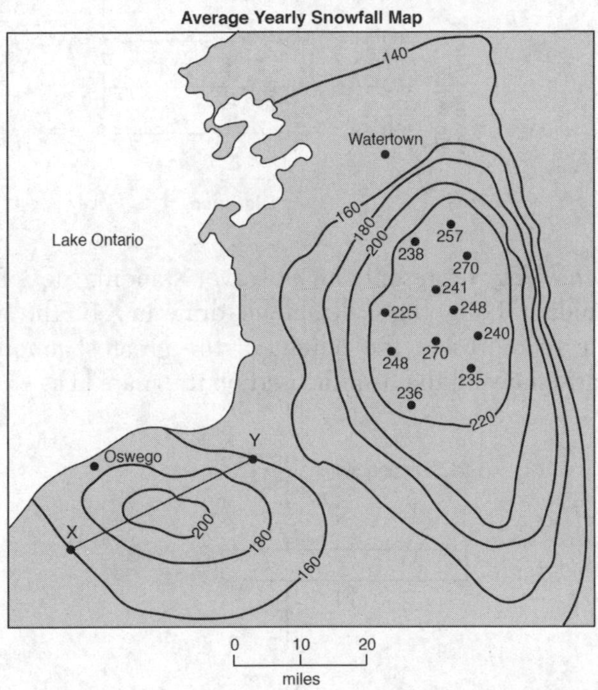

73 On the map *above*, draw the 240-inch average yearly snowfall isoline. [1]

74 On the grid *below*, construct a profile of the average annual snowfall along line *XY* by plotting the value of each isoline that crosses line *XY*. Connect *all six* plots with a line to complete the profile. [1]

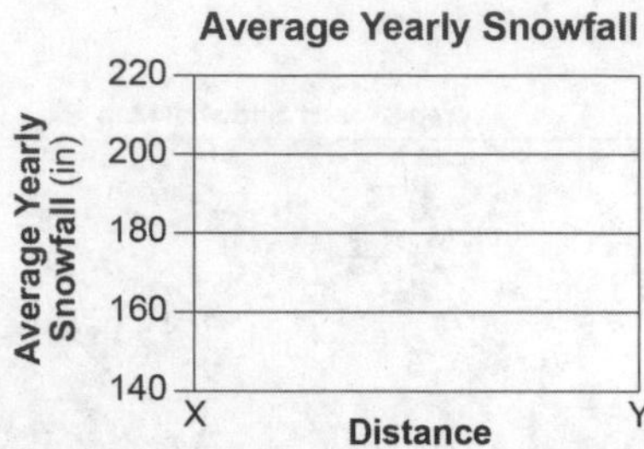

75 The diagram *below* represents an observer standing next to the side of a building. Using the scale shown, draw an **X** on the side of the building to represent the height of the greatest amount of average yearly snowfall that is indicated on the map. [1]

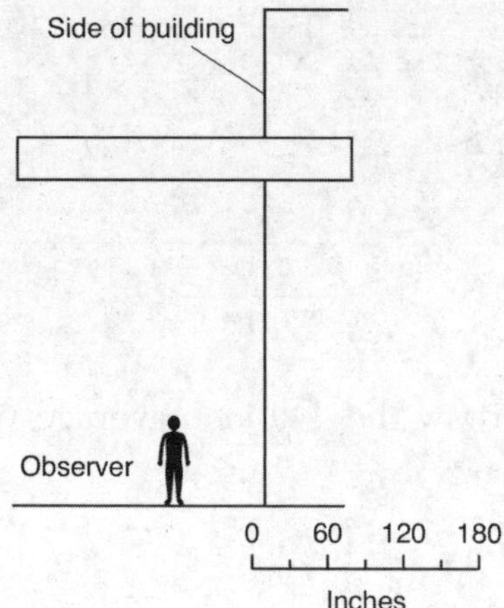

Base your answers to questions 76 through 78 on the diagram below and on your knowledge of Earth science. The diagram represents a cutaway view of a flat-plate solar collector used to heat water at a New York State location.

Solar Collector

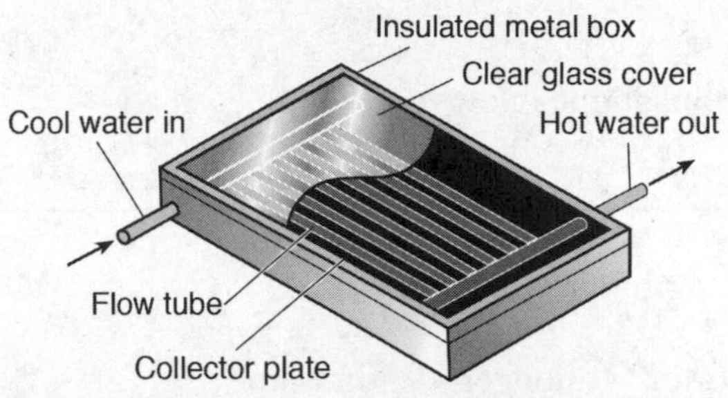

76 Identify the energy transfer process by which light travels through space from the Sun to the solar collector. [1]

__________________________ ++

77 Explain why the flow tubes and collector plate inside the solar collector are black in color. [1]

____

____

____

78 The glass cover on this solar collector allows visible light to enter the collector. Identify the type of electromagnetic energy emitted by the flow tubes and collector plate that is trapped inside the collector by the glass cover. Also, circle the relative wavelength of this trapped electromagnetic energy compared to wavelengths of visible light. [1]

Emitted electromagnetic energy:

____

Relative wavelength (circle one):

shorter longer the same

Base your answers to questions 79 through 82 on the passage and diagram below, on the data table on the next page, and on your knowledge of Earth science. The diagram compares the inner planets of our solar system to the planetary system surrounding the star *Kepler-62*, which is located in our galaxy. The data table shows some data for the planets in the *Kepler-62* system.

Kepler-62 Planetary System

Five planets orbit a seven-billion-year-old star, *Kepler-62*, which has a surface temperature of approximately 4900 Kelvin. Two of these planets are located within the habitable zone, which is the region around a star where life may exist due to the possible presence of water in the liquid phase. The shaded areas in the orbital diagrams below indicate the habitable zone of each system.

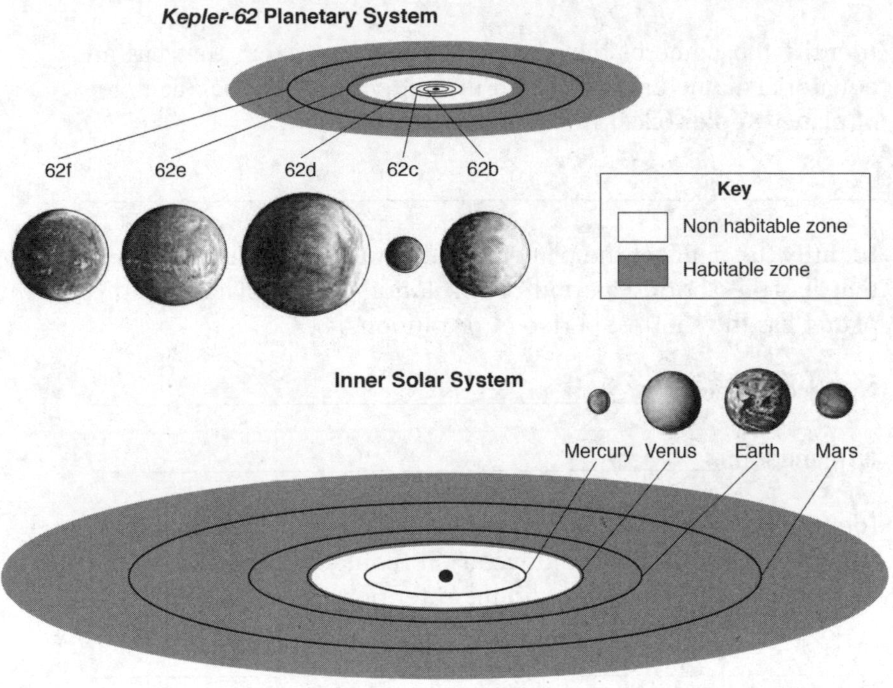

(Planets and orbits are drawn to scale)

Data Table

Name of Planet	Distance from *Kepler-62* (million kilometers)	Equatorial Diameter (compared to Earth's diameter)
62b	8.23	1.31
62c	13.76	0.54
62d	17.95	1.95
62e	63.88	1.6
62f	107.41	1.4

79 Identify the name of the galaxy where the *Kepler-62* planetary system is located. [1]

____

80 Identify the name of the planet in our solar system that has an equatorial diameter most similar in size to the equatorial diameter of planet Kepler-62c. [1]

____

81 Identify the name of the planet in the *Kepler-62* planetary system that has the shortest period of revolution, and explain why this planet has the shortest period of revolution. [1]

Name of planet: ______________________________________

Explanation: ______________________________________

82 Identify the names of the *two* planets in the *Kepler-62* planetary system that may have liquid water on their surfaces, and explain why these planets may have liquid water on their surfaces. [1]

____________________ and ____________________

Explanation: ______________________________________

____

Base your answers to questions 83 through 85 on the block diagram below and on your knowledge of Earth science. The diagram represents glacial features formed by a continental glacier and its melt water.

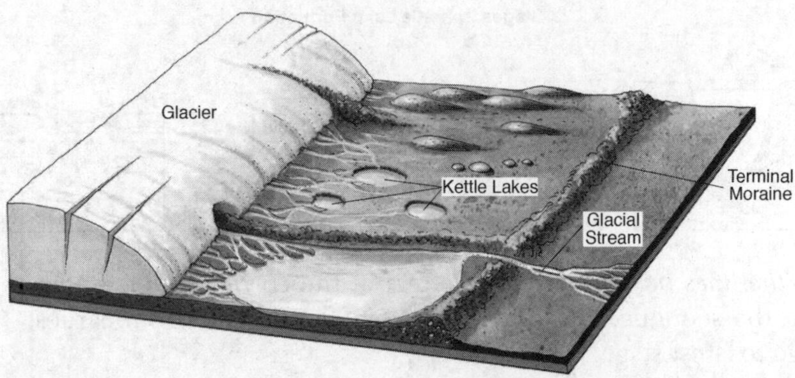

83 Describe the arrangement of the sediments found within the terminal moraine, which marks the farthest advance of the glacier. [1]

____

____

84 The cross sections below, labeled *A*, *B*, *C*, and *D*, represent four stages in the development of a kettle lake. The stages are *not* shown in the correct order.

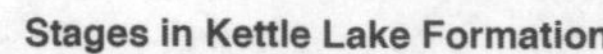

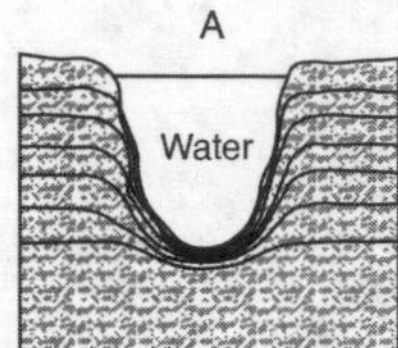

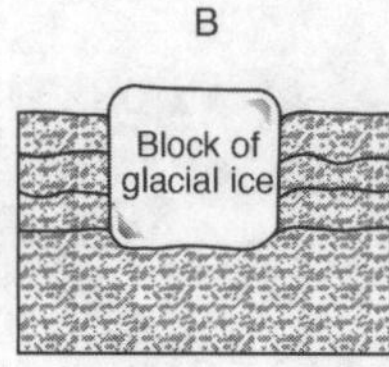

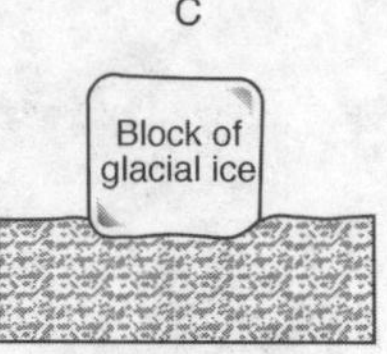

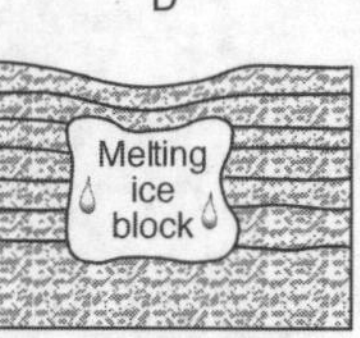

On the lines below, place the letters in the correct order to indicate the sequence of development of a kettle lake from earliest stage to latest stage. [1]

______ → ______ → ______ → ______

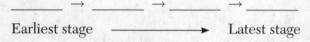

Earliest stage ⟶ Latest stage

85 Terminal moraines found on Long Island were deposited during the advance and retreat of glacial ice during the last ice age. Identify, by name, the geologic epoch during which these moraines were deposited. [1]

______________________________ **epoch**

Answers
June 2018
Physical Setting/Earth Science

Answer Key

PART A

1. 3	**8.** 1	**15.** 2	**22.** 2	**29.** 4
2. 4	**9.** 4	**16.** 1	**23.** 3	**30.** 1
3. 1	**10.** 2	**17.** 1	**24.** 1	**31.** 1
4. 1	**11.** 2	**18.** 4	**25.** 4	**32.** 1
5. 3	**12.** 3	**19.** 3	**26.** 2	**33.** 3
6. 4	**13.** 2	**20.** 4	**27.** 3	**34.** 1
7. 2	**14.** 4	**21.** 2	**28.** 4	**35.** 2

PART B–1

36. 2	**39.** 3	**42.** 1	**45.** 4	**48.** 2
37. 1	**40.** 1	**43.** 3	**46.** 2	**49.** 2
38. 4	**41.** 3	**44.** 3	**47.** 3	**50.** 4

PART B–2 and **PART C**. *See* **Answers Explained**.

Answers Explained

PART A

1. **3** A solar eclipse occurs when the Moon passes directly between Earth and the Sun, casting a shadow on Earth. An observer on Earth would see the Moon blocking the Sun. The Moon does not produce visible light of its own. The Moon is visible only due to light from the Sun that is reflected from the Moon's surface. When the Moon passes between Earth and the Sun, the side of the Moon facing the Sun is illuminated and the side of the Moon facing Earth is in darkness. Thus, it is reasonable to infer that the dark object *B* with the circular surface is the Moon and the bright object *A* in the partial solar eclipse photograph is the Sun.

The Sun is surrounded by a layer of hot gases called its atmosphere. The outermost part of the Sun's atmosphere is called the corona. The corona is usually hidden by the bright light of the Sun's surface. During a total solar eclipse, though, the Moon blocks the light of the Sun's surface. So the corona can be seen as a glowing aura of light around the dark surface of the Moon. Therefore, object *A* in the total solar eclipse photograph is the corona, which is a part of the Sun, and object *B* is the Moon. Thus, in both photographs, *A* is the Sun and *B* is the Moon.

2. **4** The planets can be divided by mass and density into the terrestrial (Earth-like) and the Jovian (Jupiter-like) planets. The terrestrial planets are small, dense, and rocky. They are composed mostly of metal silicates and iron. The terrestrial planets include the four innermost planets: Mercury, Venus, Earth, and Mars. The Jovian planets are large, low in density, and gaseous. They are composed mainly of hydrogen and helium. The Jovian planets include the outer planets: Jupiter, Saturn, Uranus, and Neptune.

Find the Solar System Data table in the *Reference Tables for Physical Setting/Earth Science.* Locate the column labeled "Period of Rotation at Equator." Note that the periods of rotation for the terrestrial planets range from 23 h 56 min 4 s to 243 d. Note that the periods of rotation of the Jovian planets range from 9 h 50 min 30 s to 17 h 14 min. Thus, compared to the terrestrial planets, the Jovian planets have shorter periods of rotation.

WRONG CHOICES EXPLAINED:

(1) Find the Solar System Data table in the *Reference Tables for Physical Setting/Earth Science.* Locate the column labeled "Mass (Earth = 1)." Note that the terrestrial planets range in mass from 0.06 to 1.00 Earth masses. Note that the Jovian planets range from 14.54 to 317.83 Earth masses. Thus, compared to terrestrial planets, Jovian planets are more massive, not less massive.

(2) Find the Solar System Data table in the *Reference Tables for Physical Setting/Earth Science.* Locate the column labeled "Density (g/cm^3)." Note that the densities of the terrestrial planets range from 3.9 g/cm^3 to 5.5 g/cm^3 while the densities of the Jovian planets range from 0.7 g/cm^3 to 1.8 g/cm^3. Thus, compared to terrestrial planets, Jovian planets are less dense, not more dense.

(3) Increasing the distance between a star and a planet orbiting the star decreases the gravitational attraction between the two objects. This, in turn, causes the orbital velocity of the planet to decrease. Find the Solar System Data table in the *Reference Tables for Physical Setting/Earth Science.* In the column labeled "Mean Distance from Sun (million km)," note that from Mercury to Neptune, mean distance from the Sun increases. Thus, Jovian planets are farther from the Sun than terrestrial planets. Therefore, compared to terrestrial planets, the Jovian planets have lesser, not greater, orbital velocities.

3. **1** According to the Big Bang theory, the universe started out with all of its matter and energy in a small volume and then expanded outward in all directions, a motion very much like an explosion. The explosion associated with the Big Bang theory and the formation of the universe occurred *before* the formation of Earth and the solar system. In other words, the universe is older than either Earth or the solar system. Find the Geologic History of New York State chart in the *Reference Tables for Physical Setting/Earth Science.* Locate the column labeled "Eon." Note that the entry "Estimated time of origin of Earth and solar system" corresponds to 4600 million years ago (4.6 billion years ago) on the time scale. Thus, the explosion associated with the Big Bang theory took place more than 4,6 billion years ago. Based on the current expansion rate of the universe, scientists extrapolate back to when the universe was all concentrated in a small volume and infer that the Big Bang occurred approximately 13.8 billion years ago. Thus, of the choices given, the event that occurred more than 10 billion years ago is the Big Bang.

WRONG CHOICES EXPLAINED:

(2) Earth had to exist before life on Earth could originate. Therefore, the origin of life occurred after the estimated time of origin of Earth and the solar system 4.6 billion years ago. Thus, the origin of life on Earth occurred less than 10 billion years ago, not more than 10 billion years ago.

(3) Find the Geologic History of New York State chart in the *Reference Tables for Physical Setting/Earth Science.* Locate the column labeled "Important Geologic Events in New York." Note the entry "Pangaea begins to break up." Trace horizontally to the left to the time scale labeled "Million years ago," and note that Pangaea began to break up about 200 million years ago. Thus, Pangaea began to break up less than, not more than, 10 billion years ago.

(4) The explosion associated with the Big Bang theory and the formation of the universe occurred before the formation of Earth and the solar system 4.6 billion years ago. Thus, the origin of Earth and its Moon occurred less than, not more than, 10 billion years ago.

4. **1** In 1851, French physicist Jean Foucault suspended a heavy iron ball on a long steel wire from the top of the dome of the Pantheon in Paris. As this pendulum swung back and forth, it appeared to change direction slowly, passing over different lines on the floor until eventually coming full circle to its original position after 24 hours. Since Foucault knew that the path of a freely swinging pendulum would not change on its own, he concluded that the apparent shift in the direction of swing of the pendulum was due to the floor (Earth's surface) rotating beneath the pendulum. Thus, the apparent change in direction of swing of a Foucault pendulum provides evidence that Earth is rotating.

WRONG CHOICES EXPLAINED:

(2) The apparent direction of swing of a Foucault pendulum changes cyclically every 24 hours. Earth takes 365 days to revolve around the Sun. If the cyclic change in a Foucault pendulum was due to Earth revolving, the path would change over a period of 365 days, not over a period of 24 hours.

(3) The shape of the surface beneath a freely swinging pendulum, whether that surface is flat or spherical, has no effect on the pendulum's direction of swing because there is no contact between the pendulum and the surface. Thus, a Foucault pendulum provides no evidence that Earth has a curved surface.

(4) The swing of any pendulum is the result of Earth's gravity acting on the suspended weight. However, Earth's gravity would act on a pendulum's suspended weight causing it to swing even if Earth was not rotating. Foucault used the cyclic change in direction of a swinging pendulum to demonstrate Earth's rotation, not that Earth had a gravitational pull.

5. **3** Earth revolves around the Sun once every 365.26 days, or about 1° per day. A month contains approximately 30 days. Therefore, in one month Earth travels approximately 30° in its orbit.

6. **4** It is given that the dry-bulb temperature is 16°C and the wet-bulb temperature is 10°C. Thus, the difference between the wet-bulb and dry-bulb temperatures is 6°C. Find the Relative Humidity (%) table in the *Reference Tables for Physical Setting/Earth Science.* Locate the column headed "6" in the "Difference Between Wet-Bulb and Dry-Bulb Temperatures (°C)" scale along the top of the chart. Find the "Dry-Bulb Temperature (°C)" scale along the left side of the chart. Locate the row labeled "16," trace to the right until it

intersects the column headed "6," and note the value of 45. Thus, when the dry-bulb temperature is 16°C and the wet-bulb temperature is 10°C, the relative humidity is 45%.

7. **2** People board up windows to protect the windows from being broken by very high winds and windblown debris. If windows break and high winds can penetrate a house, these winds can put pressure onto the roof, causing it to lift off the building. Of the natural disasters listed as choices, only a hurricane produces very high winds. Thus, boarding up windows is a common emergency action taken to prepare for a hurricane.

WRONG CHOICES EXPLAINED:

(1) The motions of an earthquake shake a building and can cause the building to break and, in turn, can cause windows to shatter. However, boarding up windows would not prevent the building from shaking and thus would not prevent the windows from breaking.

(3) Windows are typically placed above floor level. Rising water entering through openings at floor level would cause extensive damage before it reached high enough to penetrate a window. Thus, boarding up windows would not prevent a home from being damaged by flooding and is not an emergency action taken to prepare for a flood.

(4) Tsunamis are waves caused by undersea movements of Earth's crust. Tsunamis may be only a few meters high, but they have very long wavelengths and travel much more rapidly than ordinary ocean waves. When tsunamis reach shallow water, they are slowed by friction with the ocean bottom and begin to pile up as large waves. In bays and narrow channels, their high speed and long wavelength may be funneled into huge breaking waves more than 40 m high. The force exerted by such a huge, fast-moving mass of water can do extensive damage and can cause sudden, widespread coastal flooding. Most of the deaths caused by tsunamis are drownings. Boarding up windows would do little to prevent flooding and drowning and is therefore not an emergency action taken to prepare for a tsunami.

8. **1** When two different air masses meet, the boundary between them is called a front. Find the Planetary Wind and Moisture Belts in the Troposphere diagram in the *Reference Tables for Physical Setting/Earth Science.* Locate the polar front jet stream indicated by the symbol ⊗ at the boundary between the convection cell located between the North Pole and 60° N latitude and the convection cell located between 30° N and 60° N latitude. The polar front marks the boundary along which cold air masses of the polar convection cell meet warm tropical air masses carried north by the convection cell between 30° N and 60° N.

The sharp temperature gradient at this boundary results in high winds. Combined with the effects of Earth's rotation, these high winds form jet streams. The polar jet stream that tracks through the United States moves from west to east. Cold fronts pushing against the polar front from the north and warm air masses pushing against the polar front from the south cause bends to form in the polar front and its jet stream. In the winter, extremely cold air masses pushing against the boundary from the north can cause this boundary to dip to the south of its usual position at 60° N.

Now, find the Tectonic Plates map in the *Reference Tables for Physical Setting/Earth Science.* Locate 60° N along the right-hand edge of the map, and trace horizontally to the left across North America. Note that the Great Lakes are located south of latitude 60° N. Note that in diagram (1), the polar front jet stream dips below the Great Lakes and south of 60° N latitude. Thus, diagram (1) best represents the general position and flow of the polar front jet stream in the Northern Hemisphere during the winter months.

WRONG CHOICES EXPLAINED:

(2) Diagram (2) shows the polar front jet stream located near the equator. Although on rare occasions the polar front jet stream has dipped as far south as northern Florida, the polar front jet stream has never been observed to reach as far south as the equator. Thus, the general position of the polar front jet stream during the winter months is not near the equator.

(3) The polar jet stream that tracks through the United States moves from west to east. The arrows on diagram (3) show the direction of flow of the polar front jet stream to be from east to west, not from west to east.

(4) Diagram (4) shows the polar front jet stream near the equator with a direction of flow from east to west. As explained above for choices (2) and (3), this does not represent the general position and direction of flow of the polar front jet stream.

9. **4** An observer on Earth can see only that portion of the Moon that is illuminated by the Sun's rays and only from the side of Earth that is in darkness. In the diagram, the part of the Moon that is *not* shaded represents the illuminated portion of the Moon. According to the diagram, when the Moon is in position *D*, its entire illuminated portion faces Earth. An observer in New York State would see the full moon phase. When the Moon is in position *C*, most, but not all, of the illuminated portion would be visible to the observer's right. The observer would see the new gibbous moon phase. When the Moon is in position *B*, half of the illuminated portion would be visible to the observer's right. The observer would see the first quarter moon phase. When the Moon is in position *A*, only a thin crescent of the illuminated portion of the Moon's surface would be visible to the observer's right. The observer

would see the new crescent moon phase. Thus diagram (4), which shows the full moon at position *D* and the illuminated portions at positions *A*, *B*, and *C* to the observer's right, best represents the sequence of Moon phases as seen by an observer in New York State.

WRONG CHOICES EXPLAINED:

(1) and (2) As explained above, an observer in New York State would see the illuminated portions of the Moon to the observer's right. Both of these choices show the illuminated portions of the Moon to the observer's left.

(3) This choice shows the full moon at position *A*, not at position *D*.

10. **2** According to the diagrams, compared to the spectral lines of hydrogen observed in a laboratory, the spectral lines of hydrogen observed in the light from a distant star are shifted toward the end of the spectrum labeled "longer wavelength." Find the Electromagnetic Spectrum chart in the *Reference Tables for Physical Setting/Earth Science.* Note that visible light at the red end of the spectrum has a longer wavelength than light at the blue end of the spectrum. Thus, a shift toward the longer wavelength end of the spectrum is a shift toward the red end of the spectrum. If a source of electromagnetic waves is moving *away* from an observer at the same time as the source is emitting light of a particular wavelength, fewer wave crests will reach the eye of the observer each second. The eye will interpret this as meaning that the light has a longer wavelength than it actually has. In other words, the light will appear shifted toward the red end of the spectrum. The fact that the spectral lines of hydrogen from a distant star show a red shift indicates that the distant star is moving away from Earth. Therefore, compared to the spectral lines observed in the laboratory, the spectral lines observed in the light from the distant star have shifted toward the red end of the spectrum, indicating the star's movement away from Earth.

WRONG CHOICES EXPLAINED:

(1) If a source of electromagnetic waves is moving toward an observer at the same time as it is emitting light of a particular wavelength, more wave crests will reach the eye of the observer each second. The eye will interpret this as meaning that the light has a shorter wavelength than it actually has. In other words, the light will appear shifted toward the blue end of the spectrum. The diagrams in this question show the opposite.

(3) and (4) According to the diagrams, compared to the spectral lines of hydrogen observed in a laboratory, the spectral lines of hydrogen observed in the light from a distant star are shifted toward the red end of the spectrum, not toward the blue end of the spectrum.

11. **2** Earth's rotational axis is tilted at an angle of 23.5° from a perpendicular to Earth's orbital plane.

12. **3** A constellation is an imaginary pattern formed by a group of stars in an area of the sky as seen by an observer on Earth. As Earth revolves around the Sun, the side of Earth facing the Sun experiences day and the side of Earth facing away from the Sun experiences night. Stars are visible only at night. Therefore, the stars (and constellations) seen by an observer on Earth vary as Earth revolves around the Sun and as Earth's night side faces different portions of the universe. During winter nights in January, an observer in New York State sees constellations in the region of the universe facing Earth's night side. Six months later during summer nights in July, Earth has revolved to a position in its orbit directly opposite the Sun from its winter position. Earth's night side now faces the region of the universe directly opposite the region it faced in winter. So different constellations are visible. Therefore, the best explanation for why these two constellations are visible in the night sky in the months identified is that the nighttime side of Earth is facing different parts of our galaxy as Earth orbits the Sun.

WRONG CHOICES EXPLAINED:

(1) Earth's daily rotation causes a daily apparent motion of constellations, not a seasonal change in constellations visible from Earth.

(2) The rate at which Earth spins on its axis affects the daily apparent motion of constellations, not the seasonal change in constellations visible from Earth.

(4) The idea that the stars in constellations revolve around Earth is part of the geocentric theory, which has been proven to be incorrect. Furthermore, Earth is too small and too far from the stars in constellations to exert enough gravitational attraction to hold the stars in orbit around Earth and cause them to revolve around Earth.

13. **2** Find the Inferred Properties of Earth's Interior chart in the *Reference Tables for Physical Setting/Earth Science*. At the bottom of the chart, find the horizontal scale labeled "Depth (km)" and locate the value "4000." Trace the vertical line corresponding to 4000 km upward until it intersects the dashed line labeled "Melting Point." From this intersection, trace horizontally left to the scale labeled "Temperature (°C)." Note that the melting point corresponds to "5400." Now, trace the vertical 4000 km depth line upward until it intersects the solid line labeled "Interior Temperature." From this intersection, trace horizontally left to the scale labeled "Temperature (°C)." Note that the interior temperature corresponds to "5700." Thus, the temperature of the materials at a depth of 4000 km is

5700°C and their melting point is 5400°C. Since the temperature of the materials exceeds their melting point, they are in the liquid state (phase) of matter. Thus, the table that correctly shows the interior temperature, melting point, and state (phase) of matter is shown in choice (2).

14. **4** Deforestation is the removal of a forest or a stand of trees to clear the land for other uses such as farms, ranches, and construction projects. Trees absorb carbon dioxide (CO_2) and emit oxygen (O_2) as they carry out photosynthesis. Photosynthesis ties up CO_2 in the form of sugars, starches, and cellulose. Therefore, deforestation decreases the amount of carbon dioxide removed from the atmosphere by trees, causing a net increase in carbon dioxide in the atmosphere. Thus, the greenhouse gas that has increased in Earth's atmosphere partly as a result of deforestation over the last 100 years is carbon dioxide.

15. **2** Find the Surface Ocean Currents map in the *Reference Tables for Physical Setting/Earth Science.* Locate Africa. Note that the continent of Africa is split, with the western half on the right side of the map and the eastern half on the left side of the map. Locate the southeastern tip of Africa on the left side of the map. Note that the ocean current along the southeastern tip of southern Africa is the Agulhas Current, which flows to the south. According to the key, the Agulhas Current is a warm current. Thus, the ocean current that brings warm water to the southeastern coast of Africa is the Agulhas Current.

16. **1** Find the Geologic History of New York State chart in the *Reference Tables for Physical Setting/Earth Science*. In the column labeled "Life on Earth," find the entry "Humans, mastodonts, mammoths." Trace left to the scale labeled "Million years ago," and note that humans appeared 1.8 million years ago. In the column labeled "Era," find the statement "Estimated time of origin of Earth and the solar system." Trace left to the scale labeled "Million years ago," and note that the origin of Earth occurred 4600 million years ago. Thus, the approximate percentage of geologic time that humans have existed on Earth since its origin is 0.039% ($1.8/4600 \times 100 = 0.039\%$), or less than 1%. Thus, the pie graph that is shaded to best represent the approximate percentage of time that humans have existed during Earth's entire history is shown in choice (1).

17. **1** Volcanic ash is ejected high into the atmosphere during a volcanic eruption. The ash is then carried by global winds, quickly spreads out, and finally settles to the ground, forming a volcanic ash layer over a wide area. Once the volcanic eruption is over, the source of volcanic ash is cut off and deposition quickly tapers off. Thus, a volcanic ash layer can be used as a time marker for correlating rock layers because the ash is deposited rapidly over a large area.

WRONG CHOICES EXPLAINED:

(2) A buried erosional surface represents a gap in the sequence of rock layers due to uplift and erosion, causing some of the rock layers to be removed. A layer of volcanic ash is the result of deposition, not erosion.

(3) Intrusive igneous rock forms when molten magma cools beneath Earth's surface. Volcanic ash consists of tiny particles of rock that form from lava ejected from a volcano and is a type of extrusive igneous rock. Thus, volcanic ash forms an extrusive igneous rock, not an intrusive igneous rock.

(4) Volcanic ash is deposited from the air and forms a layer on top of Earth's exposed surface, not across rock layers.

18. **4** According to the cross section, locations *A* and *B* are at the same elevation on different sides of the mountain. The arrow labeled "Prevailing wind" indicates that location *A* is on the windward side of the mountain and that location *B* is on the leeward side of the mountain. As moist air is carried upward and over the windward side of the mountain by the prevailing wind, it expands and cools. When the air temperature reaches the dewpoint, moisture begins to condense and clouds form, causing precipitation at location *A*. As the air descends on the leeward side of the mountain, it is compressed and the air temperature increases. Thus, location *B* has a warmer climate than location *A*. The increase in air temperature causes a decrease in relative humidity, resulting in less precipitation at location *B*. Thus, location *B* has a drier climate than location *A*. Therefore, compared to the climate at location *A*, the climate at location *B* is warmer and drier.

19. **3** Find the Scheme for Sedimentary Rock Identification in the *Reference Tables for Physical Setting/Earth Science*. In the "Rock Name" column, locate "conglomerate." Trace left to the "Texture" column, and note that conglomerate has a "clastic (fragmental)" texture. Clastic sedimentary rocks are composed of fragments of rock, or sediments. Thus, sediments are rocks that already existed and that were broken into fragments by weathering and erosion. Sediments become a clastic sedimentary rock through the process of lithification. Lithification begins when the sediments are buried and become compacted. Then certain minerals begin to recrystallize in the fluids trapped in the pores among the rock fragments that make up the sediments, cementing the fragments together. Thus, the rock particles of the sediments had to exist already in order to be compacted and cemented together to form the clastic sedimentary rock. Therefore, compared to the age of the calcite cement and the conglomerate, the relative age of the pebbles is older than both the calcite cement and the conglomerate.

20. **4** All processes of the water cycle involve the movement of water. The arrows at letter *A* extend from the surface of the ocean, lake, and Earth's surface to the atmosphere. Water on land surfaces and in oceans and lakes exists mainly as liquid water. Water in the atmosphere exists mainly as water vapor. Therefore, the process at letter *A* involves liquid water entering the atmosphere as water vapor. The process by which water changes from a liquid to a vapor is called evaporation, or vaporization. Therefore, water cycle process *A* is evaporation.

The arrows at letter *B* extend from trees to the atmosphere. Trees are plants. Water in plants exists mainly as liquid water. Water in the atmosphere exists mainly as water vapor. Therefore, letter *B* represents a process by which liquid water enters the atmosphere as water vapor from plants. Plants remove liquid water from the ground and release it into the atmosphere in the form of water vapor by a process called transpiration. In transpiration, liquid water absorbed by plant roots from the soil increases the water pressure inside the lower parts of the plant. Simultaneously, evaporation of water through plant stems or leaf stomata decreases the water pressure in the upper parts of a plant. This difference in pressure causes the liquid water to move upward from the roots and toward the leaves. When the liquid water reaches the leaves, it evaporates. Then more water is drawn upward from the roots, and the process of transpiration continues. Thus, water cycle process *B* best represents the process of transpiration because it shows water moving into the atmosphere from plants. The arrows at letter *C* extend from a cloud toward Earth's surface. Clouds are composed of tiny droplets of liquid water. Therefore, it is a process by which the liquid water droplets in the cloud fall to the ground. The process by which water in the form of rain, snow, hail, or sleet falls to the ground is called precipitation. Therefore, water cycle process *C* is precipitation. The process labeled with the letter *D* is shown taking place in a cloud and in fog. Clouds and fog are composed of tiny droplets of liquid water that form when water vapor in air cools to its dewpoint and changes from the gas phase to the liquid phase. The change from a gas to a liquid is called condensation. Therefore, water cycle process *D* is condensation. Thus, the table in choice (4) correctly matches each letter with the process it represents: *A*—evaporation, *B*—transpiration, *C*—precipitation, and *D*—condensation.

21. **2** Arid climates are dry, and humid climates are wet. Find the Planetary Wind and Moisture Belts in the Troposphere diagram in the *Reference Tables for Physical Setting/Earth Science*. Note the zones labeled "wet" and "dry." Note that dry (arid) zones exist at 90° N and 90° S (the poles) and at 30° N and 30° S. Note that wet (humid) zones exist at 0° (the equator) and at 60° N and 60° S. Thus, table (2) best shows the relationship between latitude and general climate conditions on Earth.

22. **2** Wave action is turbulent. The turbulence of waves crashing against the shore roils up sediments. As sediments are thrown about by wave action, they bounce off of and rub against one another and bounce, roll, and scrape against the bottom. These collisions cause smaller pieces to break off of sharp corners and edges on the surface of the sediments, particularly at corners that protrude—a process called abrasion. As a result, the sediments become rounded. Thus, table (2) shows the most likely process (abrasion) and agent of erosion (wave action) responsible for the rounded sediments shown in the photograph.

WRONG CHOICES EXPLAINED:

(1) Sandblasting is the process by which sand carried by the wind collides with and wears away solid objects. When the sand particles collide with the solid object, the impact causes small pieces of the object to break off. This can cause the object being hit with particles to become smooth or worn. The agent of erosion responsible for sandblasting is wind, not running water.

(3) Mass movements involve the downhill movement of sediments under the direct influence of gravity. When pieces of bedrock are deposited by a landslide, the particles move downhill rapidly through air. Air resistance is so small that particles of all sizes move through air at virtually the same speed. Thus, the particles do not become sorted by size as they fall. During the landslide, turbulence causes the pieces of bedrock material to collide and break apart into angular fragments. Thus, the sediments deposited by a landslide are best described as angular and unsorted. Thus, a mass movement such as a landslide is unlikely to be responsible for this rounded sediment.

(4) By definition, deposition is the process by which transported sediment is dropped in a new place. Deposition by wind would cover the sediments shown in the photograph and protect them from abrasion, not cause them to become rounded.

23. **3** Note that basaltic intrusion *B* cuts across both rock unit *C* and basaltic intrusion *A*. In order for basaltic intrusion *B* to cut across both of these rocks, *A* and *C* already had to exist when *B* formed. Therefore, basaltic intrusion *B* is the youngest rock. Note that basaltic intrusion *A* cuts across rock unit *C*. Therefore, rock unit *C* already had to exist when basaltic intrusion *A* formed. Therefore, rock unit *C* is older than basaltic intrusion *A*. Thus, the relative age of these three rock units from oldest to youngest is $C \rightarrow A \rightarrow B$.

24. **1** A rift valley is an elongated, low-lying region that forms where Earth's tectonic plates move apart, or rift, allowing the land to drop down between parallel faults. These valleys also form when oceanic plates are moving apart and forming a divergent boundary. Find the Tectonic Plates map in the *Reference Tables*

for Physical Setting/Earth Science. On the map, locate the positions corresponding to surface locations *A*, *B*, *C*, and *D*. Location *A* corresponds to the East African Rift, and the arrows indicate that the plates are diverging in this location, or moving apart. Location *B* corresponds to a convergent plate boundary. Location *C* corresponds to the Hawaii Hot Spot. Location *D* corresponds to a transform plate boundary. Thus, the location that is on a major rift valley is *A*.

WRONG CHOICES EXPLAINED:

(2) Rift valleys form at divergent boundaries. According to the Tectonic Plates map in the *Reference Tables for Physical Setting/Earth Science*, location *B* corresponds to a convergent boundary.

(3) Rift valleys form at divergent boundaries. Location *C* is in the center of the Pacific Plate, far from any plate boundaries. According to the Tectonic Plates map in the *Reference Tables for Physical Setting/Earth Science*, location *C* corresponds to the Hawaii Hot Spot, which is associated with the formation of volcanic mountains, not rift valleys.

(4) Rift valleys form at divergent boundaries. According to the Tectonic Plates map in the *Reference Tables for Physical Setting/Earth Science*, location *D* is along the northern boundary between the Scotia Plate with the South American Plate. Location *D* corresponds to a transform boundary, not a divergent boundary.

25. **4** It is given that the *P*-wave travel time was 11 minutes. Find the Earthquake P-Wave and S-Wave Travel Time graph in the *Reference Tables for Physical Setting/Earth Science*. Locate 11 minutes on the vertical axis labeled "Travel Time (min)." From this value, trace horizontally to the right until you intersect the bold line labeled "P." From the intersection, trace vertically downward to the horizontal axis labeled "Epicenter Distance ($\times 10^3$ km)" at the bottom of the graph and note the value 7.6×10^3, or 7600 km. Thus, the distance to the epicenter is 7600 km. To determine the *S*-wave travel time, trace vertically upward from 7600 km until you intersect the bold curve marked "S," and then trace horizontally to the left to the vertical scale labeled "Travel Time (min)." Note that this corresponds to 20 min. Thus, the distance to the epicenter was 7600 km, and it took the first *S*-wave 20 minutes to travel that distance.

26. **2** It is given that the Catskills are part of the Alleghany Plateau. They are commonly called mountains because of their high elevation and hilly surface. Mountains are typically formed by folding or faulting due to tectonic motions or by volcanism. In contrast, plateaus generally have an underlying structure of horizontal layers of rock. Find the Generalized Bedrock Geology of New York State

map in the *Reference Tables for Physical Setting/Earth Science.* Locate the region corresponding to the Catskills, and note the map symbol for the surface bedrock. Locate this map symbol in the key labeled "Geologic Periods and Eras in New York," trace right, and note the statement "limestones, shales, sandstones, and conglomerates." Find the Scheme for Sedimentary Rock Identification in the *Reference Tables for Physical Setting/Earth Science.* Locate the column labeled "Rock Name," and note that limestone, shale, sandstone, and conglomerate are all sedimentary rocks. Streams and other agents of erosion can cut deep valleys into the surface of a plateau, and the walls of these valleys can be eroded into steep hills. This is most likely how the Catskills formed. Thus, although they look like small mountains, the Catskills are classified as a plateau because of their bedrock structure.

WRONG CHOICES EXPLAINED:

(1) Plateaus are located at high elevations, not at low elevations.

(3) A region is classified as a plateau because of its elevation, relief, and underlying bedrock structure. Bedrock age alone does not determine how a landscape region was formed, its structure, or the processes that shape it. For example, bedrock of many different ages may have been formed by volcanic activity, metamorphism, or deposition in water, giving rise to very different types of landscape regions.

(4) As explained above, the Catskills consist of limestones, shales, sandstones, and conglomerates—all sedimentary rocks. If the Catskills had undergone a high degree of metamorphism, these sedimentary rocks would have been changed into metamorphic rocks. That is not the case.

27. **3** Find the Relationship of Transported Particle Size to Water Velocity graph in the *Reference Tables for Physical Setting/Earth Science*. Locate 0.1 cm on the vertical axis labeled "Particle Diameter (cm)." Trace horizontally to the right until you intersect the bold curve. At this point, trace down to the horizontal axis labeled "Stream Velocity (cm/s)." Read the minimum stream velocity that can maintain the transportation of a particle 0.1 cm in diameter. That minimum stream velocity is 5.5 cm/s.

28. **4** Evaporites are sediments chemically precipitated due to evaporation of an aqueous solution (such as seawater). Find the Scheme for Sedimentary Rock Identification in the *Reference Tables for Physical Setting/Earth Science*. In the "Comments" column, note the statement "crystals from chemical precipitates and evaporites." Trace this section to the right to the column labeled "Rock Name." Note that rock salt, rock gypsum, and dolostone can form from evaporites. Thus, the rock that is classified as an evaporite is crystalline rock salt.

29. **4** Find the Average Chemical Composition of Earth's Crust, Hydrosphere, and Troposphere chart in the *Reference Tables for Physical Setting/Earth Science*. Locate the column labeled "Percent by volume" in the section labeled "Crust." In this column, find the row corresponding to the greatest percent by volume (94.04). Trace this row left to the column labeled "Element (symbol)." Note that the element that makes up the greatest percent by volume of Earth's crust is oxygen. Now locate the row corresponding to the second greatest percent by volume (1.42). Trace this row left to the column labeled "Element." Note that the element that makes up the second greatest percent by volume of Earth's crust is potassium. Thus, the pair of elements that makes up most of Earth's crust by volume is potassium and oxygen.

30. **1** According to the diagram, soil zone *A* is located just beneath the surface and contains plant roots. The surface of zone *A* shows grass, plants, and trees. Leaves and other plant materials that die fall to the ground and begin to decompose. Rain washes these organic materials into the pore spaces in zone *A*. Thus, zone *A* contains both fresh and partially decomposed organic matter. Note that the number of larger particles decreases as you move upward from the underlying bedrock in zone *D* to zone *A*. This is because the particles near the surface have been exposed to the weathering process for a longer period of time and have been broken down into smaller and smaller pieces by weathering. Thus, the zone having the most organic and weathered material is zone *A*.

31. **1** Find the Generalized Bedrock Geology of New York State map in the *Reference Tables for Physical Setting/Earth Science*. Locate Utica, New York. Note the rock symbol for the bedrock in that area—lines slanting downward to the right. Now find this symbol in the "Geologic Periods and Eras in New York" key at the lower left of the map. Note that the symbol corresponds to Ordovician-age rocks consisting of limestones, shales, sandstones, and dolostones. Find the Scheme for Sedimentary Rock Identification in the *Reference Tables for Physical Setting/Earth Science*. Note that limestone, shale, sandstone, and dolostone are all sedimentary rocks. Thus, the type of surface bedrock most commonly found in the Utica, New York, area is sedimentary, with limestone, shale, sandstone, and dolostone.

32. **1** Water flows from higher elevations to lower elevations. Therefore, water will flow to either side of the tops of the circular ridges shown in the block diagram and then into the valleys between the ridges, forming a stream drainage pattern, as shown in map (1).

WRONG CHOICES EXPLAINED:

(2) On a flat plain, the slope is gentle and streams flow downhill slowly. Streams typically form a dendritic stream drainage pattern. In a dendritic stream drainage pattern, numerous small streams join together to form larger streams in a treelike pattern, as shown in map (2). The geologic landscape shown in the diagram in the question shows a series of circular ridges, not a flat plain.

(3) The trellis stream pattern in map (3) forms in tilted or folded rock layers of unequal resistance to erosion. Major streams run through long, parallel valleys following belts of weak rock between parallel ridges of stronger rock. Tributary streams enter major streams at right angles. The geologic landscape shown in the diagram in the question shows a series of circular ridges, not a series of long, parallel ridges.

(4) The geologic landscape shown in the diagram in the question shows a series of circular ridges with the highest point in the center. Map (4) shows a stream pattern in which the streams are flowing toward the center from larger streams into smaller streams. In order for this to occur, water would have to flow uphill, which it does not.

33. **3** Note the symbol for ice in the key. Find the area corresponding to ice in all three maps. Note that the maps indicate that the average area covered by sea ice is decreasing. In order for the amount of ice to decrease, some of the ice must have melted. Ice must absorb heat in order to melt. Thus, the maps best support the inference that Earth's climate is warming because the average area covered by sea ice is decreasing.

34. **1** Photograph *A* shows a large, fan-shaped deposit where a stream flows into a large body of water. When a stream flows into a standing body of water, the stream slows down and most of the stream's sediment is deposited. The resulting landform is a large, flat, fan-shaped pile of sediment at the mouth of the stream called a delta. Thus, depositional feature *A* is a delta.

Photograph *B* shows an elongated island running parallel to the coastline. Note the inlets at the northeast and southwest ends of the island. Barrier bars are long, narrow, offshore deposits of sand that run parallel to the coastline. Barrier bars form when waves deposit enough sand onto a sandbar so that the sandbar is above water. When tidal inlets break a barrier bar into elongated islands, the islands are called barrier islands. Thus, depositional feature *B* is a barrier island. Therefore, depositional features *A* and *B* are correctly identified as *A*—delta; *B*—barrier island.

35. **2** In diagram *A*, the mineral crystals of the rock are randomly oriented. In diagram *B*, the crystals occur in a pattern of parallel layers of flat, elongated crystals separated into light- and dark-colored bands. The type of mineral crystal arrangement shown in diagram *B* is called a foliated texture; it shows mineral alignment and banding. Foliation occurs when mineral grains recrystallize, or are flattened, under heat and pressure. Grains that may have been randomly oriented form parallel layers or become aligned when deformed under high heat and pressure. Banding occurs when minerals of different densities recrystallize under heat and pressure. As they recrystallize, minerals of different densities separate into layers like a mixture of oil and water. Since light-colored minerals tend to be less dense than dark-colored ones, the layers that form are alternating light and dark. The changes that occur in solid rock due to heat, pressure, and chemical activity are called metamorphism. The rock in diagram *A* has been changed by recrystallizing under heat and pressure into the metamorphic rock shown in diagram *B*. Thus, the geologic processes most likely responsible for the banding and alignment of mineral crystals represented in diagram *B* are heating and increased pressure.

WRONG CHOICES EXPLAINED:

(1) Melting and solidification would lead to the formation of a new igneous rock, not a change in an existing rock.

(3) Compaction and cementation are processes that transform loose particles of sediment into solid sedimentary rocks, not processes that change an existing rock into a metamorphic rock.

(4) Weathering and erosion are processes that break down rock into smaller fragments called sediments and transport them to a new location, not processes that change an existing rock into a metamorphic rock.

PART B–1

36. **2** Locate the point corresponding to 12:00 p.m. (noon) on January 24 on the horizontal axis labeled "Time" at the bottom of the graph. From this point, trace vertically upward until you intersect the bold curved line. From this intersection, trace horizontally to the left until you intersect the vertical axis labeled "Height Above Average Low Tide Level (ft)." Note the value of 1.6. Thus, the height of the water above average low tide level at noon on January 24 was 1.6 ft.

37. **1** Note that the line representing water levels at Montauk rises and falls in a repeating pattern. In other words, the changes in water level are cyclic. Note, too, that all of these changes repeat in a specific period of time. Therefore, they are predictable. For example, three low tides are shown on the graph. The first

low tide occurs at 3:00 a.m. on January 24, the second at 3:30 p.m. on January 24, and the third at 4:00 a.m. on January 25. Each of the low tides are 12 hours 30 minutes apart. Thus, the changing water levels at Montauk can best be described as cyclic and predictable.

38. **4** Both the Sun and Moon exert a force of gravity on Earth that causes bulges in the oceans. As Earth rotates on its axis, the positions of the bulges remain fixed in line with the Moon and Sun. The water levels at a particular location on Earth rise and fall as that location rotates into and out of these bulges. The repeated pattern of Earth's rotation and the orbital motions of the Moon and Earth cause ocean water levels to change in a cyclic pattern called tides. Thus, the water-level variation pattern shown by the graph is caused by Earth's rotation and the gravitational pull of the Moon.

WRONG CHOICES EXPLAINED:

(1) The changes in water levels at Montauk, New York, are cyclic and predictable. The changes in wind velocity produced by coastal storms are neither cyclic nor predictable. Therefore, it is unlikely that a noncyclic, nonpredictable event would produce a cyclic and predictable pattern.

(2) Water is not a magnetic substance. So changes in the magnetic orientation of the North American Plate would not affect the water or change its depth.

(3) The distance between Montauk and the equator is constant, not cyclic. Therefore, this distance is unlikely to produce a cyclic change in water levels. Earth's revolution is cyclic, but each cycle is one year. A yearly cyclic change is unlikely to result in a 12 hour 30 minute cyclic change.

39. **3** It is given that the tide gauge is located at Montauk, New York, at the easternmost end of Long Island. Find the Generalized Bedrock Geology of New York State map in the *Reference Tables for Physical Setting/Earth Science.* Locate the easternmost end of Long Island, and note that it is very close to the coordinates 41°, 72°. Note the compass rose in the lower right-hand corner of the map. Note that the latitude values increase from 41° to 45° as you move northward on the map in New York State. Therefore, these values represent north latitude. Note that the longitude values increase from 72° to 79° as you move westward on the map in New York State. Therefore, these values represent west longitude. Thus, the approximate latitude of the tide gauge is 41°00′ N 72°00′ W.

40. **1** The question refers to the "most massive stars." In the diagram, begin at "Star formation." Follow the arrow labeled "Massive" to the main sequence, and note that supergiant stars are formed. Supergiants then undergo supernova explosions.

The most massive stars have a very high mass. Follow the arrow labeled "Very high mass," and note that black holes are the result. Thus, the final stage in the life cycle of the most massive stars is a black hole.

41. **3** On the diagram, locate the star labeled "Sun." Note that the next stage in the Sun's life cycle is a giant star. After becoming a giant, the next stage is a white dwarf, which then ends its life cycle as a black dwarf. Thus, giants, white dwarfs, and black dwarfs may once have been similar to our Sun in mass and luminosity. Find the Characteristics of Stars chart in the *Reference Tables for Physical Setting/Earth Science.* Locate the four stars listed as choices. Note that *Deneb* is a supergiant, *Spica* is a main sequence star, *Procyon B* is a white dwarf, and *Proxima Centauri* is a main sequence star. Thus, *Procyon B,* which is the white dwarf, may once have been similar to our Sun in mass and luminosity.

42. **1** In the cores of stars, gravity is strong enough to overcome the force of repulsion between atomic nuclei, allowing the nuclei to combine in a process called nuclear fusion. By definition, nuclear fusion involves the combining of several atoms of a lighter element to form a single atom of a heavier element. The single heavier atom typically has less mass than the lighter atoms from which it formed. The mass "missing" from the heavier atom is not lost. Instead, it is converted into energy according to Einstein's formula $E = mc^2$. This formula states that if mass is converted into energy, the amount of energy released, E, is equal to the mass, m, times the speed of light, c, squared. The speed of light squared is a very large number. Therefore, the conversion of even a small amount of mass into energy during nuclear fusion results in the release of a very large amount of energy. Thus, energy is produced in the cores of main sequence stars when lighter elements combine and form heavier elements during the process of nuclear fusion.

WRONG CHOICES EXPLAINED:

(2) By definition, fusion is the process or result of joining two or more things together to form a single thing. Nuclear fusion involves the combining of several atoms of a lighter element to form a single atom of a heavier element, not combining heavier elements to form lighter elements.

(3) Cosmic background radiation is radiation coming from all directions in the universe. It is believed to be radiation left over from the Big Bang that formed the universe. In the mid-1960s, this radiation was detected by Arno Penzias and Robert Wilson at Bell Laboratories in New Jersey. It was found to fill the universe uniformly in every direction and had a temperature of about 3 kelvins (K). Fusion requires temperatures of about 100 million kelvins. Thus, absorbing cosmic radiation would not increase the temperature of a star's core enough to support fusion.

(4) Releasing energy would decrease the energy in the star, not produce energy in the star's core.

43. **3** Radioactive isotopes break apart, or decay, at a steady, constant rate that is not affected by outside factors such as changes in temperature, pressure, or chemical state. Therefore, the decay process may be used as a clock to determine the actual age of the rocks in which the isotopes are found. Radioactive isotopes start to decay when a rock forms. The decay of a radioactive isotope occurs at a statistically predictable rate known as its half-life. Half-life is the time required for one-half of the unstable radioisotope to change into a stable decay product. According to the graph, at time = 0 there were 1000 atoms of radioactive Isotope *X* in the sample. After 1 half-life, half of those atoms would have decayed into a stable daughter product and only half of the original 1000 atoms (500 atoms) would still be radioactive. Locate "500" on the vertical axis labeled "Number of Radioactive Atoms Present," and trace right until you intersect the bold curve. From this intersection, trace vertically downward to the horizontal axis labeled "Time (h)." Note that the value is 3. Thus, based on the graph, the half-life of this radioactive isotope is 3 h.

44. **3** On the horizontal axis labeled "Time (h)," locate the point corresponding to "8" between the values 6 and 9. From this point, trace vertically upward until you intersect the bold curve. From this point, trace horizontally to the left until you intersect the vertical axis labeled "Number of Radioactive Atoms Present." Note that this point corresponds to approximately 155 atoms. Thus, the approximate number of radioactive atoms of Isotope *X* present when 8 hours of decay has occurred is 155.

45. **4** From point *A* on the diagram, trace horizontally until you intersect the vertical axis labeled "Altitude (km)." Note that point *A* is located at an altitude of about 9 km. Find the Selected Properties of Earth's Atmosphere chart in the *Reference Tables for Physical Setting/Earth Science*. On the vertical axis labeled "Altitude," locate the point corresponding to 9 km. Note that km are shown on the left side of the axis and miles on the right side. So be sure to use the values on the left side of the axis to locate 9 km. From 9 km, trace horizontally to the right to the portion of the chart labeled "Temperature Zones." Note that the temperature zone at an altitude of 9 km is called the troposphere. Thus, point *A* is located in the troposphere.

46. **2** Air is a fluid. Heated fluids are less dense than cooler fluids. When air is warmed, it becomes less dense than the surrounding cooler air and rises upward. Similarly, when air is cooled, it becomes denser than the surrounding warmer air

and sinks downward. Note that the arrow labeled “Updraft” starts in warm air. Note that the arrow labeled “Downdraft” starts in cold air. The updraft is therefore warm air that is rising, and the downdraft is cold air that is sinking. Thus, the updrafts and downdrafts are primarily caused by differences in air density.

WRONG CHOICES EXPLAINED:

(1) Note that the updrafts and downdrafts are occurring at the same altitude above sea level. Therefore, it is unlikely that altitude causes the updrafts and downdrafts.

(3) Relative humidity is a measure of the concentration of water vapor in the atmosphere. When air temperature reaches the dewpoint, the relative humidity is 100%, and water vapor in the air begins to condense and form clouds. Thus, the relative humidity in a cloud is 100%. Note that both the updrafts and the downdrafts are occurring in the cloud, where relative humidity is 100%. Therefore, it is unlikely that relative humidity causes the updrafts and downdrafts.

(4) Specific heat is a measure of the heat energy needed to raise the temperature of 1 gram of a substance by 1 degree Celsius. It is a characteristic of matter, not a force that causes motion.

47. **3** It is given that the diagram shows a thunderstorm cloud. Thus, this weather event is a thunderstorm. Find the Key to Weather Map Symbols in the *Reference Tables for Physical Setting/Earth Science.* Find the section labeled “Present Weather,” and locate the symbol for a thunderstorm. Note that this symbol corresponds to the symbol in choice (3).

48. **2** There are 5 contour intervals between the bold 250-foot index contour line and the bold 500-foot index contour line. Thus, the contour interval on this map is 50 feet ($250 \div 5 = 50$).

49. **2** Find the Equations section of the *Reference Tables for Physical Setting/Earth Science* and note the equation for gradient:

$$\text{Gradient} = \frac{\text{change in field value}}{\text{change in distance}}$$

The map shown is a topographic map. The field value on a topographic map is elevation. Note that point A is located directly on the 250-foot contour line and that point B is located directly on the 500-foot contour line. Thus, the change in field value (elevation) from A to B is 250 feet.

On a piece of scrap paper, mark off the distance along the straight line between points *A* and *B*. Compare the distance marked off on the scrap paper to the scale printed beneath the map to determine the distance between points *A* and *B*. The distance is about 10 miles. Substitute these values in the equation and solve:

$$\text{Gradient} = \frac{500 \text{ ft} - 250 \text{ ft}}{10 \text{ mi}}$$

$$\text{Gradient} = \frac{250 \text{ ft}}{10 \text{ mi}} = 25 \text{ ft/mi}$$

Thus, the gradient between location *A* and location *B* is approximately 25 ft/mi.

50. **4** Note that the bold index contour to the west of location *C* is labeled "250" and the bold index contour east-northeast of location *C* is labeled "500." Thus, the elevation increases between these two index contours. From location *B* to location *D*, the elevation continues to increase. As explained earlier, the contour interval on this map is 50 feet. Thus, the elevation of location *C* is between 300 and 350 feet.

Note that the trail leads through the center of a series of concentric contour lines within the oval-shaped 500-foot index contour. On a topographic map, concentric contour lines typically represent a hill. Thus, the elevation increases within these contours. The innermost concentric contour has an elevation of 600 feet. Therefore, the top of the hill lies within this 600-foot contour at an elevation of more than 600 feet but less than 650 feet. Thus, a person walking along the trail from location *C* to the center of the 600-foot contour would be walking uphill. From the top of the hill, the trail continues back to the 500-foot contour. So a person would now be walking downhill. Note that the trail then crosses the 500-foot contour and continues across the area between it and the next 500-foot contour line to the east. The elevation between the two 500-foot contours is lower than 500 feet but greater than 450 feet. Thus, a person walking along the trail would walk downhill to about the midpoint in this area and then begin walking uphill again to cross the next 500-foot contour. Note that location *D* lies on the fourth contour line from the 500-foot contour, so the elevation of location *D* is 700 feet. Therefore, a person walking from the 500-foot contour line to location *D* would be walking uphill. When taken together, a person walking along the trail from location *C* to location *D* would walk uphill, then downhill, then uphill again.

PART B–2

51. Find the Properties of Common Minerals chart in the *Reference Tables for Physical Setting/Earth Science.* In the column labeled "Mineral Name," locate "Muscovite mica" and locate "Biotite mica." Trace left to the column labeled "Composition*." Note that the chemical formula for muscovite mica is $KAl_3Si_3O_{10}(OH)_2$ and the chemical formula for biotite mica is $K(Mg,Fe)_3AlSi_3O_{10}(OH)_2$. Refer to the key labeled "Chemical symbols" at the bottom of the chart. Note that both muscovite mica and biotite mica contain potassium (K), aluminum (Al), silicon (Si), oxygen (O), and hydrogen (H). However, biotite mica also contains iron (Fe) and/or magnesium (Mg), while muscovite mica does not. Thus, two chemical elements present in biotite mica that are not present in muscovite mica are magnesium (Mg) and iron (Fe).

One credit is allowed for **magnesium (Mg) and iron (Fe)**.

52. The passage states that all members of the mica family are similar in density, luster, hardness, streak, type of breakage, and crystal shape. Thus, identifying these characteristics for two members of this family will provide information about all members of this family. Find the Properties of Common Minerals chart in the *Reference Tables for Physical Setting/Earth Science.* In the column labeled "Mineral Name," locate "Muscovite mica" and locate "Biotite mica." In the row for each of these minerals, trace left to the column labeled "Cleavage/Fracture." Note that both micas display cleavage. Recall that cleavage is the tendency of a mineral to split along definite crystalline planes, yielding smooth surfaces. Note in the column labeled "Distinguishing Characteristics" that both micas are "flexible in thin sheets." Thus, both muscovite mica and biotite mica cleave into thin sheets.

Now trace left to the column labeled "Hardness." Note that muscovite mica has a hardness of 2–2.5 and that biotite mica has a hardness of 2.5–3. Next, trace left to the column labeled "Luster" and note that both micas have a nonmetallic luster. Nonmetallic lusters can be described by a variety of terms, such as vitreous/glassy or pearly/silky. However, these terms are somewhat subjective.

One credit is allowed for a correct luster, hardness, and dominant form of breakage. Acceptable responses include but are not limited to:

Luster:

- **nonmetallic**
- **vitreous/glassy**
- **pearly**
- **silky**

Hardness:

- **any value or range from 2.0 to 3.0**

Dominant form of breakage:

- **cleavage/basal cleavage**
- **breaks into thin sheets**
- **one direction of cleavage**

53. Find the Scheme for Igneous Rock Identification in the *Reference Tables for Physical Setting/Earth Science*. In the column labeled "Crystal Size," locate the row corresponding to "10 mm or larger." Trace this row to the left, and note that the only igneous rock listed in this row is pegmatite. From the column containing pegmatite, trace vertically down to the section labeled "Mineral Composition (relative by volume)." Note that one of the sections representing various minerals that the pegmatite intersects is the black section labeled "Biotite (black)." Thus, pegmatite is an igneous rock in which crystals of biotite mica are larger than 10 millimeters in diameter.

One credit is allowed for **pegmatite**.

54. Large crystals of mica that weigh several hundred tons are clearly larger than 10 millimeters in diameter. Thus, as explained in answer 53, the only igneous rock with mica crystals in this size range is pegmatite. Find the Scheme for Igneous Rock Identification in the *Reference Tables for Physical Setting/Earth Science*. Locate "Pegmatite," and trace this row left to the column labeled "Environment of Formation." Note that the row corresponding to pegmatite is in the range labeled "Intrusive (Plutonic)" igneous rock. Thus, pegmatite is an intrusive igneous rock. Intrusive igneous rocks form by the cooling and solidification of molten magma deep underground where the magma is insulated by surrounding rock. Therefore, the magma cools slowly and takes a long time to solidify to form intrusive rock. Very slow cooling over a long period of time allows very large crystals to form before the rock is completely solid. Thus, an igneous rock with very large crystals formed deep underground in an intrusive environment and its relative rate of cooling was slower than that of other igneous rocks.

One credit is allowed if *both* responses are correct. Acceptable responses include but are not limited to:

Environment of formation:

- **intrusive**
- **plutonic**
- **deep underground**

Relative rate of cooling:

- **slower/slowly**
- **took a longer time/long time**

Note: Credit is *not* allowed for "long" alone, as this does not accurately describe a rate of cooling but, instead, describes a duration.

55. Note that the Sun's position at solar noon increases in altitude from December 21 to March 21 to June 21. From the diagram, it can be inferred that the position of the Sun at solar noon on April 21 will lie between its position on March 21 and its position on June 21 because April occurs between March and June. From March 21 to June 21 is three months. From March 21 to April 21 is one month, or about one-third the time between March 21 and June 21. Therefore, the position of the Sun at solar noon on April 21 will be about one-third of the distance between its position on March 21 and its position on June 21. Therefore, the **X** marking the Sun's position at noon on April 21 should be drawn, as shown below.

One credit is allowed if **the center of the X is within or touches the boxed area shown below**.

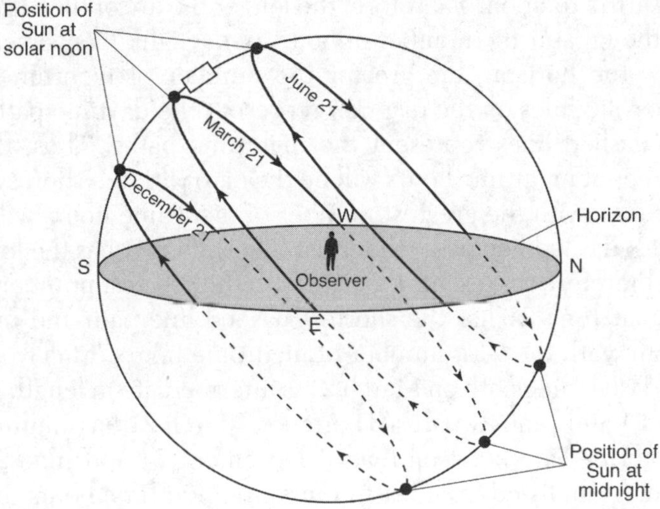

[**NOTE:** It is recommended that an overlay of the same scale as the student answer booklet be used to ensure reliability in rating.]

56. Due to Earth's rotation, all celestial objects, including the Sun, appear to move across the sky from east to west along a path that is an arc, or part of a circle. Earth completes one complete 360° rotation every day (including December 21). Thus, Earth rotates from west to east at a constant rate of

15°/hour (360° day ÷ 24 hr/day = 15°/hr). As Earth rotates on its axis from west to east at 15° per hour, the Sun appears to move through the sky from east to west at 15° per hour.

One credit is allowed for **15°/h**.

57. To determine the direction in which the observer's shadow would point at solar noon on March 21, do the following. Draw a straight line connecting north and south on the horizon. Note that this line passes beneath the observer's feet. Now draw a straight line connecting the position of the Sun at solar noon on March 21 and the tip of the observer's head. Extend that line to the line beneath the observer's feet. This intersection represents the tip of the observer's shadow. Now draw a bold horizontal line connecting the observer's feet to the tip of the observer's shadow; this represents the observer's shadow. Note that the observer's shadow lies directly on the north-south line with the tip of the shadow to the north. Thus, the observer's shadow at solar noon on March 21 points north.

One credit is allowed for **North/N**.

58. The Sun appears to move across the sky at a constant rate of 15° of arc per hour due to Earth's rotation. Therefore, the longer the arc of the Sun's path above the horizon, the greater the number of hours of daylight. Conversely, the longer the arc below the horizon, the greater the number of nighttime hours. It is given that the solid lines on the diagram represent the daytime paths of the Sun and that the dashed lines represent the nighttime paths. Thus, the date with the least number of nighttime hours will be the date with the shortest dashed line path, and the date with the greatest number of nighttime hours will be the date with the longest dashed line path. Note that December 21 has the longest dashed line path and therefore represents the date with the greatest number of nighttime hours. Note that June 21 has the shortest dashed line path and therefore represents the date with the least number of nighttime hours. Finally, note that the length of the dashed line path on March 21 is intermediate in length between that of December 21 and June 21 paths. Therefore, March 21 has a number of nighttime hours intermediate between that of December 21 and June 21. Thus, the dates on the diagram listed from least number of nighttime hours to the greatest number of nighttime hours is June 21 → March 21 → December 21.

One credit is allowed for **a correct sequence, as shown below**.

June 21	**March 21**	**December 21**
Least number of nighttime hours	⟶	Greatest number of nighttime hours

59. It is given that the map in your answer booklet shows a portion of the Tectonic Plates map in the *Reference Tables for Physical Setting/Earth Science.* Find the Tectonic Plates map in the *Reference Tables for Physical Setting/Earth Science.* On the map, locate the plate boundary corresponding to the one shown on the map in your answer booklet. According to the key at the bottom of the map, a bold arrow represents relative motion at a plate boundary. Note that the Tectonic Plates map in the *Reference Tables for Physical Setting/Earth Science* shows the North American Plate moving to the WSW at this plate boundary and the Caribbean Plate moving to the ENE. Thus, an arrow pointing WSW should be drawn in the box on the North American Plate, and an arrow pointing ENE should be drawn in the box on the Caribbean Plate, as shown below.

One credit is allowed for **the correct direction of both arrows, as shown below**.

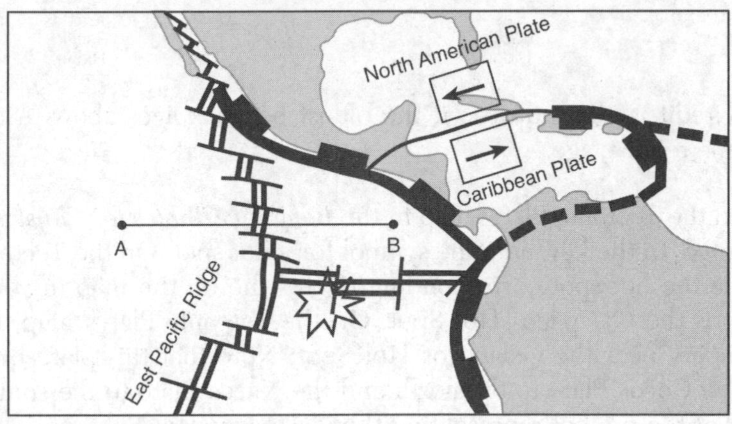

60. Find the Tectonic Plates map in the *Reference Tables for Physical Setting/ Earth Science.* Locate the plate boundary corresponding to the one labeled "East Pacific Ridge" on the map in your answer booklet. Note that this corresponds to the boundary between the Pacific Plate and the Cocos Plate. Thus, location *A* is on the Pacific Plate, and location *B* is on the Cocos Plate. According to the key at the bottom of the map, the boundary between these two plates is a divergent boundary. Along divergent boundaries, adjacent plates move apart and open rifts through which magma can rise to the surface and solidify to form new ocean floor. As the plates continue to move apart, this process is repeated, constantly creating new ocean floor and pushing the older ocean floor on either side of the ridge away from the ridge. Thus, the age of the ocean floor rock increases with distance away from the East Pacific Ridge. Therefore, the line representing the relative age of the ocean floor bedrock from location *A* to location *B* should indicate the oldest bedrock ages at locations *A* and *B* and the youngest at the East Pacific Ridge, as shown below.

One credit is allowed for **a V-shaped or U-shaped line showing the oldest bedrock ages above *A* and *B* and the youngest bedrock age at the East Pacific Ridge**.

Examples of 1-credit responses:

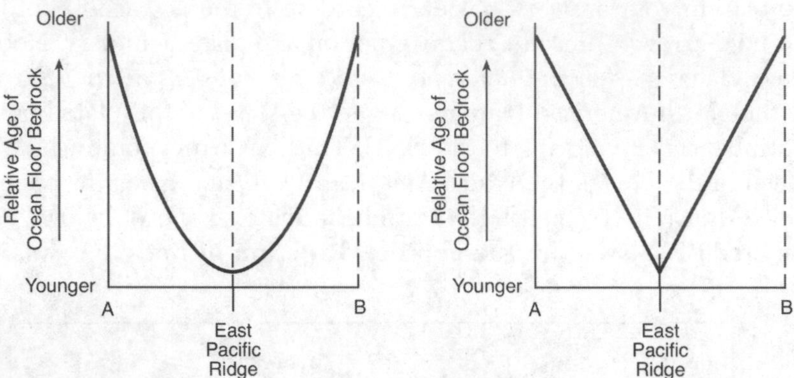

Note: Credit is allowed even if the oldest bedrock ages above *A* and *B* are not the same relative age.

61. Find the Tectonic Plates map in the *Reference Tables for Physical Setting/Earth Science.* In the key, note the symbol for a hot spot. On the Tectonic Plates map, locate the hot spot corresponding to the one on the map in your answer booklet. It is the Galapagos Hot Spot. On the Tectonic Plates map, locate the plate boundary near the Galapagos Hot Spot. Note that this plate boundary is between the Cocos Plate to the north and the Nazca Plate to the south. Locate the center of the symbol representing the Galapagos Hot Spot, and note that it lies south of the boundary on the Nazca Plate. Thus, the hot spot on the map is the Galapagos Hot Spot, and the center of this hot spot is located on the Nazca Plate.

One credit is allowed for **Galapagos Hot Spot and Nazca Plate**.

62. The bedrock at location *A* formed when magma rose to Earth's surface beneath the ocean through cracks at the plate boundary and then solidified to form new ocean floor. Thus, the ocean floor at location *A* is considered an extrusive igneous rock, because it cooled at Earth's surface. Find the Scheme for Igneous Rock Identification in the *Reference Tables for Physical Setting/Earth Science.* Locate the column labeled "Environment of Formation." In that column, locate the rows labeled "Extrusive (Volcanic)." It is given that bedrock at location *A* is mafic igneous rock. Locate the arrow labeled "Composition" in the "Characteristics" section in the middle of the chart. Note that the arrow extends from "Felsic (rich in Si, Al)" on the left to "Mafic (right in Fe, Mg)" on the right. Trace the

columns on the mafic side of the scale upward until they intersect with the rows corresponding to an extrusive environment of formation. Thus, the extrusive igneous rocks that are mafic in composition are basaltic glass, scoria, vesicular basalt, basalt, and diabase. Collectively, these are considered to be basaltic rocks. Thus, the bedrock at location *A* is either basalt and/or other basaltic rocks. Find the Inferred Properties of Earth's Interior chart in the *Reference Tables for Physical Setting/Earth Science*. Locate the "Density (g/cm^3)" scale along the right side of the cross section. Note that oceanic crust is listed as basaltic and has a density of 3.0 g/cm^3. Thus, the mafic igneous bedrock that is most likely to make up the oceanic crust at location *A* is basalt or basaltic igneous rocks, and the average density of oceanic crust is 3.0 g/cm^3.

One credit is allowed for **basalt/basaltic bedrock and a density of 3.0 g/cm^3 *or* 3 g/cm^3**.

63. Porosity is a measure of the amount of empty space among the particles in a rock or sediment sample. It is often expressed as the percentage of the total volume of a rock or a sediment sample that is open pore space. It is given that the mixture consists of pebbles, sand, silt, and clay. Find the Relationship of Transported Particle Size to Water Velocity graph in the *Reference Tables for Physical Setting/Earth Science.* Note the axis labeled "Particle Diameter (cm)" along the left side of the graph, and note the names of sediments corresponding to size ranges along the right side of the graph. From largest to smallest, the particles in the mixture are pebbles, sand, silt, and clay. In a mixture of different particle sizes, the smaller particles can fit into and fill the open pore spaces among the larger particles. This decreases the total amount of open pore space, or porosity, of the mixture. For example, sand, silt, and clay particles are small enough to fit into and fill the open pore spaces among pebbles. Silt and clay can fill the empty pore spaces among grains of sand. Clay can fill the empty pore spaces among silt. Thus, the particle sizes fit more closely in the mixture. This results in the lowest porosity because smaller particles fill the open pore spaces among larger particles, decreasing the total amount of open pore space.

One credit is allowed. Acceptable responses include but are not limited to:

- **The smaller particles filled the pore spaces among the larger particles and decreased the total amount of open pore space.**
- **Pore spaces among the pebbles were filled by sand, silt, or clay.**
- **Small particles take up spaces between larger particles.**

Note: Credit is *not* allowed for "it has less pores" or "less pore space" because that is restating the question.

Note: Credit is *not* allowed for "particles fit together more closely" or "particles are more closely packed" alone because this is stated in the question.

64. Infiltration is the downward movement of liquid water into and through interconnected openings among soil particles due to gravity. It is given that the rate of infiltration is measured in centimeters per second. Thus, the infiltration rate is the distance water moves downward through a material in a given amount of time.

Find the Equations chart in the *Reference Tables for Physical Setting/Earth Science.* Locate the equation for rate of change:

$$\text{Rate of change} = \frac{\text{change in value}}{\text{time}}$$

The change in value is the distance from the top of the column of sand to the bottom. It is given that this distance is 28 centimeters. On the table in the question, find "Sand" in the column labeled "Particle Type." Trace this row right to the column labeled "Time for Water to Infiltrate (s)," and note the value of 7 seconds. Thus, the time the water takes to infiltrate the 28 centimeters from the top to the bottom of the column is 7 seconds. Substitute these values in the rate of change equation and solve:

$$\text{Rate of change} = \frac{28 \text{ cm}}{7 \text{ s}} = 4\frac{\text{cm}}{\text{s}}$$

One credit is allowed for **4 cm/s**.

65. On the table in the question, find "Plastic beads" in the column labeled "Particle Type." Trace this row right to the column labeled "Particle Diameter (cm)," and note the value of 0.4. Find the Relationship of Transported Particle Size to Water Velocity graph in the *Reference Tables for Physical Setting/Earth Science.* On the axis labeled "Particle Diameter (cm)" along the left side of the graph, locate 0.4 cm. Note that 0.4 lies between 0.1 and 1.0. Note that on this segment of the axis, each subdivision represents 0.1 cm, so 0.4 is the third subdivision above 0.1. From 0.4 cm, trace horizontally to the right side of the graph. Note that 0.4 cm falls into the range of particle sizes called "Pebbles." Thus, based on particle diameter, the plastic beads represent pebbles.

One credit is allowed for **pebbles**.

PART C

66. Surface wind speed is directly related to the air pressure gradient. The closer together the isobars on the map, the greater the pressure gradient and, therefore, the greater the wind speed. On the given map, the isobars are closest together near Watertown, New York. Therefore, the highest wind speed was most

likely recorded near Watertown, New York. Thus, the evidence shown on the map that indicates that the highest wind speeds occurred near Watertown, New York, is that the isobars there are closely spaced, showing a steep pressure gradient.

One credit is allowed. Acceptable responses include but are not limited to:

- **The isobars are closest together.**
- **The steepest pressure gradient occurs at Watertown.**
- **The closer the isolines/lines are, the faster the wind speed.**

67. Find the Key to Weather Map Symbols in the *Reference Tables for Physical Setting/Earth Science*, and locate the section labeled "Station Model Explanation." Note the positions in which the values for dewpoint, amount of cloud cover, and barometric pressure are placed on a station model. Note that dewpoint (°F) is shown to the lower left of the station model. Thus, the dewpoint in Albany, New York, is 36°F. Note that the amount of cloud cover is represented by how much of the circle at the center of the station model is shaded. One-quarter of the circle at the center of the station model is shaded. Thus, the cloud cover at Albany, New York, is 25%. Finally, note that barometric pressure is shown at the upper right of the station model and that the three digits listed represent the last three digits of the air pressure in millibars. Since air pressure rarely rises above 1050.0 mb or drops below 950.0 mb, the general rule is to place a 9 in front of the three digits on the station model if they are greater than 500 and place a 10 in front of the three digits if they are less than 500. Then place a decimal point between the last two digits. According to the station model, the barometric pressure is represented by the digits 051. Since the three digits are less than 500, place a 10 in front of the digits and a decimal point between the last two digits. Therefore, the barometric pressure at the weather station is 1005.1 mb. Record these values in the table provided in your answer booklet, as shown below.

One credit is allowed if ***all three* weather variables are correct, as shown in the table below**.

Albany, New York

Weather Variable	**Weather Data**
Dewpoint	**36°F**
Cloud cover	**25%**
Actual barometric pressure	**1005.1 mb**

68. Find the Generalized Bedrock Geology of New York State map in the *Reference Tables for Physical Setting/Earth Science.* Note that New York State lies between 40° N latitude and 45° N latitude.

Find the Planetary Wind and Moisture Belts in the Troposphere diagram in the *Reference Tables for Physical Setting/Earth Science*. Note that New York State is located in the belt of southwest planetary winds that lie between 30° N latitude and 60° N latitude. As these winds travel northward, they curve more and more to the east. Thus, if this low-pressure system follows a normal storm track over the next two days, it will move in a northeasterly direction and curve toward the east.

One credit is allowed. Acceptable responses include but are not limited to:

- **East/E**
- **East northeast/ENE**
- **Northeast/NE**
- **North northeast/NNE**

69. To plot the average body volume for each of the geologic ages listed in the Brachiopod Data Table, proceed as follows. In the column labeled "Geologic Age (mya)," locate 480. Trace this row right to the column labeled "Average Body Volume Including the Shell (mL)" and note the value of 0.1. Thus, 480 mya, the average body volume of brachiopods was 0.1 mL. On the graph provided in your answer booklet, locate the point corresponding to 0.1 mL on the vertical axis labeled "Average Brachiopod Body Volume (mL)." Now locate the point corresponding to 480 mya along the horizontal axis labeled "Geologic Age (mya)." Trace horizontally right from 0.1 and vertically upward from 480 until the lines intersect. Plot this point. Repeat this procedure for each of the geologic ages listed in the data table. When completed, connect the plots with a solid line, as shown below.

One credit is allowed if **the centers of *all five* plots are within or touch the circles shown and the plots are correctly connected with a line that passes within or touches the circles**.

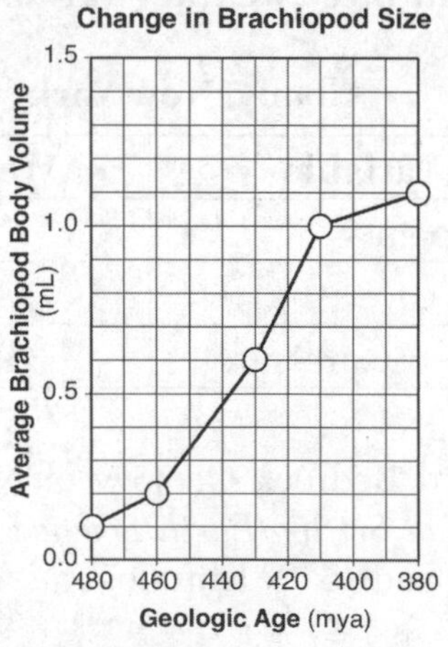

Note: Credit is allowed if the student-drawn line does *not* pass through the student plots but is still within or touches the circles. It is recommended that an overlay of the same scale as the student answer booklet be used to ensure reliability in rating.

70. The brachiopods represented on the Brachiopod Data Table existed between 380 and 480 mya. Find the Geologic History of New York State chart in the *Reference Tables for Physical Setting/Earth Science*. Locate the time scale labeled "Million years ago" along the right edge of the column labeled "Epoch." On this time scale, locate the positions corresponding to 380 and 480 million years ago. From each of these times, trace left to the column labeled "Period." Note that 380 mya corresponds to the Devonian period and 480 mya corresponds to the Ordovician period. Note that this time range encompasses the Ordovician, Silurian, and Devonian periods. Record the names of any two of these three periods in your answer booklet.

One credit is allowed for **Ordovician period, Silurian period, and/or Devonian period**.

Note: Credit is *not* allowed for Early, Middle, or Late Ordovician, for Early or Late Silurian, or for Early, Middle or Late Devonian because these are epochs, *not* periods.

71. Find the Geologic History of New York State chart in the *Reference Tables for Physical Setting/Earth Science.* Locate the column headed "Time Distribution of Fossils (including important fossils of New York State)." Note the bold black lines labeled with names and circled letters. The names are of types of fossil organisms found in New York State bedrock. The circled letters are keyed to the illustrations of specific important fossils printed along the bottom of the chart. Along the black line labeled "Brachiopods," note the circled letters "Y" and "Z." Locate drawings "Y" and "Z" at the bottom of the chart. Note that brachiopod "Y" is named *Eospirifer* and that brachiopod "Z" is named *Mucrospirifer*. Thus, two brachiopod index fossils found in New York State bedrock are *Eospirifer* and *Mucrospirifer.*

One credit is allowed for ***Eospirifer* and *Mucrospirifer***.

72. According to the reading passage, "Cope's Rule states that the average size of animals preserved in the fossil record tends to increase as each group evolves from a previous group." It is given that the earliest horses appeared in the Eocene epoch and were about the size of a large dog today. Find the Geologic History of New York State chart in the *Reference Tables for Physical Setting/Earth Science*. Locate Eocene in the column labeled "Epoch." Note along the right edge of the

column that the Eocene extended from about 33.9 to 55.8 million years ago. It is logical to infer that since the Eocene epoch, horses evolved into the horses of today. The average horse today is a much larger animal than the earliest horses were in the Eocene when they were only the size of a large dog today. Therefore, the evolution of horses supports Cope's Rule because the fossil record shows that horses have become larger over geologic time.

One credit is allowed. Acceptable responses include but are not limited to:

- **The fossil record shows that horses have become larger over geologic time.**
- **Large dogs are much smaller than horses of today.**
- **The average horse of today is a large animal, so horses must have become larger since Eocene time.**

73. An isoline connects points of equal field value, and the field values shown on this map are average yearly snowfall measured in inches. The 240-inch average yearly snowfall isoline connects all points that experienced 240 inches of average yearly snowfall. To draw the 240-inch isoline, do the following. Note that only one of the points specified on this map received exactly 240 inches of snowfall. However, that does not mean that there are no other places on the map with that value. Note that there are values greater than 240 adjacent to values less than 240. For example, note the point labeled "235" just south and slightly east of a point labeled "270." It is logical to infer that somewhere between the locations that recorded 235 and 270 inches of snowfall is a location that recorded 240 inches of snowfall. Therefore, the 240-inch isoline should pass between these two points. Similarly, as the 240-inch isoline moves west from that point, it should pass between the points labeled "270" and "236." Further west, it should pass between the point labeled "248" and the 220-inch isoline. Then the isoline should swing northeast to between the points labeled "225" and "241" and then north between the points labeled "241" and "238." Continue the isoline north between the point labeled "257" and the 220-inch isoline. From there, the isoline should turn south and run between the 220-inch isoline and the points labeled "257," "270," and "248." The isoline should end up back at the point labeled "240."

One credit is allowed for **correctly drawing the 240-inch isoline. The isoline must pass through or touch the dot labeled 240**.

Note: If additional lines are drawn, all must be drawn correctly to receive credit.

Example of a 1-credit response:

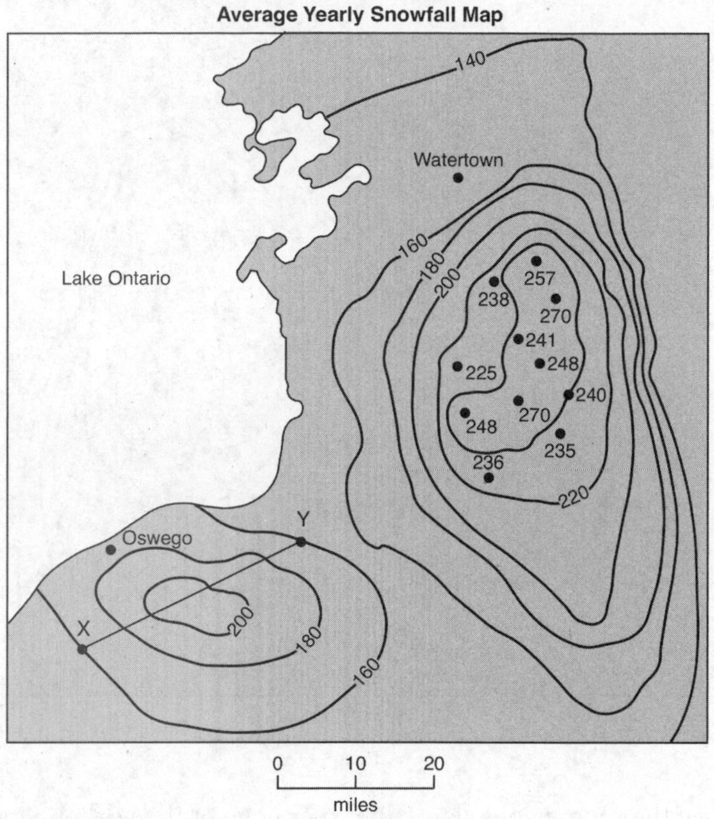

74. To construct a profile of the average annual snowfall along line *XY*, proceed as follows. Place the straight edge of a piece of scrap paper along the solid line connecting point *X* to point *Y*. Mark the edge of the paper at points *X* and *Y* and wherever the paper intersects an average yearly snowfall isoline. Note that the interval between isolines is 20 inches. Wherever the paper intersects an isoline, label the mark with the value of the average yearly snowfall of that isoline, as shown below. (You may want to label the isolines on the Oswego side of line *XY*. If your paper is not on the Oswego side, it will cover the labels.)

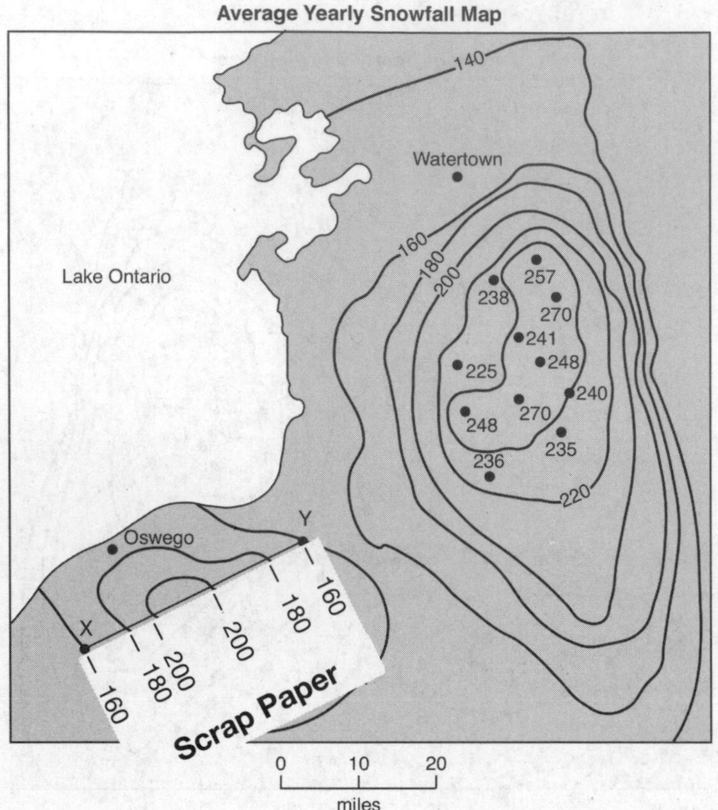

Now, place the scrap paper along the lower edge of the grid provided on your answer sheet so that points *X* and *Y* on the paper align with points *X* and *Y* on the grid, as shown below.

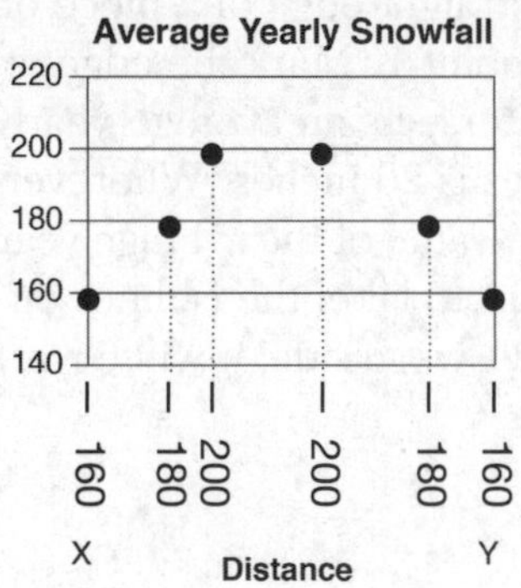

At each point where an isoline crossed the edge of the paper, draw a dot on the grid at the appropriate number of inches of average yearly snowfall directly above it. Finally, connect all of the dots in a smooth curve to form the finished profile, as shown below. Note that the top of the curve should extend above 200 inches. Since no additional isoline exists within the 200-inch isoline, and the interval between isolines is 20 inches, the average yearly snowfall in that area is more than 200 inches but less than 220 inches. Therefore, the highest point of the profile must be greater than 200 inches but less than 220 inches.

One credit is allowed if **the centers of *all six* plots are within or touch the rectangles shown and are correctly connected with a line that passes within or touches each rectangle. The line must extend above the 200-inch line but remain below the 220-inch line**.

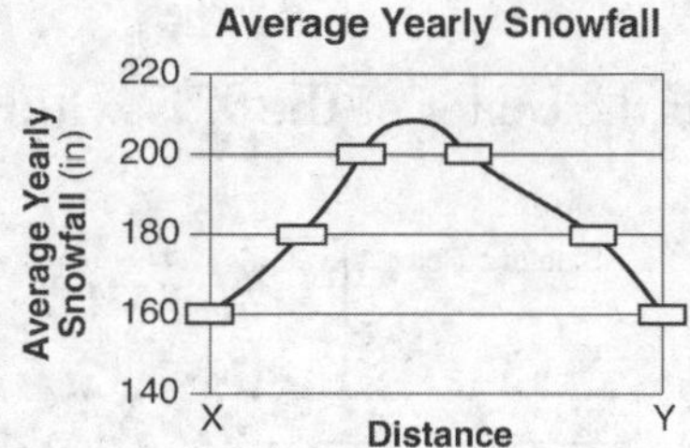

Note: Credit is allowed if the student-drawn line does *not* pass through the student plots but is still within or touches the rectangles. It is recommended that an overlay of the same scale as the student answer booklet be used to ensure reliability in rating.

75. Note that the greatest amount of average annual snowfall indicated on the map is 270 inches. Locate the scale beneath the diagram. Note that the scale is marked in increments of 30 inches and labeled every 60 inches. Note, too, that the scale only extends to 180. On the straight edge of a piece of scrap paper, mark off a distance equivalent to 180 inches. Then move the 180 mark so that it lines up with 0 on the scale. Mark off an additional 90 inches, and label this as 270 (180 + 90 = 270.) Now place the scrap paper next to the wall. Mark off the distance equivalent to 270 inches above the ground on the wall, as shown below. Mark the point on the wall that is 270 inches above the ground with an **X**.

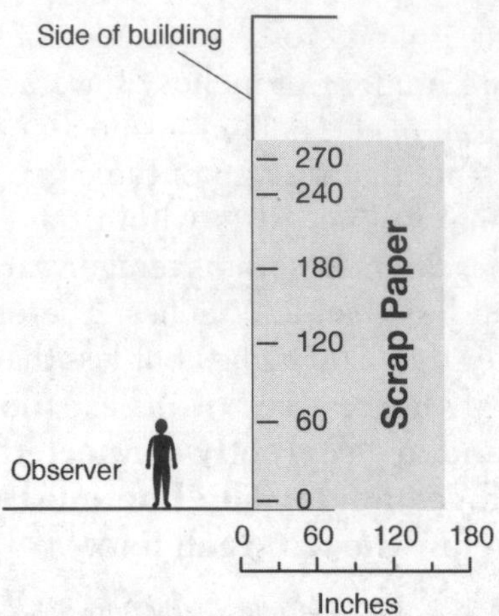

One credit is allowed if the center of **the X is within or touches the edge of the box shown**.

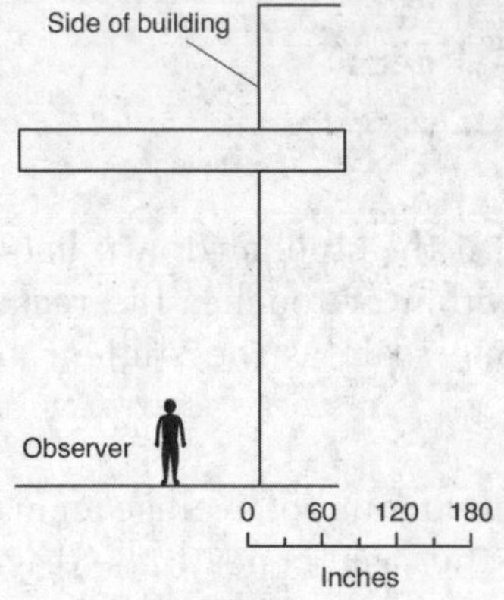

Note: Credit is allowed if a symbol other than an **X** is used. It is recommended that an overlay of the same scale as the student answer booklet be used to ensure reliability in rating.

76. The solar collector is located on Earth. Energy from the Sun must travel through a vacuum to reach Earth. Energy cannot be transmitted through a vacuum by either conduction (molecular collisions) or convection (movements of fluids due to density differences) because both of these methods require a medium. Therefore, energy is transferred from the Sun to Earth mainly in the form of radiation consisting of electromagnetic waves, many of which are light waves. Thus, the energy transfer process by which light travels through space from the Sun to the solar collector is radiation.

One credit is allowed for **radiation**.

77. It is given that the solar collector is used to heat water. Sunlight consists of electromagnetic waves of many wavelengths, but peaks occur in the range of visible light. Clear glass is transparent to visible light but blocks most other wavelengths. Thus, most of the sunlight that passes through the clear glass is visible light. When the light that passes through the clear glass strikes the collector plate and flow tubes, some of the light is absorbed and some of it is reflected.

The energy that is absorbed by the collector plate and flow tubes causes their molecules to move faster, increasing their temperature. The faster-moving molecules of the hot collector plate and flow tubes then transfer some of their energy to molecules of the air by molecular collisions (conduction). The collector plate and flow tubes also reradiate energy in the form of infrared (heat) waves, which is trapped by the clear glass cover. These infrared rays are then absorbed by the molecules of the air inside the collector, causing that air to heat up even more. The heat energy in the air is then transferred first to the flow tubes and then to the water inside the flow tubes by conduction, causing the water to heat up.

Substances of all colors absorb light energy. In general, though, dark colors absorb more light energy than light colors. Dark colors are dark because so much of the light energy that strikes them is absorbed and so little is reflected back to our eye. Dark colors are also better radiators than light colors. The more energy that is radiated by the collector plate and flow tubes, the more the air in the collector heats up. Black is such a dark color because it is a particularly good absorber (and radiator). Therefore, by painting the collector plate and the flow tubes black, the amount of energy absorbed and reradiated by these objects is increased, which, in turn, increases the temperature of the air in the collector. The hotter the air is in the collector, the greater the amount of energy that is transferred to the flow tubes and then to the water inside them. The more energy that is transferred to the water, the more the water heats up.

Therefore, the collector plate and flow tubes are painted black because the color black is a good absorber and is a good radiator of electromagnetic energy such as sunlight.

One credit is allowed. Acceptable responses include but are not limited to:

- **Black is a good absorber of electromagnetic energy/sunlight/insolation.**
- **Black is a good absorber and a good radiator.**
- **Dark colors take in radiation better than light colors.**
- **Black absorbs more energy.**

Note: Credit is *not* allowed for "black absorbs energy" alone because all colors absorb energy. Black is just a better absorber of that energy.

78. Temperature is a measure of the speed at which the molecules in a substance are moving. The faster the molecules in the substance are moving, the higher is the temperature of that substance. The speed at which molecules are moving (temperature) also determines the wavelength of the electromagnetic energy emitted. The faster the molecules of a substance are moving, the shorter the wavelength of the electromagnetic waves it emits. Conversely, the slower the molecules in a substance are moving, the lower its temperature and the longer the wavelength of the electromagnetic waves it emits. Most substances begin to radiate electromagnetic waves in the red visible range when they get "red hot." The light energy that is absorbed by the collector plate and flow tubes raises their temperature. However, that light energy is never high enough to cause the plate and tubes to get red hot and to emit visible light. Thus, the molecules of the collector plate and flow tubes are moving slower than molecules that would emit visible light. So the plate and tubes emit electromagnetic waves that have a longer wavelength than visible light.

Find the Electromagnetic Spectrum chart on the *Reference Tables for Physical Setting/Earth Science*. Locate the segment labeled "Visible light." Note the arrow labeled "Increasing wavelength" on the right on the spectrum. Note that the electromagnetic waves to the right of visible light are labeled "Infrared." Thus, the type of electromagnetic energy emitted by the flow tubes and collector plate is infrared, or IR. Infrared waves cannot pass through glass and are trapped inside the collector by the glass cover. As explained above, infrared waves have a longer wavelength than visible light waves. Therefore, "longer" should be circled in your answer booklet.

One credit is allowed for **infrared/IR and circling longer**.

79. It is given that the star *Kepler-62* is located in our galaxy. Our solar system is located in the Milky Way galaxy. Therefore, *Kepler-62* and its planetary system are located in the Milky Way galaxy.

One credit is allowed for **Milky Way**.

80. In the Data Table in the question, find the column labeled "Name of Planet" and locate "62c." Trace this row right to the column labeled "Equatorial Diameter (compared to Earth's diameter)." Note the value of 0.54. Thus, the equatorial diameter of Kepler-62c is 0.54 Earth diameters. Find the Solar System Data chart in the *Reference Tables for Physical Setting/Earth Science.* In the column labeled "Celestial Object," locate Earth. Trace this row right to the column labeled "Equatorial Diameter (km)," and note the value of "12,756." Thus, the equatorial diameter of Earth is 12,756 km. Since the equatorial diameter of Kepler-62c is 0.54 Earth diameters, multiply 12,756 km by 0.54 and solve to find

the equatorial diameter of Kepler-62c: 12,756 km × 0.54 = 6888.24 km. Thus, Kepler-62c has an equatorial diameter of 6888.24 km. On the Solar System Data chart, locate the column labeled "Equatorial Diameter (km)." Note that the value nearest to 6888.24 is 6794. From this value, trace left to the column labeled "Celestial Object." Note that this value is the equatorial diameter of the planet Mars. Thus, the planet in our solar system that has an equatorial diameter most similar in size to the equatorial diameter of planet Kepler-62c is Mars.

One credit is allowed for **Mars**.

81. The smaller the distance between a planet and the star it orbits, the greater the force of gravitational attraction between the planet and the star. The greater the force of gravitational attraction between a planet and a star, the faster the planet has to move in order to remain in orbit and not be drawn into the star. In addition, the closer a planet is to the star it orbits, the smaller the planet's orbit is and the less distance the planet has to travel to complete one revolution. The combination of a small orbit and high speed means that the planet closest to the star has the shortest period of revolution.

In the Data Table in the question, locate the column labeled "Distance from *Kepler-62* (million kilometers)." Note that the smallest value (closest to *Kepler-62*) is 8.23. From this value, trace left to the column labeled "Name of Planet." Note that the planet located closest to the star *Kepler-62* is Kepler-62b. Thus, planet Kepler-62b has the shortest period of revolution. Kepler-62b has the shortest period of revolution because it is the closest planet to its star. Kepler-62b also orbits at a faster speed and in a smaller orbit than planets farther from the star, resulting in a shorter period of revolution.

One credit is allowed for Kepler-62b *or* 62b and for an acceptable explanation. Acceptable explanations include but are not limited to:

- **Kepler-62b is closest to *Kepler-62*.**
- **Kepler-62b; the closer a planet is to a star, the shorter its period of revolution.**
- **62b; closer planets orbit faster.**
- **62b has the shortest orbital path.**
- **Kepler-62b is closest to its sun.**

82. The reading passage states, "The habitable zone, which is the region around a star where life may exist due to the possible presence of water in the liquid phase." Thus, the two planets that may have liquid water on their surfaces are located within the habitable zone around the star *Kepler-62*. According to the key to the right of the diagram of the *Kepler-62* Planetary System, the habitable zone in the diagram is shaded. According to the diagram, the two planets that

orbit *Kepler-62* in the habitable zone and therefore may have liquid water on their surfaces are Kepler-62e and Kepler-62f.

In order to have liquid water on its surface, a planet must have a surface temperature at which water can exist in the liquid phase. The surface temperature of a planet depends on its distance from the star it orbits. The intensity of radiation a planet receives from its star decreases with distance from the star. If a planet is too close to the star, the radiation is so intense that the planet's surface reaches temperatures above those at which liquid water can exist. If a planet is too far from a star, the radiation is so weak in intensity that the planet's surface does not reach a temperature high enough for liquid water to exist. Thus, Kepler-62e and Kepler-62f may have liquid water on their surface because they are in the habitable zone, the zone in which planets are at a distance from the star that allows them to have surface temperatures at which liquid water can exist.

One credit is allowed for Kepler-62e and Kepler-62f and for an acceptable explanation. Acceptable explanations include but are not limited to:

- **Kepler-62e and Kepler-62f are located in the habitable zone.**
- **Kepler-62e and Kepler-62f are located at a distance from *Kepler-62* to allow water to remain in the liquid phase.**
- **Kepler-62e and Kepler-62f; they have the correct temperature.**
- **Kepler-62e and Kepler-62f; these planets are at the correct distance from the star.**

Note: Credit is *not* allowed for "located farthest from the star" alone because this does not refer to the habitable zone.

83. The terminal moraine is a ridgelike accumulation of debris marking the farthest point that a glacier has advanced. A glacier stops advancing when its leading edge melts at a faster rate than new ice is formed at the glacier's source. At the very end of a melting glacier, all of the rocks, soil, and sediments that were picked up and frozen into, or fell on top of, the glacier and that were pushed to the front of the glacier are released as the ice around them melts. Air resistance is so small that particles of all sizes fall through air at virtually the same speed. Therefore, sediments deposited by a melting glacier do not become sorted by size as they fall and do not form layers due to different settling rates. Particles of mixed shapes and sizes, ranging from boulders to clay, drop to the ground in an unorganized jumble. The result is a ridgelike pile of debris running along the very end of the glacier that is unsorted and unlayered—a terminal moraine. Thus, the arrangement of the sediments found within the terminal moraine is an unsorted and unlayered mixture of sediment sizes.

One credit is allowed. Acceptable responses include but are not limited to:

- **The sediments are unsorted.**
- **The sediments are unlayered.**
- **Mixed sediment sizes ranging from boulders to clay**
- **Unorganized arrangement**

84. Kettles are bowl-shaped depressions formed when buried blocks of glacial ice melt. If the depressions fill with water, they are called kettle lakes. Kettle lakes form as follows. First, a block of glacial ice falls off the face of the glacier and comes to rest on the ground, as shown in diagram *C*. As the glacier melts, sediments that were trapped in or on the glacier fall to the ground and are deposited around the block of glacial ice, as shown in diagram *B*. Eventually, the block of glacial ice is buried beneath sediments and begins to melt, as shown in diagram *D*. Once the ice has melted, it no longer supports the surrounding sediments and they collapse. This forms a circular depression called a kettle, which can fill with water to form a kettle lake, as shown in diagram *A*. Thus, the correct order in which the letters should be placed to indicate the sequence of development of a kettle lake from earliest stage to latest stage is $C \rightarrow B \rightarrow D \rightarrow A$.

One credit is allowed for $\boldsymbol{C \rightarrow B \rightarrow D \rightarrow A}$.

C	***B***	***D***	***A***
Earliest Stage	⟶		Latest stage

85. Find the Geologic History of New York State chart in the *Reference Tables for Physical Setting/Earth Science*. In the column labeled "Important Geologic Events in New York," locate the statement "Advance and retreat of last continental ice." Thus, this advance and retreat occurred during the last ice age. Trace this row to the left to the column labeled "Epoch." Note that the advance and retreat of the last continental ice occurred during the Pleistocene epoch. Thus, the terminal moraines found on Long Island were deposited during the Pleistocene epoch.

One credit is allowed for **Pleistocene epoch**.

Topic	Question Numbers (Total)	Wrong Answers (x)	Grade
Standards 1, 2, 6, and 7: Skills and Application			
Skills			
Standard 1 Analysis, Inquiry, and Design	2–6, 8, 10, 13, 15, 16, 21, 22, 24, 25, 27–31, 35, 36, 38, 39, 41–49, 51–57, 59–71, 74, 76–80, 84, 85		$\frac{100(60 - x)}{60} = \%$
Standard 2 Information Systems	7, 68		$\frac{100(2 - x)}{2} = \%$
Standard 6 Interconnectedness, Common Themes	1, 5, 8–12, 14, 18–20, 22–24, 30, 32–37, 40–45, 47–50, 52, 54–59, 61, 66–68, 70, 71, 73–75, 80–84		$\frac{100(52 - x)}{52} = \%$
Standard 7 Interdisciplinary Problem Solving	7		$\frac{100(1 - x)}{1} = \%$
Standard 4: The Physical Setting/Earth Science			
Astronomy			
The Solar System (MU 1.1a, b; 1.2d)	1, 9, 81		$\frac{100(3 - x)}{3} = \%$
Earth Motions and Their Effects (MU 1.1c, d, e, f, g, h, i)	4, 5, 11, 12, 36–39, 55–57		$\frac{100(11 - x)}{11} = \%$
Stellar Astronomy (MU 1.2b)	40–42, 79		$\frac{100(4 - x)}{4} = \%$
Origin of Earth's Atmosphere, Hydrosphere, and Lithosphere (MU 1.2e, f, h)			
Theories of the Origin of the Universe and Solar System (MU 1.2a, c)	2, 3, 10, 80, 82		$\frac{100(5 - x)}{5} = \%$

Topic	Question Numbers (Total)	Wrong Answers (x)	Grade
Meteorology			
Energy Sources for Earth Systems (MU 2.1a, b)	45, 46		$\frac{100(2-x)}{2} = \%$
Weather (MU 2.1c, d, e, f, g, h)	6–8, 47, 66–68		$\frac{100(7-x)}{7} = \%$
Insolation and Seasonal Changes (MU 2.1i; 2.2a, b)	58, 76–78		$\frac{100(4-x)}{4} = \%$
The Water Cycle and Climates (MU 1.2g; 2.2c, d)	14, 15, 18, 20, 21, 33, 63, 64		$\frac{100(8-x)}{8} = \%$
Geology			
Minerals and Rocks (MU 3.1a, b, c)	29, 31, 35, 51–54		$\frac{100(7-x)}{7} = \%$
Weathering, Erosion, and Deposition (MU 2.1s, t, u, v, w)	22, 27, 28, 30, 34, 65, 83, 84		$\frac{100(8-x)}{8} = \%$
Plate Tectonics and Earth's Interior (MU 2.1j, k, l, m, n, o)	13, 24, 25, 59–62		$\frac{100(7-x)}{7} = \%$
Geologic History (MU 1.2i, j)	16, 17, 19, 23, 43, 44, 69–72, 85		$\frac{100(11-x)}{11} = \%$
Topographic Maps and Landscapes (MU 2.1p, q, r)	26, 32, 48–50, 73–75		$\frac{100(8-x)}{8} = \%$
ESRT			
2011 Edition Reference Tables for Physical Setting/Earth Science	2, 3, 5, 6, 8, 10, 13, 15, 16, 21, 24, 25, 27–29, 31, 35, 39, 41, 45, 47, 49, 51–54, 59, 61, 62, 64, 65, 67, 68, 70, 71, 78, 80, 85		$\frac{100(38-x)}{38} = \%$

To further pinpoint your weak areas, use the Topic Outline in the front of the book.
MU = Major Understanding (see Topic Outline)

Examination August 2018

Physical Setting/Earth Science

PART A

Answer all questions in this part.

Directions (1–35): For *each* statement or question, choose the word or expression that, of those given, best completes the statement or answers the question. Some questions may require the use of the *2011 Edition Reference Tables for Physical Setting/Earth Science*. Record your answers in the space provided.

1 The diagram below represents a sundial positioned in New York State. During daylight, the shadow cast by the gnomon (pointer) moves across the disc, with the tip of the shadow pointing to the time of day.

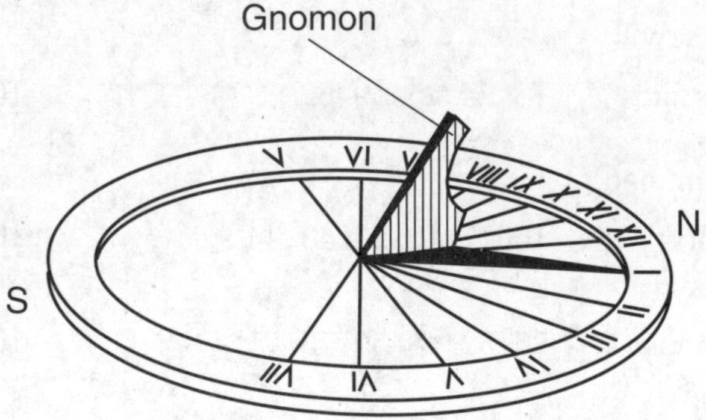

This motion of the gnomon's shadow on the sundial is mainly due to

(1) Earth's rotation
(2) Earth's revolution
(3) the Sun's rotation
(4) the Sun's revolution

1 _____

2 The formation of the planet Uranus is estimated to have occurred approximately

(1) 100,000 million years ago
(2) 2.0 billion years ago
(3) 4.6 billion years ago
(4) 13.7 billion years ago 2 _____

3 Compared to the Jovian planets in our solar system, the terrestrial planets have

(1) less mass and are less dense
(2) less mass and are more dense
(3) more mass and are less dense
(4) more mass and are more dense 3 _____

4 The diagram below represents the constellation Leo that can be seen by an observer in New York State at midnight during March.

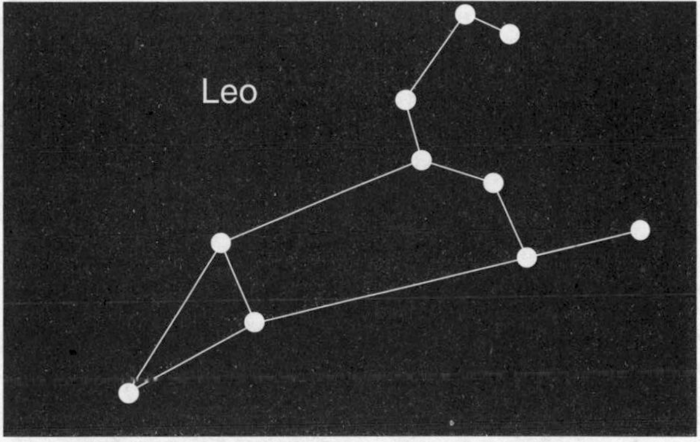

Leo is *not* visible to this observer at midnight during September because

(1) Leo has rotated on its axis
(2) Leo has revolved in its orbit around the Sun
(3) Earth has rotated on its axis
(4) Earth has revolved in its orbit around the Sun 4 _____

5 An observer in New York City measured the angle of insolation at solar noon each day. During which month did this observer see the noontime angle of insolation increase each day?

(1) April
(2) July
(3) September
(4) December

5 ____

6 The Coriolis effect occurs as a result of Earth's

(1) rotation
(2) revolution
(3) tilted axis
(4) magnetic field

6 ____

7 During the process of condensation, water vapor

(1) releases 334 J/g of heat energy
(2) releases 2260 J/g of heat energy
(3) gains 334 J/g of heat energy
(4) gains 2260 J/g of heat energy

7 ____

8 Infiltration is generally greater than runoff where the land has a

(1) gentle slope and permeable soil
(2) gentle slope and impermeable bedrock
(3) steep slope and permeable soil
(4) steep slope and impermeable bedrock

8 ____

9 Nearly 90% of the water vapor that enters Earth's atmosphere comes from the evaporation of Earth's surface waters. Most of the remaining 10% is water vapor that enters the atmosphere through

(1) precipitation from clouds
(2) transpiration from plants
(3) condensation within the troposphere
(4) melting of polar ice caps

9 ____

10 What is the relative humidity if the dry-bulb temperature is 26°C and the wet-bulb temperature is 18°C?

(1) 13%
(2) 33%
(3) 45%
(4) 51%

10 ____

11 Most of the long-wave energy radiated from Earth and lost to space on a cloudless night is

(1) ultraviolet
(2) infrared
(3) visible light
(4) gamma rays

11 ____

12 In addition to carbon dioxide, two other major greenhouse gases in Earth's atmosphere are

(1) oxygen and nitrogen
(2) oxygen and methane
(3) water vapor and nitrogen
(4) water vapor and methane

12 ____

13 The arrows in the diagram below represent the movement of air over a mountain.

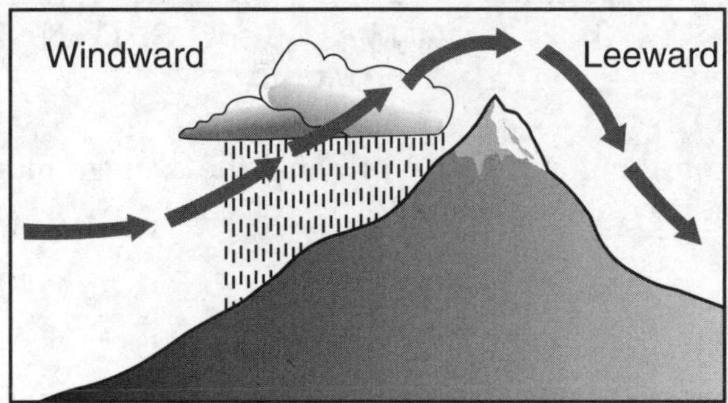

Clouds are forming on the windward side of this mountain because the air is

(1) expanding and cooling to the dewpoint
(2) expanding and warming to the dewpoint
(3) compressing and cooling to the dewpoint
(4) compressing and warming to the dewpoint

13 ____

14 The graph below shows the average monthly amount of insolation received throughout a year at four locations (*A*, *B*, *C*, and *D*) on Earth.

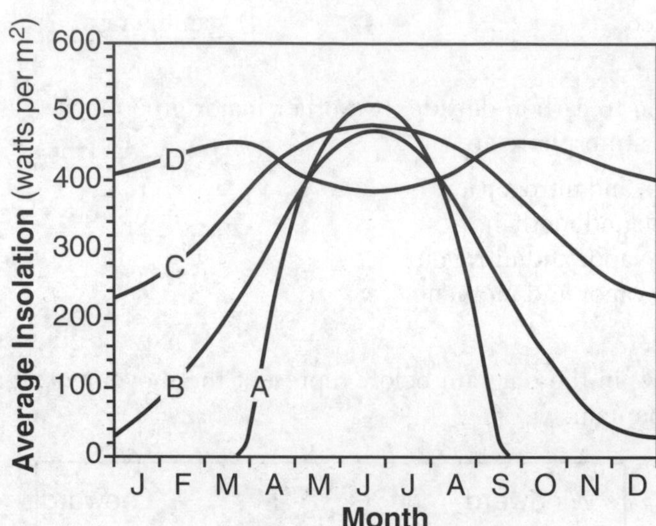

Which line on the graph best represents the average monthly insolation received at the equator?

(1) *A* (3) *C*
(2) *B* (4) *D* 14_____

15 Earth's polar regions have cold, dry climates because the Sun's rays are at a

(1) low angle, and upper atmospheric air is sinking
(2) low angle, and upper atmospheric air is rising
(3) high angle, and lower atmospheric air is sinking
(4) high angle, and lower atmospheric air is rising 15_____

16 Atmospheric transparency will increase when

(1) volcanic eruptions occur
(2) fog is produced
(3) insolation is reflected by clouds
(4) precipitation removes dust particles from the air 16_____

17 The existence of which group of organisms spans the shortest geologic time?

(1) birds
(2) humans
(3) dinosaurs
(4) placoderm fish 17 ____

18 Which New York State geologic event occurred most recently?

(1) Taconian orogeny
(2) Grenville orogeny
(3) formation of the Catskill delta
(4) dome-like uplift of the Adirondack region 18 ____

19 The only dinosaur fossils found in New York State are footprints found on 210-million-yearold bedrock. In which New York State landscape region were these dinosaur fossils found?

(1) Tug Hill Plateau
(2) Newark Lowlands
(3) Allegheny Plateau
(4) Adirondack Mountains 19 ____

20 The first *P*-wave of an earthquake travels 5600 kilometers from the epicenter and arrives at a seismic station at 10:05 a.m. At what time did this earthquake occur?

(1) 9:49 a.m.
(2) 9:56 a.m.
(3) 10:02 a.m.
(4) 10:14 a.m. 20 ____

21 The photograph below shows a New York State index fossil.

What is the best classification of this fossil, and during which geologic time period did the organism that produced this fossil exist?

(1) Classification: Coral
Geologic time period: Permian
(2) Classification: Coral
Geologic time period: Ordovician
(3) Classification: Trilobite
Geologic time period: Permian
(4) Classification: Trilobite
Geologic time period: Ordovician 21 ____

22 As a quartz pebble is transported by a stream, the pebble will become more rounded as a result of

(1) dissolving as water is running over the rock
(2) abrasion by colliding with other rocks
(3) deposition in well-sorted layers
(4) resistance to weathering and erosion 22 ____

23 Which three minerals are most likely used in the construction of a house?

(1) graphite, pyrite, and halite
(2) garnet, galena, and sulfur
(3) talc, amphibole, and fluorite
(4) selenite gypsum, dolomite, and muscovite mica 23 ____

24 The topographic map below shows a portion of a volcanic island in the Pacific Ocean. Elevations are shown in feet. Letters *A* and *B* represent locations on Earth's surface. Locations *A* and *B* are 2.5 miles apart.

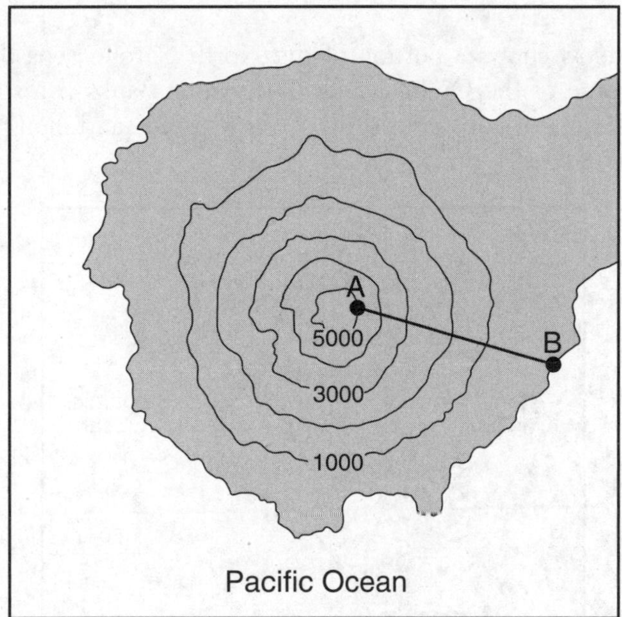

What is the approximate gradient from point *A* to point *B* on the island?

(1) 1000 ft/mi
(2) 1250 ft/mi
(3) 2000 ft/mi
(4) 2500 ft/mi 24 ____

25 Which two rocks usually consist of only one mineral, but may contain additional minerals?

(1) hornfels and diorite
(2) quartzite and dunite
(3) rock salt and basalt
(4) gabbro and bituminous coal

25 ____

26 Which rock has never melted, but was produced by great heat and pressure, which distorted and rearranged its minerals?

(1) siltstone
(2) breccia
(3) pegmatite
(4) metaconglomerate

26 ____

27 The map below shows a portion of the North Carolina coastline, including some of the Outer Banks. The Outer Banks is a string of narrow barrier islands consisting of well-sorted sand along the Atlantic Ocean coast.

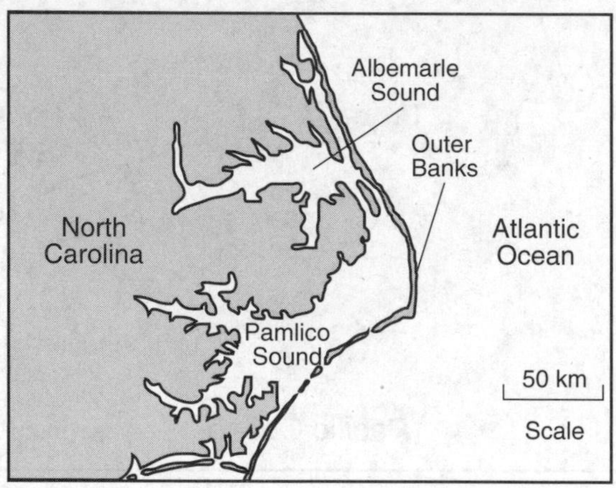

Which agent of erosion is primarily responsible for the formation of these barrier islands?

(1) wave action
(2) landslides
(3) streams
(4) glacial ice

27 ____

28 The photograph below shows a magnified view of a portion of a rock that can float if placed in water.

Which terms best describe this rock?

(1) non-crystalline and vesicular
(2) coarse and non-vesicular
(3) clastic and fragmental
(4) foliated and banded

28 _____

29 The map below shows three locations, labeled *A*, *X*, and *B*, on Earth's surface.

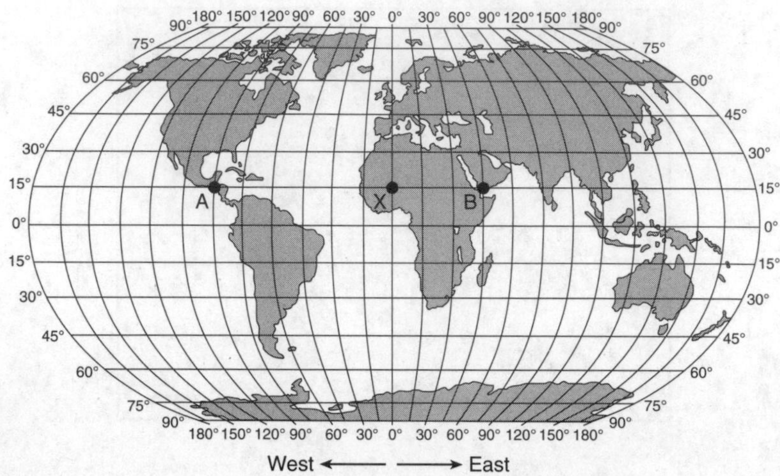

Which table correctly indicates the solar times at locations *A* and *B* when it is 12 noon at location *X*?

Location	Solar Time
A	6 a.m.
B	9 a.m.

(1)

Location	Solar Time
A	6 p.m.
B	9 a.m.

(3)

Location	Solar Time
A	6 a.m.
B	3 p.m.

(2)

Location	Solar Time
A	6 p.m.
B	3 p.m.

(4)

29 _____

30 Which equation is used to determine the approximate rate of Earth's revolution?

(1) Approximate rate of Earth's revolution $= \dfrac{365°}{360\text{ days}}$

(2) Approximate rate of Earth's revolution $= \dfrac{360°}{24\text{ hours}}$

(3) Approximate rate of Earth's revolution $= \dfrac{360°}{365\text{ days}}$

(4) Approximate rate of Earth's revolution $= \dfrac{24°}{360\text{ hours}}$

30 _____

31 Which cross section best represents a cold front?

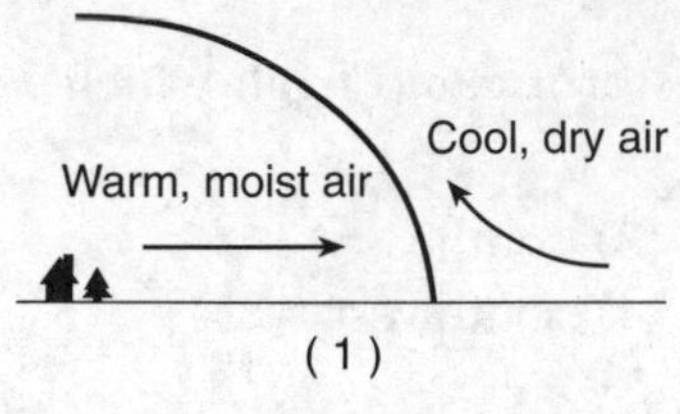

(1)

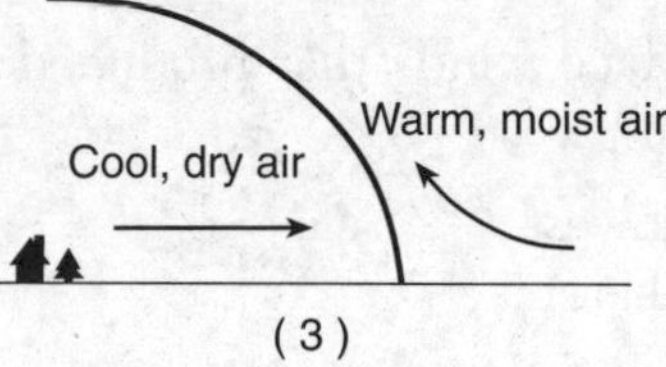

(3)

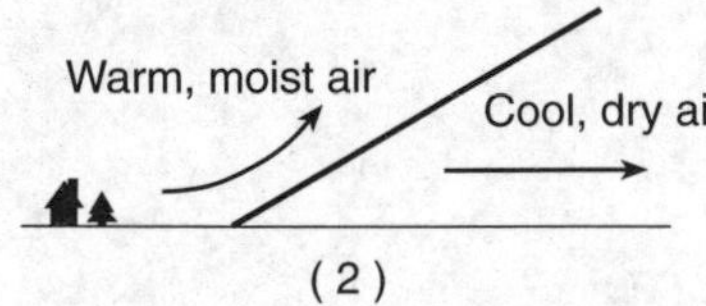

(2)

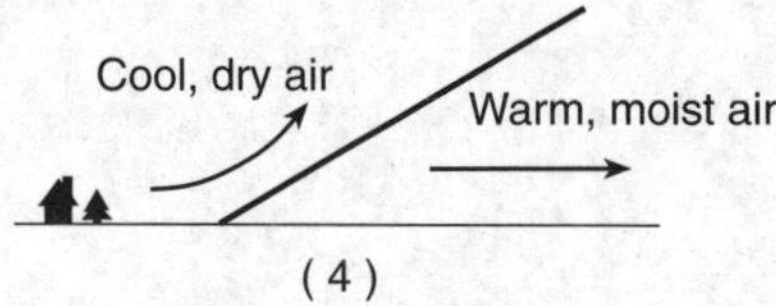

(4)

31 _____

32 The isolines on the map below show snowfall totals from a lake-effect storm that affected a portion of New York State.

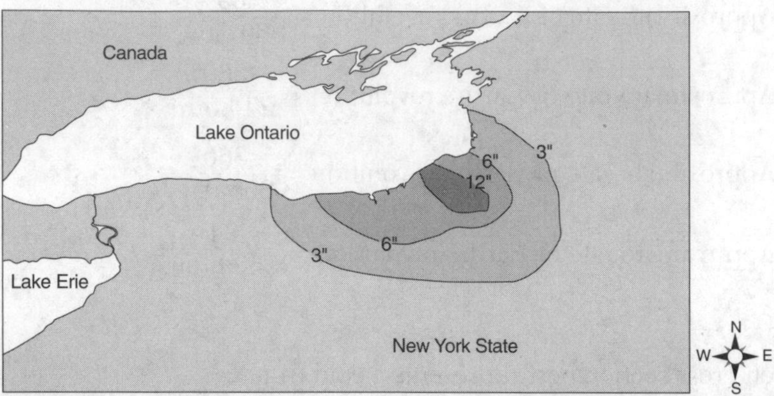

The surface winds that produced this storm came from which direction?

(1) northwest
(2) northeast
(3) southeast
(4) southwest

32 _____

33 The photograph below shows Mount Rainier, a volcano in the state of Washington.

Which map best shows the complete stream drainage pattern for this mountain?

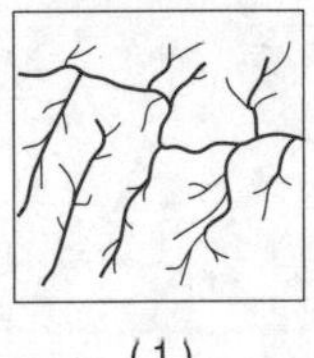

(1)

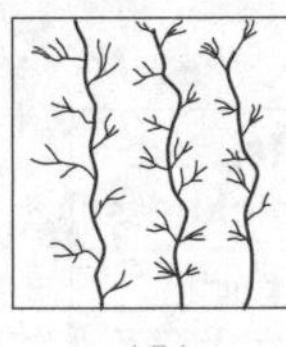

(2)

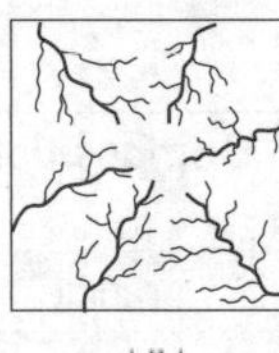

(3)

(4)

33 _____

34 Which table best represents the characteristics of the continental crust and the oceanic crust?

Type of Crust	Density (g/cm^3)	Composition	Relative Thickness
Continental	3.0	basaltic	thicker
Oceanic	2.7	granitic	thinner

(1)

Type of Crust	Density (g/cm^3)	Composition	Relative Thickness
Continental	3.0	granitic	thicker
Oceanic	2.7	basaltic	thinner

(2)

Type of Crust	Density (g/cm^3)	Composition	Relative Thickness
Continental	2.7	granitic	thinner
Oceanic	3.0	basaltic	thicker

(3)

Type of Crust	Density (g/cm^3)	Composition	Relative Thickness
Continental	2.7	granitic	thicker
Oceanic	3.0	basaltic	thinner

(4)

34 _____

35 The map below shows the locations of some oil and gas fields in New York State.

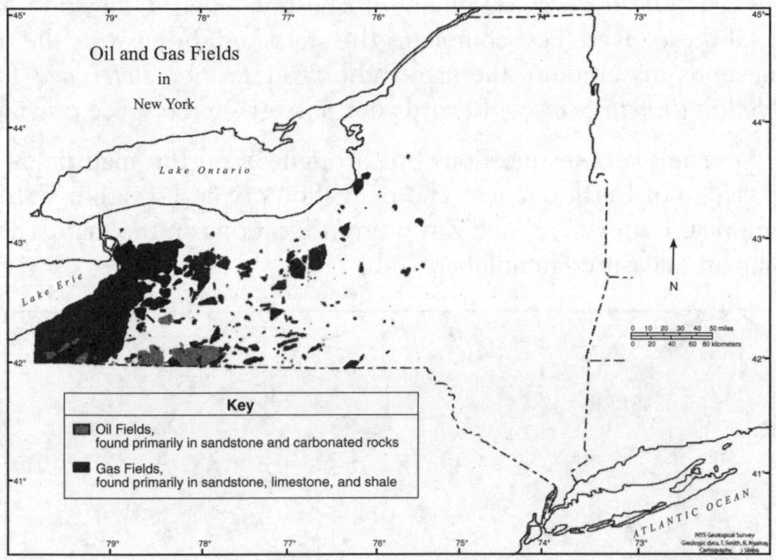

Source: New York State Museum, State Geological Survey

Which type of bedrock contains these oil and gas deposits?

(1) extrusive igneous rock
(2) intrusive igneous rock
(3) metamorphic rock
(4) sedimentary rock

35 _____

PART B–1

Answer all questions in this part.

Directions (36–50): For *each* statement or question, choose the word or expression that, of those given, best completes the statement or answers the question. Some questions may require the use of the 2011 *Edition Reference Tables for Physical Setting/Earth Science*. Record your answers in the space provided.

Base your answers to questions 36 through 38 on the map below and on your knowledge of Earth science. The map shows typical weather systems over North America. Letters *X*, *Y*, and *Z* represent locations on the map. The isobars on the map are measured in millibars (mb).

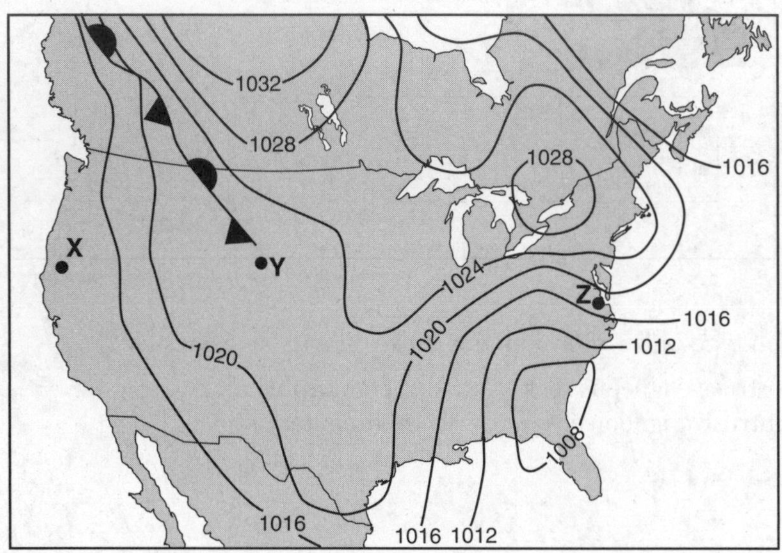

36 Which map information indicates that the wind velocity is greater at location Z than at location *X*?

(1) Location Z is closer to the ocean.
(2) The isobars are closer together at Z.
(3) The latitude of location *X* is greater.
(4) Location *X* is closer to the front. 36 _____

37 Which type of front extends northwest from location *Y*?

(1) warm front
(2) cold front
(3) occluded front
(4) stationary front 37 _____

38 Which map best shows the locations for the centers of high pressure (**H**) and low pressure (**L**)?

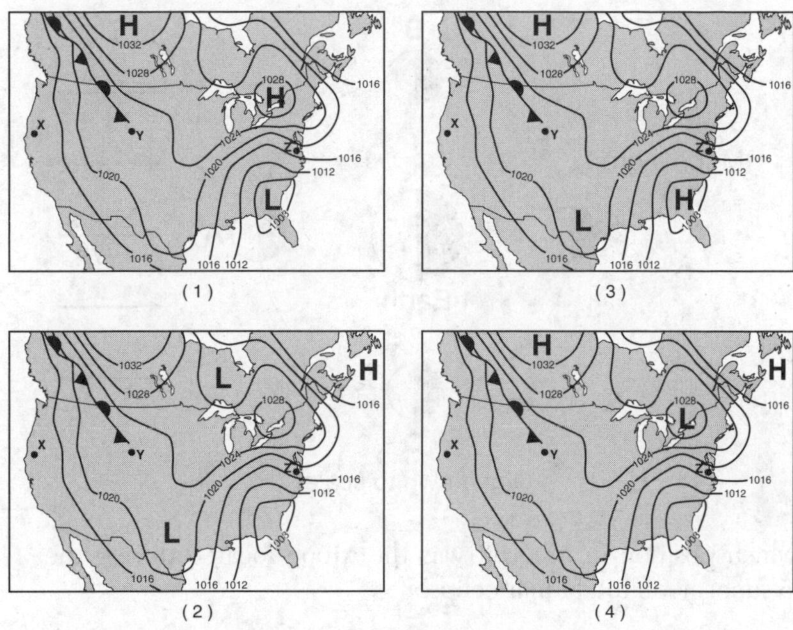

38 _____

Base your answers to questions 39 through 41 on the passage below and on your knowledge of Earth Science.

Supermoon Eclipse

On September 27, 2015, a rare total lunar eclipse of a supermoon occurred. A supermoon occurs when the entire lighted half of the Moon faces Earth (full Moon phase) and the Moon is at its closest point to Earth in its orbit. At this time, the Moon will appear 14% larger and 30% brighter than normal. Supermoon events are rare, but a total lunar eclipse during a supermoon is even more rare. There have been only six total supermoon lunar eclipses since 1900. The next one will not happen until 2033.

39 Supermoon total lunar eclipses are celestial events that

(1) are random occurrences
(2) are predictable
(3) will never happen again after 2033
(4) will happen every full Moon

39 _____

40 The diagram below represents the Moon in four positions, *A* through *D*, in its orbit around Earth.

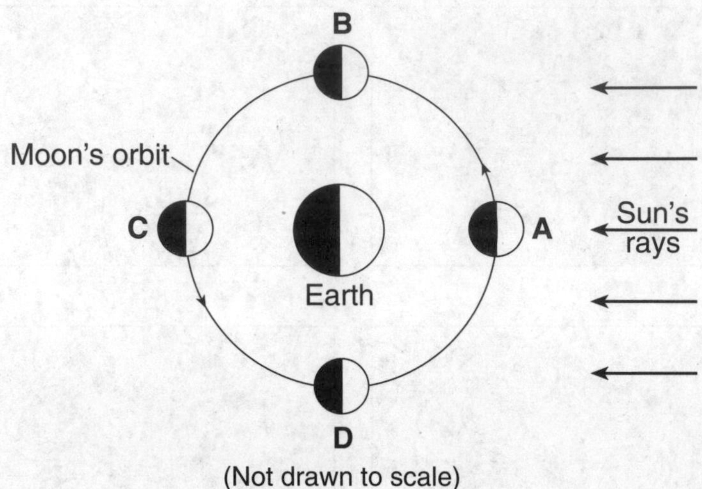

At which position in its orbit was the Moon located during the 2015 supermoon total lunar eclipse?

(1) *A* (3) *C*
(2) *B* (4) *D* 40 ____

41 The time it took for the Moon to go from this supermoon to the next full moon phase was

(1) 15 days (3) 29.5 days
(2) 27.3 days (4) 365 days 41 ____

Base your answers to questions 42 through 44 on the diagram and graph below and on your knowledge of Earth science. The diagram represents a portion of Earth's interior. Letters *A*, *B*, and *C* represent interior layers. The graph shows the velocity of *P*-waves and *S*-waves at various depths in Earth's interior.

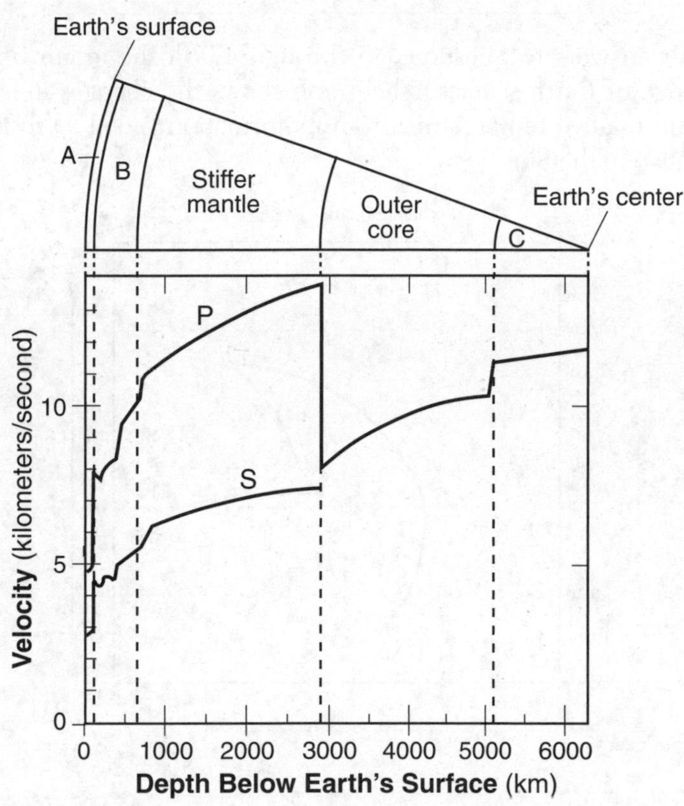

42 Which layers of Earth's interior are represented by letters *A* and *B*?

(1) *A* is the crust and *B* is the rigid mantle.
(2) *A* is the lithosphere and *B* is the asthenosphere.
(3) *A* is the asthenosphere and *B* is the crust.
(4) *A* is the rigid mantle and *B* is the lithosphere. 42 _____

43 What is the approximate velocity in kilometers/second of the *P*-waves at a depth of 1000 kilometers?

(1) 6.2 km/s
(2) 7.2 km/s
(3) 11.3 km/s
(4) 13.8 km/s 43 _____

44 Some locations within layer *C* have an inferred density of

(1) 3.4 g/cm^3 (3) 11.5 g/cm^3
(2) 5.6 g/cm^3 (4) 12.9 g/cm^3 44 ____

Base your answers to questions 45 through 47 on the graph below and on your knowledge of Earth science. The graph shows the changes in a single star's luminosity and relative temperature from its formation (point 1) to its late stage (point 4) relative to the Sun.

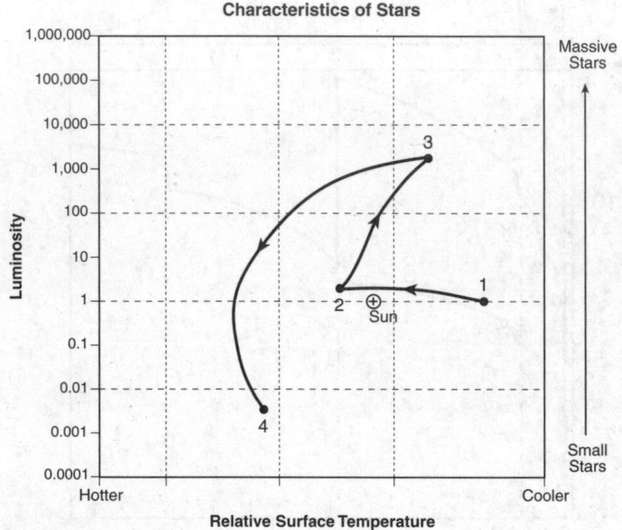

45 Which is a possible surface temperature of this star at point 2?

(1) 3000 K (3) 7000 K
(2) 5000 K (4) 10,000 K 45 ____

46 Between points 1 and 3, this star is visible to observers on Earth because it emits light energy. This energy is released by the process of nuclear fusion when

(1) dust collides with the star
(2) dust is broken apart by radiation
(3) lighter elements combine to form heavier elements
(4) heavier elements are broken down to form lighter elements 46 ____

47 Which table correctly classifies this star at points 3 and 4?

Point	Classification
3	Giant
4	White Dwarf

(1)

Point	Classification
3	Supergiant
4	Main Sequence

(3)

Point	Classification
3	White Dwarf
4	Supergiant

(2)

Point	Classification
3	Giant
4	Main Sequence

(4)

47 ____

Base your answers to questions 48 through 50 on the photograph below and on your knowledge of Earth science. The photograph shows a meandering stream in a wooded area. Points *A* and *B* represent locations on the streambanks. Letter *X* represents a flat area near the stream.

48 The streambank at location *B* is steeper than the streambank at location *A* because the water near location *B* is moving

(1) slower than the water near location *A*, causing more erosion
(2) slower than the water near location *A*, causing more deposition
(3) faster than the water near location *A*, causing more erosion
(4) faster than the water near location *A*, causing more deposition 48 ____

49 The area labeled letter *X* represents a portion of a

(1) delta (3) finger lake
(2) sand dune (4) floodplain 49 _____

50 Most of the particles deposited where the stream velocity *decreases* from 50 centimeters per second to 5 centimeters per second are

(1) small cobbles and large pebbles
(2) small pebbles and large sand
(3) small sand and large silt
(4) small silt and large clay 50 _____

PART B–2

Answer all questions in this part.

Directions (51–65): Record your answers in the spaces provided. Some questions may require the use of the *2011 Edition Reference Tables for Physical Setting/Earth Science*.

Base your answers to questions 51 through 54 on the diagram below and on your knowledge of Earth science. In the diagram, letters *A*, *B*, *C*, and *D* represent Earth's location on the first day of the four seasons as it orbits the Sun. Aphelion (Earth's farthest distance from the Sun) and perihelion (Earth's closest distance to the Sun) are labeled to show the approximate positions where they occur in Earth's orbit. The dashed lines represent Earth's axis, and the North Pole is labeled N.

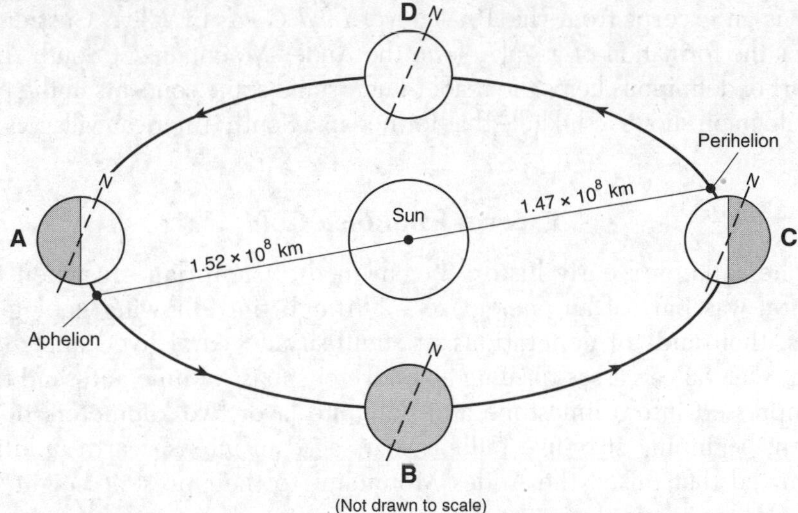

(Not drawn to scale)

51 State the number of degrees that Earth's axis is tilted from a line perpendicular to the plane of its orbit at each lettered location. [1]

__________°

52 Identify the season in New York State during which Earth is at perihelion. [1]

____

53 State the number of hours of daylight that an observer in New York State will experience when Earth is at position *D*. [1]

______________ **h**

54 Identify the name of the star that is aligned with Earth's axis above the North Pole. [1]

____

Base your answers to questions 55 through 58 on the passage, chart of definitions, and photograph below, and on your knowledge of Earth science. The passage is an excerpt from the 1994 novel *Inca Gold*, by Clive Cussler, which describes the formation of a sinkhole in the Andes Mountains of South America. The chart of definitions helps the reader understand some concepts in the passage. The photograph shows a sinkhole that formed in a South American village.

Excerpt from *Inca Gold*

...The sinkhole's early history began in the Cambrian era when the region was part of an ancient sea. Through the following geological eras, thousands of generations of shellfish and coral lived and died, their skeletal carcasses forming an enormous mass of lime and sand that compressed into a limestone and dolomite layer two kilometers thick. Then, beginning sixty-five million years ago, an intense earth uplifting occurred that raised the Andes Mountains to their present height. As the rain ran down from the mountains it formed a great underground water table that slowly began dissolving the limestone. Where it collected and pooled, the water ate upward until the land surface collapsed and created the sinkhole....

Chart of Definitions

Sinkhole	A natural depression or hole in the ground surface caused by some form of collapse of the bedrock beneath. Most are caused when slightly acidic groundwater chemically breaks down the carbonate rocks or the carbonate cement holding the rock particles together. The removal of the carbonates by groundwater gradually forms a hollow space or cavern under the surface layer. As the roof of the cavern weakens, it sometimes collapses, forming a sinkhole.
Lime	A general term for material containing the element calcium that combined with oxygen.

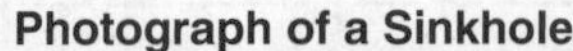

Photograph of a Sinkhole

55 In the first line of the passage, referring to Cambrian as an "era" is scientifically incorrect. State the unit of geologic time that should be substituted for the word "era." [1]

____

56 Identify *one* group of animals that became extinct at the time the passage states that intense uplifting of the Andes Mountains began. [1]

____

57 Using chemical symbols, state the chemical composition of the mineral found in limestone. [1]

___

58 Describe the chemical weathering that contributes to the formation of sinkholes. [1]

___

___

Base your answers to questions 59 through 61 on the geologic cross section below and on your knowledge of Earth science. The cross section represents rock units, labeled *A* through *K*, that have *not* been overturned. Two unconformities and a volcanic ash layer are indicated.

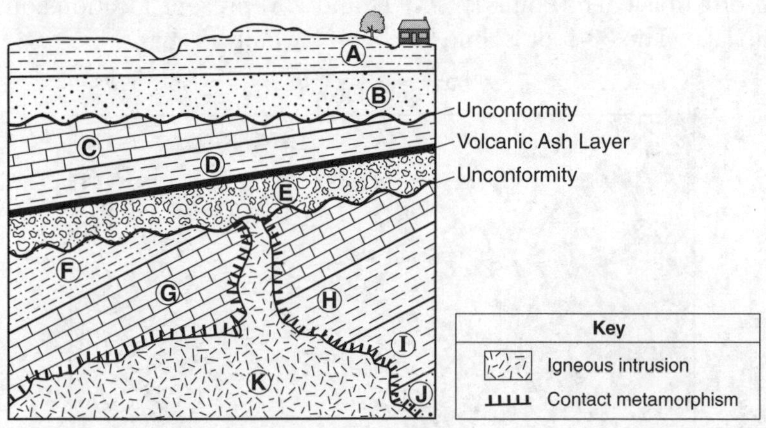

59 List the letters *E, H,* and *K* to indicate the correct order of rock unit formation, from oldest to youngest, that formed this portion of Earth's crust. [1]

Letters: ______________, ______________, ______________

Oldest ————————————————→ Youngest

60 Identify *two* processes that most likely caused the formation of both unconformities. [1]

(1) ____

(2) ____

61 Identify *one* metamorphic rock that most likely formed within rock unit *G* at the boundary of rock unit *K*. [1]

____

Base your answers to questions 62 through 65 on the field map below and on your knowledge of Earth science. The map shows the depth of Cuba Lake, located in New York State at latitude 42°14′ N, longitude 78°18′ W. Isoline values indicate water depth, in feet. Points *A* and *B* represent locations on the shoreline of Cuba Lake. Points *W*, *X*, *Y*, and *Z* represent locations on the bottom of the lake. The 30-foot isoline has been partially drawn.

Cuba Lake

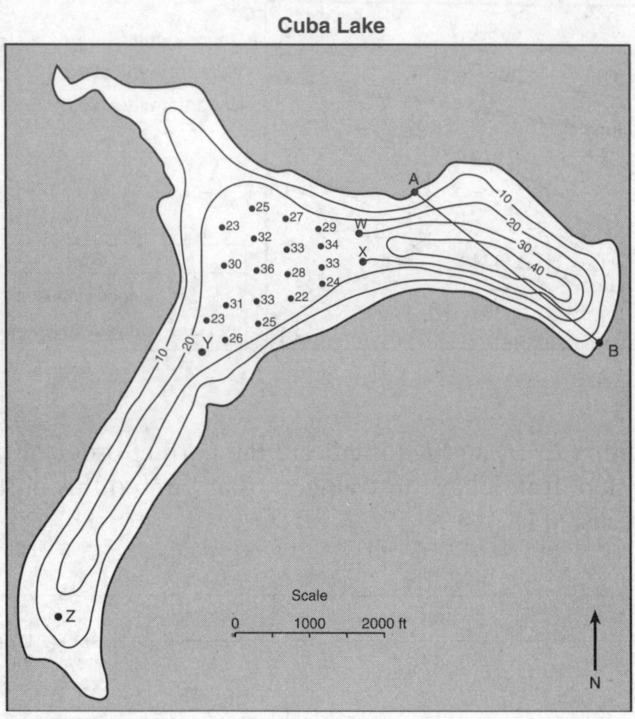

62 On the map above, complete the 30-foot water depth isoline from point *W* to point *X*. [1]

63 On the grid below, construct a profile of the bottom of Cuba Lake from point *A* to point *B*. Plot each point where an isoline showing depth is crossed by line *AB*. Connect the plots with a line, starting at *A* and ending at *B*, to complete the profile. [1]

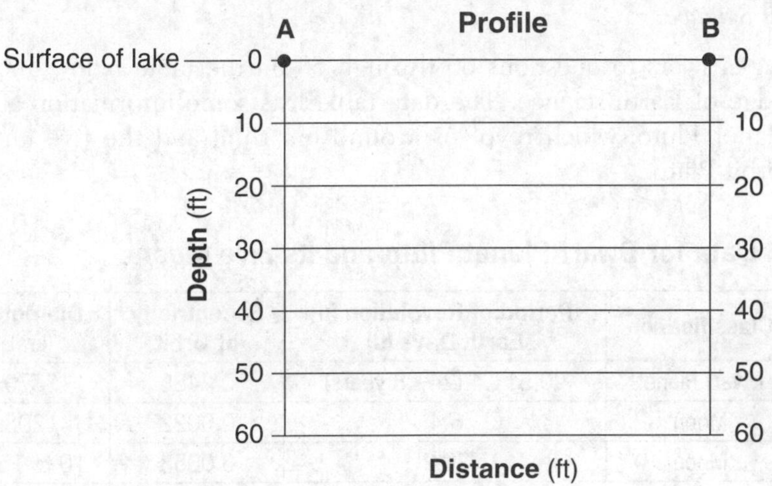

64 State the compass direction and distance in feet (ft) from point *Y* to point Z. [1]

Direction: ____________

Distance: ____________ **ft**

65 Identify the New York State landscape region where Cuba Lake is located. [1]

____

PART C

Answer all questions in this part.

Directions (66–85): Record your answers in the spaces provided. Some questions may require the use of the *2011 Edition Reference Tables for Physical Setting/Earth Science*.

Base your answers to questions 66 through 68 on the table below and on your knowledge of Earth science. The data table lists some information about the dwarf planet Pluto, which revolves around our Sun, and the five known moons that orbit Pluto.

Data for Dwarf Planet Pluto and Its Five Moons

Object Name	Classification	Period of Revolution in Earth Days (d)	Eccentricity of Orbit	Diameter (km)
Pluto	Dwarf Planet	90,511.4 (247.8 years)	0.2488	2370
Charon	Moon	6.4	0.0022	1208
Styx	Moon	20.2	0.0058	10 to 15 *
Nix	Moon	24.9	0.0020	40
Kerberos	Moon	32.2	0.0033	13 to 34 *
Hydra	Moon	38.2	0.0059	33 to 43 *

* There is a range in diameters for these moons due to their irregular shapes.

66 Identify the name of Pluto's moon that most likely has an orbit farthest from Pluto. Explain how the data indicate that this moon's orbit has the greatest distance from Pluto. [1]

Moon of Pluto: ______________________________

Explanation: ______________________________

____

67 Describe the shape of the orbit of Pluto and the orbits of its five moons. [1]

____

68 Explain why Pluto and its five moons are considered to be part of our solar system. [1]

____

____

____

Base your answers to questions 69 through 73 on the passage and map below and on your knowledge of Earth science. The map shows a satellite image of a nor'easter that influenced the weather of the northeastern United States. The white areas represent clouds associated with this storm system. The locations of North Carolina and Albany, New York, are labeled on the map. The storm's low-pressure center is represented by letter L. Letters cP and mT represent two air masses.

Nor'easters

A nor'easter is a large, low-pressure storm system that moves along the east coast of the United States. The wind over the land blows generally from the northeast as the center of the low passes by a location, hence the name nor'easter. Due to the circulation of winds around the center of the low-pressure system, large amounts of precipitation occur as moist air is carried from the ocean to the land. These storms usually intensify off of the North Carolina coast as they track toward the northeast.

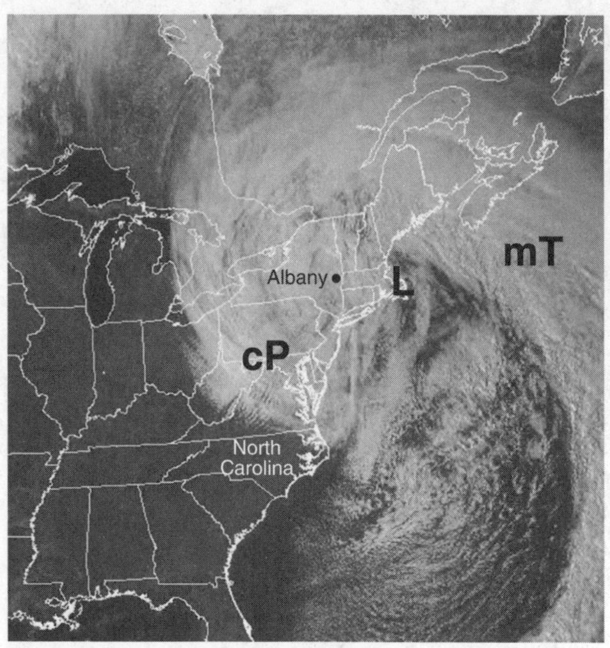

69 Describe *two* characteristics of the circulation pattern of the surface winds around the center of the low-pressure area represented on the map. [1]

Characteristic 1: ______________________________

Characteristic 2: ______________________________

70 Circle the terms that best describe the relative moisture and relative temperature characteristics of the mT air mass compared to the cP air mass shown on the map. [1]

Relative moisture of mT air mass (circle one): **more humid** **less humid** **the same**

Relative temperature of mT air mass (circle one): **cooler** **warmer** **the same**

71 The map below shows some of the principal storm tracks across the United States and the names of these storm tracks.

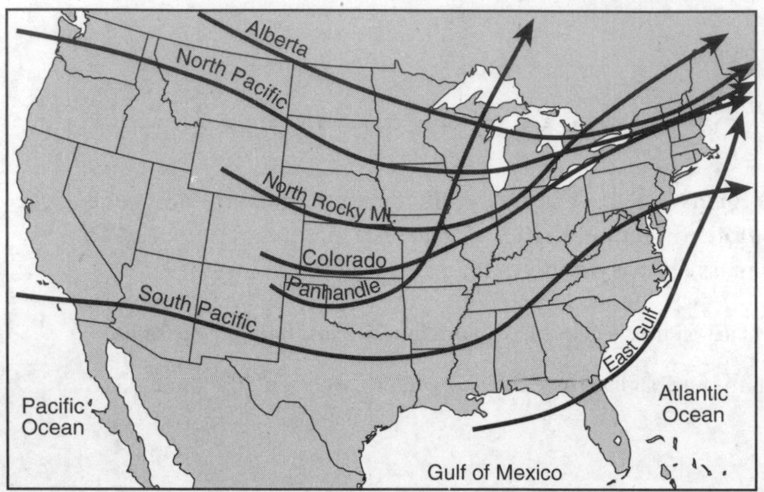

Identify the name of the storm track that this nor'easter most closely followed. [1]

____

72 The table below shows weather conditions recorded in Albany, New York, at the time that the satellite image was taken.

Weather Conditions

Dewpoint	22°F
Barometric Pressure	988.0 mb
Cloud Cover	100%
Present Weather	Snow

On the station model below, use the correct symbols and proper format to indicate the four conditions in the table. [1]

73 Identify *one* weather instrument that was most likely used to determine the dewpoint at Albany, New York. [1]

____

Base your answers to questions 74 through 76 on the data table below and on your knowledge of Earth science. The data table shows the mass of a sample of radioactive carbon-14 remaining after each half-life.

Data Table

Number of Half-lives	**Mass of Carbon-14** (g)
0	64
1	32
2	16
3	8
4	4
5	2
6	1

74 On the grid below, construct a line graph by plotting the data for the mass of carbon-14 in the sample for *each* half-life shown on the data table. Connect the plots with a line. [1]

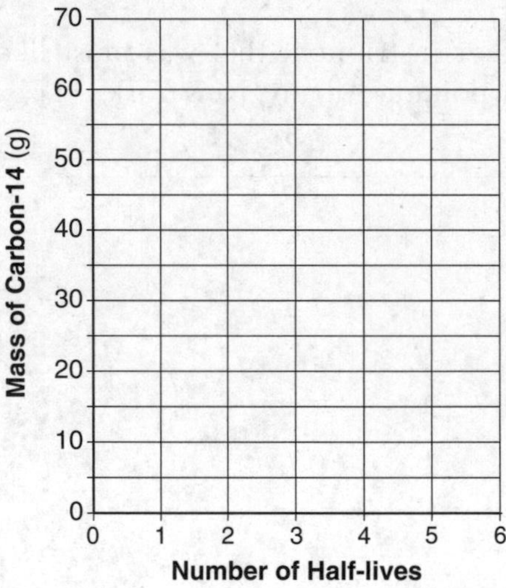

75 Identify the stable disintegration product of the radioactive isotope carbon-14. [1]

____

76 Determine the number of grams of carbon-14 remaining in this sample at 17,100 years. [1]

___________________ g

Base your answers to questions 77 through 79 on the data tables below and on your knowledge of Earth science. Data table 1 shows the average maximum and average minimum water temperatures at different depths in Lake Michigan. Data table 2 shows the average maximum and average minimum land temperatures at different depths in soil at St. Paul, Minnesota (MN).

Table 1

Water in Lake Michigan, 45° N		
Water Depth Range (m)	**Average Temperature** (°C)	
	Maximum (Occurs in Summer)	**Minimum** (Occurs in Winter)
0–10	24.0	4.0
10–20	19.5	4.0
20–30	12.5	4.0
30–40	7.5	4.0
40–60	5.5	4.0
60–110	4.5	4.0
110–150	4.0	4.0

Table 2

Land in St. Paul, MN, 45° N		
Soil Depth Range (m)	**Average Temperature** (°C)	
	Maximum (Occurs in Summer)	**Minimum** (Occurs in Winter)
0–0.1	25.5	–6.0
0.1–1	23.0	–1.5
1–2	19.5	2.0
2–3	16.0	4.5
3–4	13.5	7.0
4–5	12.5	8.5
5–6	11.0	10.0

77 Identify the water depth range and the soil depth range, in meters, that have the same average maximum temperature of 12.5°C. [1]

Water depth range: ______ **m**

Soil depth range: ________ **m**

78 Describe the general relationship between depth and average temperature for both water and soil in the summer. [1]

____

____

79 Explain why the winter water surface is warmer than the winter land surface. [1]

____

____

Base your answers to questions 80 through 82 on the map below and on your knowledge of Earth science. The map shows an enlargement of a portion of the *Tectonic Plates* map from the *Physical Setting/Earth Science Reference Tables*. Arrows showing plate motion have been omitted. Points *Y* and *Z* represent locations on plate boundaries.

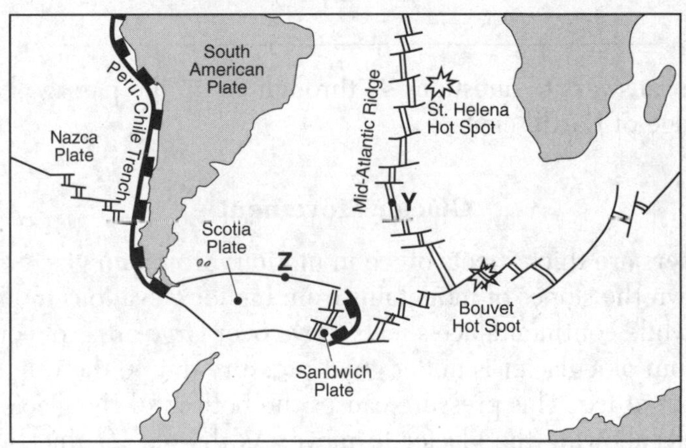

80 Identify the type of tectonic plate boundary found at location Z. [1]

____

81 State the names of the tectonic plates on *each* side of the Bouvet Hot Spot. [1]

______________ **Plate** and ______________ **Plate**

82 The cross section below represents a portion of Earth's interior beneath point *Y*. On this cross section, draw an arrowhead on *each* of the *four* bold lines, to represent the direction of the convection currents in the asthenosphere. [1]

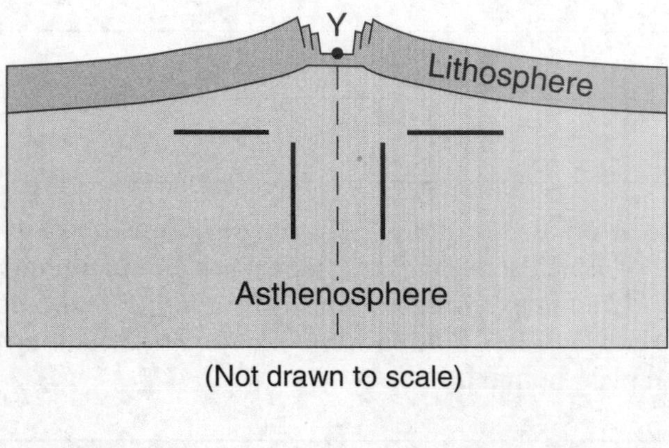

(Not drawn to scale)

Base your answers to questions 83 through 85 on the passage below and on your knowledge of Earth science.

Glacier Movement

Glaciers are thick sheets of ice in motion. Mountain glaciers tend to move down the slopes of mountains from higher elevations to lower elevations, while continental ice sheets move over large areas of continents. The bottom of a glacier is under great pressure due to the weight of the thick sheet of ice. This pressure causes the bottom of the glacier to partially melt, allowing the glacier to move. As the glacier thickens, more pressure is created and the glacier moves faster. Different parts of a glacier can move at different rates, depending on the amount of pressure and friction between the glacier and the underlying bedrock.

83 Describe the relationship between the thickness of a glacier and its rate of movement downhill. [1]

____

84 Describe the most likely shape of the cross section of a valley formed as a result of erosion by a mountain glacier. [1]

____

85 Compared to sediments deposited by meltwater from a glacier, describe the difference in the arrangement of the sediment deposited directly by a glacier. [1]

____

Answers August 2018

Physical Setting/Earth Science

Answer Key

PART A

1. 1	**8.** 1	**15.** 1	**22.** 2	**29.** 2
2. 3	**9.** 2	**16.** 4	**23.** 4	**30.** 3
3. 2	**10.** 3	**17.** 2	**24.** 3	**31.** 3
4. 4	**11.** 2	**18.** 4	**25.** 2	**32.** 1
5. 1	**12.** 4	**19.** 2	**26.** 4	**33.** 3
6. 1	**13.** 1	**20.** 2	**27.** 1	**34.** 4
7. 2	**14.** 4	**21.** 4	**28.** 1	**35.** 4

PART B–1

36. 2	**39.** 2	**42.** 2	**45.** 3	**48.** 3
37. 4	**40.** 3	**43.** 3	**46.** 3	**49.** 4
38. 1	**41.** 3	**44.** 4	**47.** 1	**50.** 2

In **PART B–2** and **PART C**, you are required to show how you arrived at your answers. For sample methods of solutions, see the *Answers Explained* section.

Answers Explained

PART A

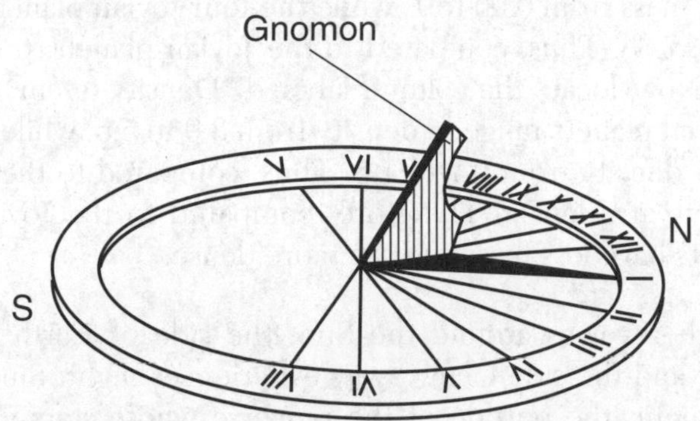

1. **1** The shadow cast by an object will always point in the direction opposite the position of the Sun in relation to the object. Therefore, a line drawn from the tip of the shadow to the tip of the object casting the shadow will point toward the Sun. It is given that the shadow cast by the gnomon moves across the disc pointing to the time of day. In order for the shadow to move, the position of the Sun in relation to the gnomon must change. Thus, when the shadow cast by the gnomon points to XI (A.M.), the Sun is located to the east of south. When the shadow cast by the gnomon points to I (P.M.), the Sun is located to the west of south. Thus, the Sun appears to be moving from east to west across the sky. The Sun's apparent path through the sky from sunrise to sunset is an arc, like all other celestial objects, and is the result of Earth's rotation. As Earth (and any object on Earth's surface) rotates on its axis from west to east, the Sun appears to move from east to west. Thus, the motion of the gnomon's shadow on the sundial is mainly due to Earth's rotation.

2. **3** The planet Uranus, like Earth, is part of the solar system. Find the Geologic History of New York State chart in the *Reference Tables for Physical Setting/Earth Science* and locate the time scale along the left edge of the column labeled "Eon." Locate 4600 million years ago (4.6 billion) on the time scale, trace right to the column labeled "Era," and note the statement "Estimated time of origin of Earth and our solar system." Thus, the formation of Earth and our solar system (which includes the planet Uranus) occurred approximately 4.6 billion years ago.

3. **2** The planets can be divided by mass and density into the terrestrial (Earth-like) and Jovian (Jupiter-like) planets. The terrestrial planets include the four innermost planets: Mercury, Venus, Earth, and Mars. The Jovian planets include the four outermost planets: Jupiter, Saturn, Neptune, and Uranus. Find the Solar System Data chart in the *Reference Tables for Physical Setting/Earth Science.* Locate the column labeled "Mass (Earth=1)" and note that the four terrestrial planets range in mass from 0.06 to 1, while the four Jovian planets range in mass from 14.54 to 317.83. Thus, compared to the Jovian planets, terrestrial planets have *less* mass. Now, locate the column labeled "Density (g/cm^3)" and note that the four terrestrial planets range in density from 3.9 to 5.5, while the four Jovian planets range in density from 0.7 to 1.8. Thus, compared to the Jovian planets, terrestrial planets are denser. Therefore, compared to the Jovian planets, the terrestrial planets have less mass and are more dense.

4. **4** As Earth revolves around the Sun, the side of Earth facing the Sun experiences day and the side facing away experiences night. Since the stars are only visible at night, the portion of the universe whose stars are visible to an observer on Earth varies cyclically as Earth revolves around the Sun. It is given that the constellation Leo is visible to an observer in New York State at midnight during March. Therefore, Earth's night side is facing Leo in March. In September, six months later, Earth would have revolved halfway through its orbit and would be on the opposite side of the Sun. Thus, its nighttime side would be facing away from Leo, and Leo would not be visible to an observer at night. See the diagram below. Therefore, the constellation Leo is not visible to this observer at midnight during September because of the revolution of Earth around the Sun.

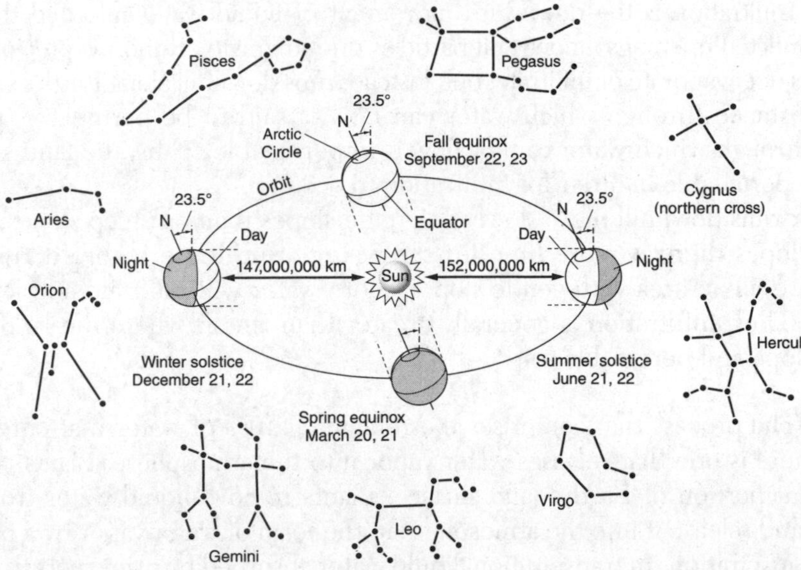

At various times of the year, different constellations are visible at night because Earth's nighttime side faces different parts of the universe.

5. **1** In New York State, the angle of insolation at solar noon is greatest on June 21 and is least on December 21. From June 21 through December 21 (which includes July, September, and December), the angle of insolation at solar noon decreases every day. From December 21 through June 21 (which includes April), the angle of insolation at solar noon increases every day. Thus, an observer in New York City would see the noontime angle of insolation increase each day during April.

6. **1** Planetary winds tend to blow in a straight line from regions of high pressure toward regions of low pressure. As these winds are blowing, Earth is turning on its axis. This rotation causes the winds to appear to be turning toward the right in the Northern Hemisphere and toward the left in the Southern Hemisphere. This phenomenon is called the Coriolis effect. Thus, the Coriolis effect occurs as a result of Earth's rotation.

7. **2** Condensation is the phase change from gas to liquid. Find the Properties of Water chart in the *Reference Tables for Physical Setting/Earth Science.* Note the entry "Heat energy released during condensation . . . 2260 J/g." Thus, during the process of condensation, when water changes phase from water vapor to liquid water, the water vapor releases 2260 J/g of heat energy.

8. **1** Infiltration is the downward movement of liquid water into and through interconnected openings among soil particles due to gravity. Runoff is precipitation that does not evaporate or infiltrate, but, instead, runs downhill along Earth's surface.

A substance through which water can flow is said to be permeable; a substance through which water cannot flow is impermeable. Thus, the land surface must be permeable in order for infiltration to occur.

Water runs downhill more slowly on gentle slopes than on steep slopes, so on gentle slopes there is more time for the water to infiltrate before it runs off. Therefore, in an area with gentle slopes, more water will infiltrate and less will run off. Thus, infiltration is generally greater than runoff where the land has a gentle slope and permeable soil.

9. **2** The process that comprises the remaining 10% of water that enters the atmosphere is one that releases water vapor into the atmosphere. Plants cover a significant portion of Earth's land surface. Plants remove liquid water from the ground and release it into the atmosphere in the form of water vapor by a process called transpiration. In transpiration, liquid water absorbed by plant roots from the soil increases the water pressure inside the lower parts of the plant. Simultaneously, evaporation of water through plant stems or leaf stomata decreases the water pressure in the upper parts of a plant. This difference in pressure causes the liquid water to move upward from the roots and toward the leaves. When the liquid water reaches the leaves, it evaporates. Then more water is drawn upward from the roots, and the process of transpiration continues. Thus, most of the remaining 10% of water vapor enters the atmosphere through transpiration from plants.

WRONG CHOICES EXPLAINED:

(1) Precipitation is the process by which water in the form of rain, snow, hail, or sleet falls to the ground. Thus, water vapor does not enter the atmosphere through precipitation.

(2) Condensation is the process by which water vapor changes phase to form liquid water. Thus, water vapor does not enter the atmosphere due to condensation; water vapor leaves the atmosphere due to condensation.

(3) Melting is the process by which solid water changes phase to form liquid water. Thus, melting does not add water vapor to the atmosphere.

10. **3** It is given that the dry-bulb temperature is 26°C and the wet-bulb temperature is 18°C. Thus, the difference between the wet-bulb and dry-bulb temperatures is 8°C. Find the Relative Humidity (%) chart in the *Reference Tables for Physical Setting/Earth Science*. Locate the column headed "8" in the "Difference Between Wet-Bulb and Dry-Bulb Temperatures (°C)" scale along the top of the chart. Find the "Dry-Bulb Temperature (°C)" scale along the left

side of the chart. Locate the row labeled "26," trace to the right until it intersects the column headed "8," and note the value of 45. Thus, when the dry-bulb temperature is 26°C and the wet-bulb temperature is 18°C, the relative humidity is 45%.

11. **2** Every object not at absolute zero radiates energy. The type of energy radiated depends on the temperature of the object. The Sun, which is at a high temperature, radiates energy of relatively short wavelengths, such as ultraviolet and visible light rays. Earth's surface, which is at a much lower temperature, radiates energy at longer wavelengths, mainly in the infrared range. Thus, most of the long-wave energy radiated and lost to space on a cloudless night is infrared.

12. **4** When sunlight strikes Earth's surface, some of it is reflected back toward space, and some of it is absorbed and then re-radiated as infrared radiation (or heat). Over time, the amount of energy received from the Sun should be about the same as the amount of energy radiated back into space, leaving the temperature of the Earth's surface roughly constant. However, many chemical compounds found in the Earth's atmosphere act as "greenhouse gases." These gases allow sunlight to enter the atmosphere and strike Earth's surface, but absorb the infrared radiation re-radiated by Earth's surface, trapping the heat in the atmosphere resulting in global warming. Some of these "greenhouse gases" occur in nature (water vapor, carbon dioxide, methane, ozone, and nitrous oxide), while others are exclusively human-made (like the chlorofluorocarbons used in aerosol sprays). Thus, in addition to carbon dioxide, only choice 4—water vapor and methane—contains two other major greenhouse gases found in Earth's atmosphere.

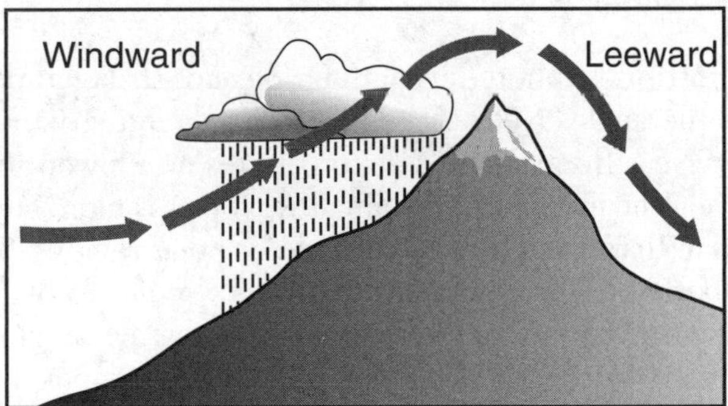

13. **1** According to the arrows in the diagram, air is moving upward on the windward side of the mountain. As air is carried upward to higher elevations, it expands and cools. When the air temperature cools to the dewpoint, moisture in the air begins to condense and clouds form. Thus, clouds are forming on the windward side of the mountain because the air is expanding and cooling to the dewpoint.

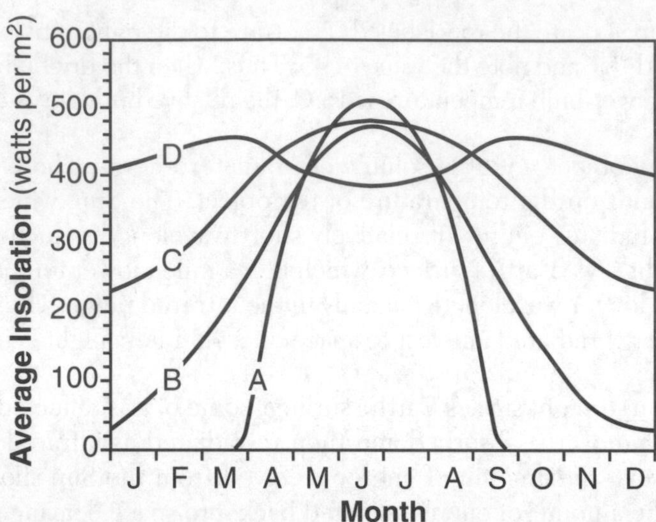

14. **4** The average monthly insolation received at the equator will be greatest when the sunlight striking Earth's surface at the equator is most direct. At the equinoxes, Earth's axis of rotation tilts neither toward nor away from the Sun, and the Sun's direct rays strike Earth's surface at the equator. Thus, the average monthly insolation received at the equator will peak at the equinoxes (March and September). At other times of the year, Earth is tilted either toward or away from the Sun and sunlight strikes Earth's surface at less than 90º, and the average monthly insolation received at the equator will be lower. Thus, the line on the graph representing the equator should show two peaks—one in March and one in September—as shown on line D.

15. **1** Since Earth is a sphere, insolation does not strike all points on Earth's surface at the same angle. Near the equator, insolation strikes Earth's surface almost vertically. Near the poles, insolation strikes at a lower angle. Therefore, throughout the year the insolation reaching the tropics is more concentrated (i.e., more intense) than that reaching mid-latitude or polar regions. The less intense the insolation, the less Earth's surface is warmed by the insolation. Therefore, polar regions have cold climates because the Sun's rays are at a low angle.

Find the Planetary Winds and Moisture Belts in the Troposphere chart in the *Reference Tables for Physical Setting/Earth Science.* Note that the arrows indicating air movements show that, near the poles, air in the upper troposphere is sinking. As the air sinks, it is compressed and warms (although it is still very cold). This increases the difference between the air temperature and the dew-point temperature, resulting in lower relative humidity. Thus, Earth's polar regions also have dry climates. Therefore, Earth's polar regions have cold, dry climates because the Sun's rays are at a low angle, and upper atmosphere air is sinking.

16. **4** Air is mainly composed of transparent gases. However, air also contains varying amounts of substances that are opaque and block light, or that reflect or scatter light. The more of these substances the air contains, the lower the atmospheric transparency. Common substances in air that block, reflect, or scatter light include water droplets, ice crystals, salt particles, or particles of dust, ash, or smoke. Removing any of these substances from the air will increase atmospheric transparency. One common process that removes dust and other particles from the air is precipitation. Precipitation is condensed moisture (such as rain, snow, hail, or sleet) in the atmosphere that falls to the ground. Precipitation removes particles of dust from the air in a number of ways. For example, dust particles are often the condensation nuclei on which water condenses. When the condensed moisture falls to the ground, these condensation nuclei are carried to the ground. Dust particles can also come into contact with the condensed moisture as they fall through the air, adhere to its surface, and be carried to the ground. Thus, atmospheric transparency will increase when precipitation removes dust particles from the air.

WRONG CHOICES EXPLAINED:

(1) Volcanic eruptions spew gases, ash, and dust into the atmosphere, which are then carried around the globe by winds in the upper atmosphere. Volcanic ash and dust are opaque and will block, reflect, or scatter incoming solar radiation. Thus, a volcanic eruption would decrease atmospheric transparency.

(2) Fog is composed of water droplets. Water droplets can reflect and scatter light. Therefore, fog decreases atmospheric transparency.

(3) When insolation is reflected by clouds, less passes through the cloud and reaches Earth's surface. Insolation consists of electromagnetic waves, and electromagnetic waves are totally transparent to other electromagnetic waves. Thus, atmospheric transparency is not affected by the quantity of insolation passing through the air. Therefore, if insolation is reflected by clouds, atmospheric transparency would neither increase nor decrease.

17. **2** Find the Geologic History of New York State chart in the *Reference Tables for Physical Setting/Earth Science*. In the column labeled "Life on Earth," locate "humans." From humans, trace left to the time scale labeled "Million years ago." Note that humans appeared less than 1.8 million years ago. Humans still exist today. Therefore, humans have existed for less than 1.8 million years.

Now find the column labeled "Time Distribution of Fossils (including important fossils of New York)." Note the vertical gray bars labeled with the names of different groups of organisms. The vertical gray bars indicate the range of time during which each group of organisms existed on Earth. The bottom of the gray bar indicates when the group first appeared on Earth, and the top of the bar indicates when the group became extinct or, if it extends to the top of the column, that the group still exists. Locate the gray bar labeled "Birds." From the bottom of this gray bar, trace left to the time scale labeled "Million years ago." Note that birds first appeared a little more than 146 million years ago. From the top of this gray bar, trace left to the time scale labeled "Million years ago," and note that birds still exist. Therefore, birds have existed for more than 146 million years. Now, locate the gray bar labeled "Dinosaurs." From the bottom of this gray bar, trace left to the time scale labeled "Million years ago." Note that dinosaurs first appeared between 200 and 251, or about 225 million years ago. From the top of this gray bar, trace left to the time scale labeled "Million years ago," and note that dinosaurs became extinct 65.5 million years ago. Thus, dinosaurs existed for about 159.5 million years (225 – 65.5 = 159.5). Finally, locate the gray bar labeled "Placoderm fish." From the bottom of this gray bar, trace left to the time scale labeled "Million years ago." Note that placoderm fish first appeared between 416 and 444, or about 430 million years ago. From the top of this gray bar, trace left to the time scale labeled "Million years ago," and note that placoderm fish became extinct 359 million years ago. Thus, placoderm fish existed for about 71 million years (430 – 359 = 71). Therefore, the group whose existence spans the shortest geologic time is humans.

18. **4** Find the Geologic History of New York State chart in the *Reference Tables for Physical Setting/Earth Science*. Locate the time scale labeled "Million years ago" to the left of the column labeled "Life on Earth." Note that, on the scale, age increases from top to bottom. Now, locate the column labeled "Important Geologic Events in New York." Note that, according to the time scale, the most recent events are nearest the top of the column. Now, locate "Taconian Orogeny," "Grenville Orogeny," "formation of Catskill delta," and "dome-like uplift of the Adirondack region." Note that, of these events, the dome-like uplift of the Adirondack region is nearest the top of the chart. Therefore, the New York State geologic event that occurred most recently is the dome-like uplift of the Adirondack region.

19. **2** It is given that the dinosaur footprints were found in 210-million-year-old bedrock. Find the Geologic History of New York State chart in the *Reference Tables for Physical Setting/Earth Science*. Locate the time scale labeled

"Million years ago" to the left of the column labeled "Life on Earth." On the time scale, locate the point corresponding to 210 million years ago. From this point, trace left to the column labeled "Period," and note that 210 million years ago corresponds to the Late Triassic period. Next, find the Generalized Bedrock Geology of New York State map in the *Reference Tables for Physical Setting/Earth Science*. In the key labeled "Geologic Periods and Eras in New York," locate the map symbol corresponding to Late Triassic and Early Jurassic conglomerates, red sandstones, red shales, basalt, and diabase (Palisades sill). On the map, locate the region shaded with this map symbol (along the border between New York and New Jersey). Now, find the Generalized Landscape Regions of New York State map in the *Reference Tables for Physical Setting/Earth Science*. Note that the region of Late Triassic rocks in New York State corresponds to the Newark Lowlands. Thus, the New York State landscape region in which these fossils were found is the Newark Lowlands.

20. **2** It is given that the first *P*-wave traveled 5600 kilometers from the epicenter of the earthquake to the seismic station. Find the Earthquake *P*-wave and *S*-wave Travel Time graph in the *Reference Tables for Physical Setting/Earth Science*. Locate 5600 kilometers (5.6×10^3 km) along the horizontal Epicenter Distance axis at the bottom of the graph. From this point, trace vertically until you intersect the bold line labeled "P." From this intersection, trace horizontally to the left to the Travel Time axis and read the *P*-wave travel time—9 minutes. If the first *P*-wave takes 9 minutes to travel the 5600 kilometers from the epicenter to the seismic station and arrives at 10:05 A.M., it must have left the epicenter 9 minutes before 10:05 A.M., or 9:56 A.M. Thus, the earthquake occurred at 9:56 A.M.

21. **4** It is given that the photograph shows a New York State index fossil. Find the Geologic History of New York State chart in the *Reference Tables for Physical Setting/Earth Science*. Locate the diagrams of index fossils along the bottom of the chart. Note that the New York State index fossil that most closely resembles the one in the photograph is B—*Cryptolithus*. In the column labeled "Time Distribution of Fossils," locate the circled letter B and note that it lies on the gray bar labeled "Trilobites." Thus, the index fossil in the photograph is a trilobite. Note, too, the explanation at the top of the column: "The center of each lettered circle indicates the approximate time of existence of a specific index fossil." From the center of the circled letter B representing *Cryptolithus*, trace left to the column labeled "Period," and note that *Cryptolithus* lived during the Ordovician period. Thus, the best classification of this fossil and geologic time period during which it existed is Classification: Trilobite; Geologic time period: Ordovician.

22. **2** As sediment particles, such as quartz pebbles, are carried along by the moving water in a stream, they bounce off of and rub against one another and bounce, roll, and scrape against the bottom of the streambed. These collisions cause smaller pieces to break off of sharp corners and edges on the surface of the sediments, particularly at corners that protrude—a process called abrasion. As a result, the sediments eroded by the water in this stream become more rounded. Thus, as a quartz pebble is transported by a stream, the pebble will become more rounded as a result of abrasion by colliding with other rocks.

WRONG CHOICES EXPLAINED:

(1) The mineral quartz is almost insoluble in water. Therefore, a quartz pebble would not become rounded by water running over it.

(3) A quartz pebble is deposited when the stream stops transporting it, not as it is transported. If it is deposited in layers, the surrounding sediments shield it from further abrasion. Therefore, it would not become rounded as a result of deposition in layers.

(4) If a quartz pebble is resistant to weathering and erosion, it is less susceptible to abrasion and likely to become less rounded as it is transported by a stream.

23. **4** Find the Properties of Common Minerals table in the *Reference Tables for Physical Setting/Earth Science.* In the column labeled "Uses," note uses that relate to the construction of a house, such as "plaster of paris, drywall," "paint, roofing," and "building stones." From each of these uses, trace right to the column labeled "Mineral Name," and note that they correspond to selenite gypsum, muscovite mica, and dolomite, as shown in choice 4.

WRONG CHOICES EXPLAINED:

(1), (2) and (3) Note that there are other minerals with uses relating to the construction of a house, such as "cement, lime," "ceramics, glass," and "construction materials," but none of the minerals in answer choices 1–3 has a construction-related use for all three of the minerals listed in the choice.

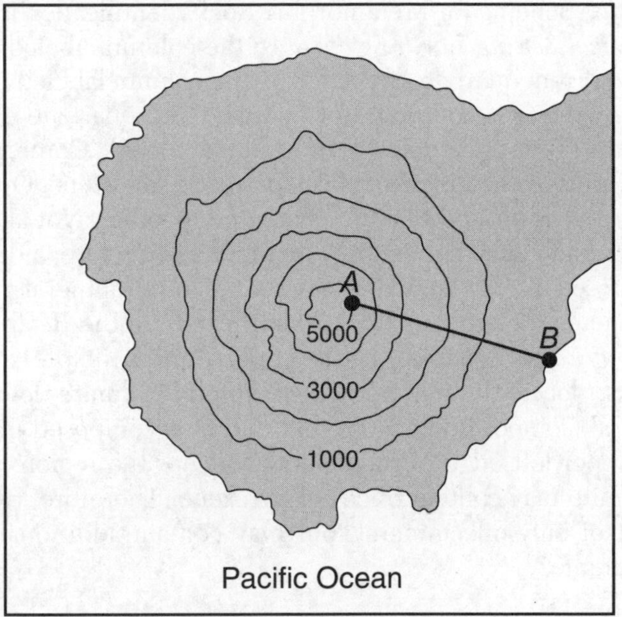

24. **3** Find the Equations section of the *Reference Tables for Physical Setting/ Earth Science* and note the equation for gradient:

$$\text{gradient} = \frac{\text{change in field value}}{\text{change in distance}}$$

The map shown is a topographic map. The field value on a topographic map is elevation. Note that point *A* is located directly on the 5000-foot contour line and that point *B* is located directly on the coastline of the island, which corresponds to sea level, or elevation 0 feet. Thus, the change in field value (elevation) from *A* to *B* is 5000 feet.

It is given that the distance between locations *A* and *B* is 2.5 miles. Substitute these values in the equation and solve:

$$\text{gradient} = \frac{500 \text{ ft}}{2.5 \text{ mi}} = 2000 \text{ ft/mi}$$

Thus, the gradient between location *A* and location *B* is approximately 2000 ft/mi.

25. **2** Find the Scheme for Metamorphic Rock Identification in the *Reference Tables for Physical Setting/Earth Science.* In the column labeled "Rock Name," locate quartzite. From quartzite, trace left to the column labeled "Composition," and note that quartzite is composed of quartz. Thus, quartzite usually contains only one mineral. Now, trace right to the column labeled "Comments," and note that quartzite forms by the metamorphism of quartz sandstone. Quartz sandstone is composed of grains of sand that are cemented together. Not all grains of sand are pure quartz, and not all cement is pure quartz. Therefore, although quartzite is usually composed of only quartz, it may contain other minerals.

Find the Scheme for Igneous Rock Identification in the *Reference Tables for Physical Setting/Earth Science*. In the upper portion of the scheme labeled "Igneous Rocks," locate dunite. Trace the column for dunite down to the graph labeled "Mineral Composition." Note that dunite is composed of olivine. Note, too, that the upper left edge of the column touches the region labeled "pyroxene." Thus, dunite may contain traces of pyroxene. Therefore, two rocks which usually consist of only one mineral, but may contain additional minerals, are quartzite and dunite.

WRONG CHOICES EXPLAINED:

(1) Find the Scheme for Metamorphic Rock Identification in the *Reference Tables for Physical Setting/Earth Science.* In the column labeled "Rock Name," locate hornfels. Trace the row for hornfels to the left to the column labeled "Composition," and note that hornfels is composed of various minerals. Thus, hornfels does not usually consist of only one mineral.

(3) Find the Scheme for Igneous Rock Identification in the *Reference Tables for Physical Setting/Earth Science*. In the upper portion of the scheme labeled "Igneous Rocks," locate basalt. Trace the column for basalt down to the graph labeled "Mineral Composition." Note that basalt is composed of plagioclase feldspar, pyroxene, biotite, amphibole, and olivine. Thus, basalt does not usually consist of only one mineral.

(4) Find the Scheme for Igneous Rock Identification in the *Reference Tables for Physical Setting/Earth Science*. In the upper portion of the scheme labeled "Igneous Rocks," locate gabbro. Trace the column for gabbro down to the graph labeled "Mineral Composition." Note that gabbro is composed of plagioclase feldspar, pyroxene, biotite, amphibole, and olivine. Thus, gabbro does not usually consist of only one mineral.

26. **4** By definition, metamorphism refers to the changes that occur in solid rock due to great heat, great pressure, and the chemical activity that sometimes results from the combination of these two factors. Metamorphism occurs while rocks are still in the solid state and does not involve molten rock in the liquid state. Thus, a rock that has never melted, but was produced by great heat and pressure that distorted and rearranged its minerals, is a metamorphic rock. Find the Scheme for Metamorphic Rock Identification in the *Reference Tables for Physical Setting/Earth Science.* In the column labeled "Rock Name," note that only answer choice 4—metaconglomerate—is a metamorphic rock.

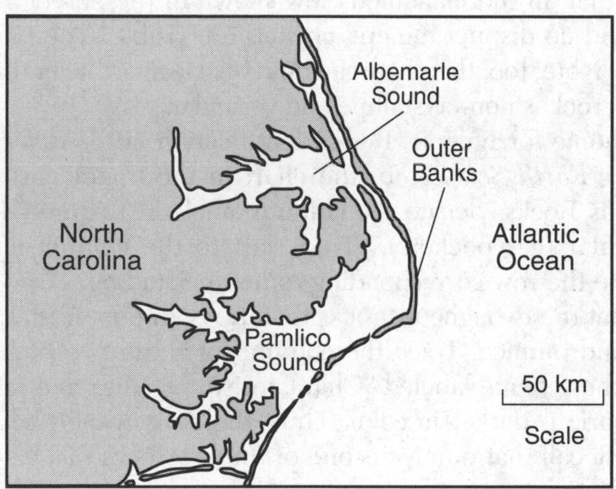

27. **1** Barrier islands form when tidal inlets break a barrier bar into elongated islands. Barrier bars are formed when waves deposit enough sand on a sandbar so that it is above water. Thus, the agent of erosion that is primarily responsible for the formation of these barrier islands is wave action.

WRONG CHOICES EXPLAINED:

(2) During mass movements such as landslides, sediment is pulled downhill by gravity and is deposited at the base of a slope when it stops moving. Since the barrier islands are located offshore, not at the base of a slope, it is unlikely that they formed as a result of landslides.

(3) When a stream flows into a standing body of water, such as the Atlantic Ocean, it stops moving, and most of the stream's sediment is deposited. The resulting landform is a large, flat, fan-shaped pile of sediment at the mouth of the stream called a delta. These barrier islands are elongated and run parallel to the coast; therefore, it is unlikely that they were formed by streams.

(4) Glaciers do not currently exist in North Carolina. At the end of the last ice age, vast amounts of water were locked up in polar ice caps, and sea level was much lower than it is today. As a result, at the time of the last ice age, the shoreline of what is now North Carolina was as much as 40 miles east of the current coastline at the edge of the North American continental shelf. In the thousands of years since the last ice age, rising sea level and the action of waves and currents would have eroded and drowned any deposits left by glaciers even if they had existed along the North Carolina coast of the Atlantic Ocean. Therefore, it is unlikely that these barrier islands were formed by glacial ice.

28. **1** Note that, in the magnified view shown in the photograph, the rock is light in color and no distinct mineral crystals are visible. Therefore, this rock is non-crystalline. Note, too, the many cavities (vesicles) visible in the surface of the rock. Thus, this rock is non-crystalline and vesicular.

Find the Scheme for Igneous Rock Identification in the *Reference Tables for Physical Setting/Earth Science*. In the chart in the upper part of the scheme labeled "Igneous Rocks," locate the column labeled "Texture." Locate the row labeled "Vesicular (gas pockets)." Trace left to the column labeled "Crystal Size," and locate the row corresponding to non-crystalline. Trace this row to the left and note that the two igneous rocks that are both non-crystalline and vesicular are scoria and pumice. Trace the column for these two rocks downward to the section of the scheme labeled "Characteristics." Note that pumice is lighter in color and scoria is darker in color. Thus, the rock in the photograph is most likely pumice. Recall that pumice is one of the few rocks that can float if placed in water. This is because pumice contains many cavities (vesicles) that formed when molten rock solidified as gases were still bubbling out of it. The presence of these vesicles decreases the overall density of the pumice sample enough that its density may be less than that of water and it will float.

Thus, the rock in the photograph that can float is most likely pumice and is best described by the terms non-crystalline and vesicular.

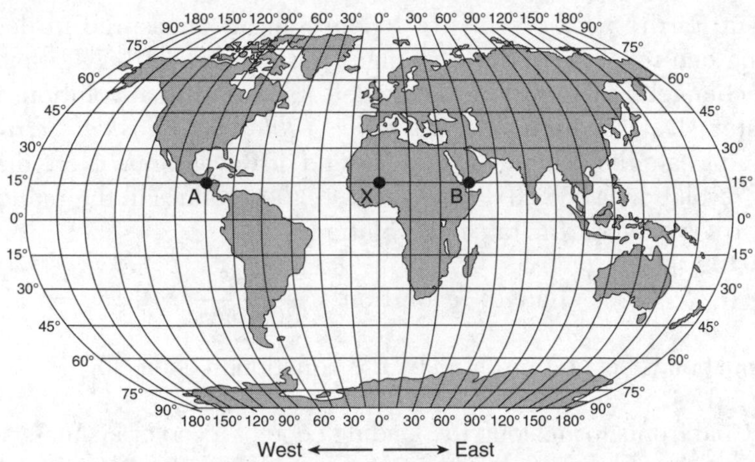

29. **2** By convention, maps are oriented with north at the top. Lines of longitude run north–south (vertically) and lines of latitude run east–west (horizontally). On the given map, longitude values are read along the top and bottom of the map; latitude values are read along the left and right side of the map. Note that location *X* is on the Prime Meridian (0° longitude). Earth rotates eastward at a rate of 15° per hour. Therefore, for every 15° longitude you travel west of the Prime Meridian, the Sun rises one hour later and solar time is one hour earlier than at the Prime Meridian. Conversely, for every 15° longitude you travel east of the Prime Meridian, the Sun rises one hour earlier and solar time is one hour later than at the Prime Meridian. It is given that it is 12 noon at location *X*. Note that location *B* is 45° of longitude east of the Prime Meridian. Therefore, when it is 12 noon at location *X*, the solar time at location *B* will be 3 hours later (45°/15°/hr = 3 hr), or 3 P.M. Note that location *A* is 90° of longitude west of the Prime Meridian. Therefore, when it is 12 noon at location *X*, the solar time at location *B* will be 6 hours earlier (45°/15°/hr = 3 hr), or 6 A.M. This corresponds to the table shown in choice 2.

30. **3** Find the Equations chart in the *Reference Tables for Physical Setting/Earth Science.* Locate the equation for rate of change:

$$\text{Rate of change} = \frac{\text{change in value}}{\text{time}}$$

$$\text{Rate of revolution} = \frac{\text{change in Earth's position relative to the Sun during one revolution}}{\text{time it takes Earth to complete one revolution}}$$

Changes in Earth's position relative to the Sun are measured in degrees. In completing one revolution, Earth changes position by 360°. The time it takes Earth to complete one revolution is called its period of revolution. Find the Solar System Data chart in the *Reference Tables for Physical Setting/Earth Science* and locate the column labeled "Period of Revolution." Note that Earth's period of revolution is 365.26 days. Substitute these values in the rate of change equation to solve for Earth's rate of revolution:

$$\text{Rate of revolution} = \frac{360°}{365 \text{ days}}$$

This equation is best represented by the equation in choice 3.

31. **3** Cold fronts form along the leading edge of a cooler air mass advancing against a warmer air mass. A cool, dry air mass is denser than the warm, moist air ahead of it. So, the cool, dry air mass bulges into the warm, moist air, pushing against and under it like a wedge. The curved slope of the front is rather steep, especially if it is moving fast. Thus, a cold front slopes downward from the cool air mass toward the warm air mass in a steep-sided, curved bulge as shown in choice 3.

WRONG CHOICES EXPLAINED:

(1) and (2) Cold fronts form along the leading edge of a cooler air mass advancing against a warmer air mass. Both of these cross sections show the opposite—warm, moist air advancing against cool, dry air.

(4) In this cross section, cool, dry air is advancing against warm, moist air. However, the boundary shown slopes upward away from the cool, dry air, indicating that the denser, cool, dry air moves up and over the less dense, warm, moist air. This will not happen because denser materials tend to sink beneath less dense materials, not rise above them.

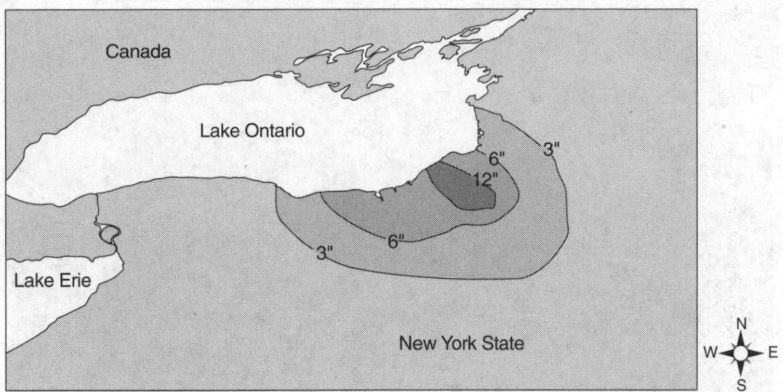

32. **1** Winter lake-effect snow storms occur when cold winds move across large stretches of warm lake water. Evaporation of the warm lake water adds water vapor to the air. The water vapor is picked up by the cold wind, freezes, and is deposited as snow on the cold land of the leeward shores of the lake. Find the Generalized Bedrock Geology of New York State map in the *Reference Tables for Physical Setting/Earth Science.* Locate Lake Ontario and note that it is located at about 43–44° N latitude. Find the Planetary Winds and Moisture Belts in the Troposphere diagram in the *Reference Tables for Physical Setting/Earth Science.* Note that, at the latitudes corresponding to Lake Ontario, the planetary winds generally blow from west to east. According to the map, the area of snowfall from the lake-effect storm is on the southeast shoreline of the lake. Note, too, that the regions encompassed by the isolines are elongated along an axis running northwest to southeast. Thus, the cold winds are moving from northwest to southeast over the lake, and the southeastern shores of Lake Ontario are the leeward shores that would be affected by lake-effect snow from this storm. Therefore, the surface winds that produced this storm came from the northwest.

33. **3** According to the photograph, Mount Rainier is a roughly cone-shaped volcano. Water flows from higher elevations to lower elevations. A cone-shaped mountain is highest at the peak and is lower in all directions outward from the peak. Thus, precipitation running off the surface of a hill will flow outward in all directions from the central, higher areas, forming a radial stream pattern as shown in diagram (3).

WRONG CHOICES EXPLAINED:

(1) The rectangular stream drainage pattern shown in map (1) forms in regions where there are faults or joints that break the surface bedrock into rectangular blocks. The main streams and their tributaries in such regions follow these faults and joints, displaying right-angle bends at the corners of the rectangular blocks and perpendicular sections of approximately the same length as they flow along their edges. The photograph in the question shows a cone-shaped volcanic mountain, not a landscape in which bedrock has broken into rectangular blocks by faulting.

(2) The trellis stream pattern in map (2) forms in tilted or folded rock layers of unequal resistance to erosion. Major streams run through long, parallel valleys following belts of weak rock between parallel ridges of stronger rock. Tributary streams enter major streams at right angles. The photograph in the question shows a cone-shaped volcanic mountain, not a series of long, parallel ridges.

(4) On a flat plain, the slope is gentle and streams will flow downhill slowly and typically form a dendritic stream drainage pattern. In a dendritic stream drainage pattern, there are numerous small streams that join together to form

larger streams in a tree-like pattern as shown in map (4). The photograph in the question shows a cone-shaped volcanic mountain with steep slopes, not a gently sloped, flat plain.

34. **4** Find the Inferred Properties of Earth's Interior chart in the *Reference Tables for Physical Setting/Earth Science.* Locate the "Density (g/cm^3)" scale along the right side of the cross section. Note the following densities: granitic continental crust—2.7 and basaltic oceanic crust—3.0.

Find the Scheme for Igneous Rock Identification in the *Reference Tables for Physical Setting/Earth Science*. Locate granite and basalt in the upper portion of the chart. Trace downward to the center section of the scheme labeled "Characteristics." Note that granite is lower in density and basalt is higher in density. A lower-density substance occupies more volume per given mass than a higher-density substance. Thus, compared to a given mass of oceanic crust, an equivalent mass of continental crust will occupy more volume. Therefore, continental crust tends to be thicker than an equal area of oceanic crust of the same mass. Thus, the characteristics of continental crust and oceanic crust are best represented by the table in choice (4.)

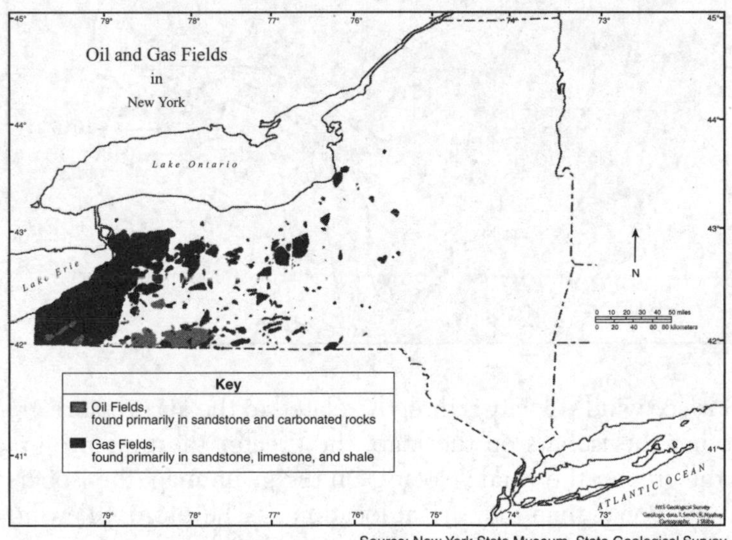

Source: New York State Museum, State Geological Survey

35. **4** According to the map key, the oil fields are found primarily in sandstone and carbonated [sic] rocks (rocks composed of carbonate minerals such as calcite and dolomite), and gas fields are found primarily in sandstone, limestone, and shale. Find the Scheme for Sedimentary Rock Identification in the *Reference*

Tables for Physical Setting/Earth Science. Locate the column labeled "Rock Name," and note that sandstone, limestone, dolostone, and shale are all sedimentary rocks. Carbonate rocks are composed of carbonate minerals. Find the Properties of Common Minerals table in the *Reference Tables for Physical Setting/ Earth Science.* In the column labeled "Composition," locate chemical formulas that indicate that the mineral is a carbonate, that is, formulas ending in CO_3. Note the two formulas $CaCO_3$ and $CaMg(CO_3)_2$. Trace left to the column labeled "Mineral Name," and note that these formulas correspond to the minerals Calcite and Dolomite, respectively. Thus, the sedimentary rocks limestone and dolostone are carbonate rocks. Thus, the type of bedrock that contains these oil and gas deposits is sedimentary rock.

PART B–1

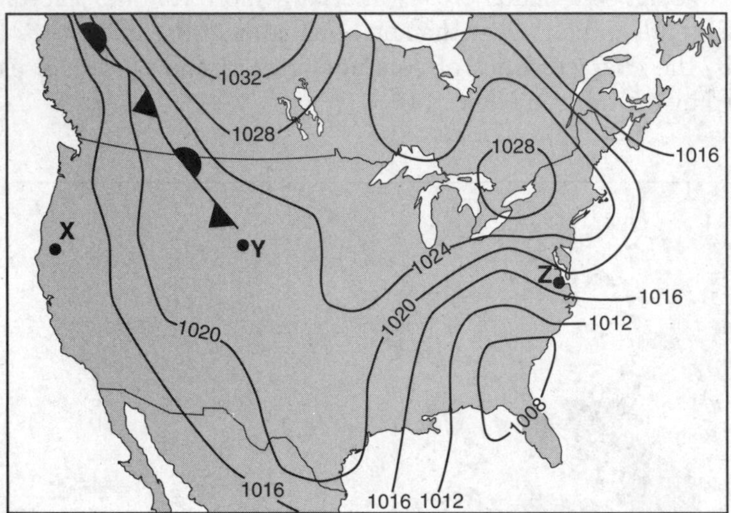

36. **2** Surface wind velocity is directly related to the air pressure gradient. The closer together the isobars on the map, the greater the pressure gradient and, therefore, the greater the wind velocity. On the given map, the isobars are closer together at location Z than they are at location *X*. Therefore, the wind velocity is greater at location Z than at location *X* because the isobars are closer together at Z.

37. **4** Find the Weather Map Symbols chart in the *Reference Tables for Physical Setting/Earth Science* and locate the key for Front Symbols. Note that the symbol shown in the weather map on the front that extends northwest from location *Y* corresponds to a stationary front.

38. **1** Isobars are lines connecting points of equal air pressure. Isobars are labeled with the air pressure in millibars of all points on that line. Low-pressure centers are surrounded by air with higher pressures. The lowest air pressure shown on the map is the circular isobar labeled "1008" just north of Florida. Note that this region is surrounded by higher-pressure air. Thus, the region within the 1008-mb isobar should be labeled as a low-pressure center (**L**.) High-pressure centers are surrounded by air with lower pressures. The highest air pressure shown on the map is the semi-circular isobar labeled "1032" in northwestern Canada. Note that this region is surrounded by lower-pressure air. Thus, the region within the 1032-mb isobar should be labeled as a high-pressure center (**H**). There is another region of high pressure surrounded by lower pressures just east of the Great Lakes centered on the circular 1028-mb isobar. Thus, the region within this circular 1028-mb isobar should also be labeled as a high-pressure center (**H**). Therefore, the map that best shows the locations for the centers of high pressure and low pressure is map (1).

39. **2** Lunar eclipses are predictable because the Moon's revolution around Earth and Earth's revolution around the Sun are cyclic; that is, they occur in a constantly repeating pattern with reference to time and space. The pattern of repetition of these cyclic motions allows predictions to be made about future positions of the Sun, Earth, and Moon relative to one another. Thus, scientists can predict the occurrence of lunar eclipses. Furthermore, the reading passage states that "The next one will not happen until 2033," indicating that supermoon eclipses can be predicted. Thus, supermoon total eclipses are celestial events that are predictable.

WRONG CHOICES EXPLAINED:

(1) By definition, random occurrences are unpredictable. Supermoon total eclipses are the result of the predictable cyclic motions of Earth and its Moon in relation to one another and the Sun. Therefore, supermoon total lunar eclipses are predictable, not random. Furthermore, the reading passage states that the next supermoon total lunar eclipse will occur in 2033. The year 2033 is in the future; therefore, this statement is a prediction—a prediction that could not be made if supermoon total lunar eclipses were random occurrences.

(3) The occurrence of supermoon total lunar eclipses is the result of the Moon's cyclic revolution around Earth and Earth's cyclic revolution around the Sun. As long as these cyclic motions continue, supermoon total eclipses will continue to occur. There is no evidence that the motions of Earth and the Moon will cease after 2033. Therefore, there is no evidence that supermoon total eclipses will never happen again after 2033.

(4) The reading passage states that "supermoon events are rare, but a total lunar eclipse during a supermoon is even rarer." A full Moon is not rare; it occurs every 29.5 days. Therefore, supermoon total lunar eclipses do not happen every full Moon.

40. **3** The reading passage states that "A supermoon occurs when the entire lighted half of the Moon faces Earth (full Moon phase) and the Moon is at its closes point to Earth in its orbit." According to the diagram, when the Moon is in position *C*, its entire lighted half faces Earth and a full Moon phase occurs. Thus, during the 2015 supermoon total eclipse, the Moon was located at position *C* in its orbit.

41. **3** Lunar phases are cyclic and predictable. Find the Solar System Data chart in the *Reference Tables for Physical Setting/Earth Science.* In the column labeled "Celestial Object," locate Earth's Moon. Trace right to the column labeled "Period of Revolution." Note that the Moon takes 27.3 days to orbit Earth once. Although the Moon makes one revolution in 27.32 days, the Moon takes 29.5 days to go through a complete cycle of phases from the full Moon position represented in the diagram to the full Moon the following month. Why the extra two days? Let's start at full Moon, when the Sun, Earth, and Moon align when the Moon is on the side of Earth directly opposite the Sun. At the same time that the Moon revolves around Earth, Earth is revolving around the Sun at a rate of about 1° per day. Thus, by the time the Moon has completed one revolution, Earth's position relative to the Sun has changed. In 27.3 days, Earth has moved about 27.3° in its orbit. The Moon moves at about 13° per day. At that rate, it takes the Moon a bit more than two more days to catch up to still moving Earth and align again with Earth and the Sun in a full Moon phase. Thus, the Moon completes a cycle of phases in 29.5 days, not 27.3 days. Therefore, the time it took for the Moon to go from this supermoon to the next full Moon phase was 29.5 days.

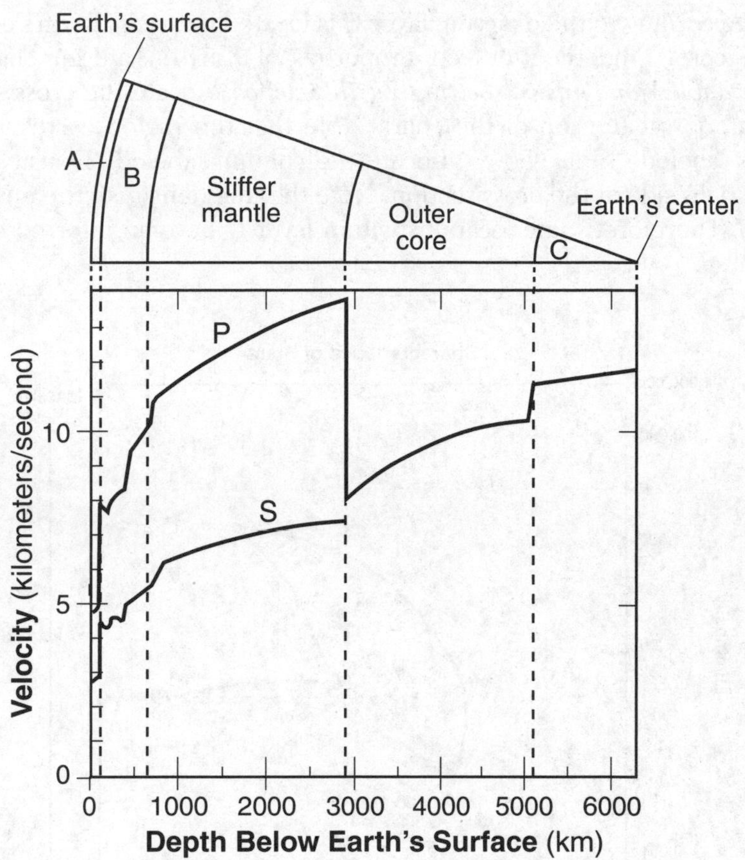

42. **2** Find the Inferred Properties of Earth's Interior chart in the *Reference Tables for Physical Setting/Earth Science*. Locate the cross section of Earth's interior at the top of the chart. Note that the region corresponding to layer *B* is labeled "Asthenosphere." Note, too, that the region corresponding to layer *A* is labeled "Lithosphere." Thus, the layers of Earth's interior represented by letters *A* and *B* are as follows: *A* is the lithosphere and *B* is the asthenosphere.

43. **3** On the horizontal axis of the graph labeled "Depth Below Earth's Surface (km)," locate 1000 kilometers. From this point, trace vertically upward until you intersect the bold curve labeled "P." At this intersection, trace horizontally left to the vertical axis labeled "Velocity (kilometers/second)," and note the value at this point—about 11.3. Thus, the approximate velocity in kilometers per second of the *P*-waves at a depth of 1000 kilometers is 11.3 km/s.

44. **4** According to the diagram, layer *C* is located between Earth's center and the outer core. Find the Inferred Properties of Earth's Interior chart in the *Reference Tables for Physical Setting/Earth Science*. Locate the cross section of Earth's interior at the top of the chart. Note that the region corresponding to layer *C* is labeled "Inner Core." Locate the column labeled "Density (g/cm^3)" along the right side of the cross section. Note that the density of the inner core is 12.7–13.0. Therefore, some locations within layer *C* have an inferred density of 12.9 g/cm^3.

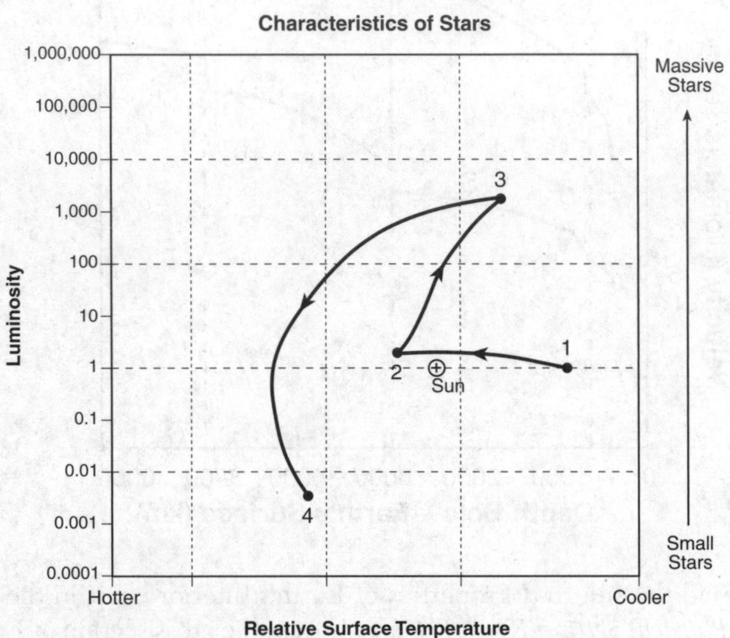

45. **3** Find the Characteristics of Stars chart in the *Reference Tables for Physical Setting/Earth Science.* On this chart, locate the position corresponding to the star at point 2 in the question. From this position, trace downward until you intersect the horizontal axis labeled "Surface Temperature (K)" and read the value 7000. Thus, a possible surface temperature of the star in the question at point 2 is 7000 K.

46. **3** In the cores of stars, gravity is strong enough to overcome the force of repulsion between atomic nuclei, allowing the nuclei to combine in a process called nuclear fusion. By definition, nuclear fusion involves the combining of several atoms of a lighter element to form a single atom of a heavier element. The single heavier atom typically has less mass than the lighter atoms from which it formed. The mass "missing" from the heavier atom is not lost. Instead, it is converted into energy according to Einstein's formula $E = mc^2$. This formula states that if mass is converted into energy, the amount of energy released, E, is equal to the mass, m, times the speed of light, c, squared. The speed of light squared is a very large number. Therefore, the conversion of even a small amount of mass into energy during nuclear fusion results in the release of a very large amount of energy. Thus, between points 1 and 3, energy is released by the process of nuclear fusion when lighter elements combine to form heavier elements.

47. **1** Find the Characteristics of Stars chart in the *Reference Tables for Physical Setting/Earth Science.* Locate the position corresponding to the star at point 3 in the chart in the question. Note that point 3 on the chart in the question corresponds to a region labeled "Giants" on the Characteristics of Stars chart. Now, locate the position on the Characteristics of Stars chart corresponding to point 4 on the chart in the question. Note that point 4 on the chart in the question corresponds to a region labeled "White Dwarfs" on the Characteristics of Stars chart. Thus, at point 3 in the life cycle, the star is a Giant, and at point 4 in its life cycle, it is a White Dwarf, as shown in table (1).

48. **3** Note that, in the photograph, the area containing A and B cuts across a section of the stream where the stream channel curves in a meander. When a stream curves, or meanders, water velocity is higher along the outside of the curve and lower along the inside of the curve because the water is forced to cover

a greater distance along the outside of the curve. Since stream erosion is greatest where the velocity of the water is highest, the outside of the curve of a meandering channel experiences more erosion than the inside of the curve. In the diagram, point *B* is located at the outside of a curve in the stream, where the velocity of the water is highest and erosion is the greatest. Therefore, the streambank at location *B* is steeper than the streambank at location *A* because the water near location *B* is moving faster than the water near location *A*, causing more erosion.

49. **4** Location *X* is a flat area adjacent to a stream. When a stream is floods, it carries much more sediment than normal because of its increased speed and volume and often overflows its banks, and the sediment-laden water spreads out over the surrounding land. Outside of the channel, the water is shallower and slower moving, so the flood water begins to deposit its sediment load. Over time, these deposits form a broad, flat area adjacent to the stream called a floodplain. Therefore, the area labeled letter *X* represents a portion of a floodplain.

50. **2** Find the Relationship of Transported Particle Size to Water Velocity graph in the *Reference Tables for Physical Setting/Earth Science.* Note the dashed lines labeled with the size ranges corresponding to clay, silt, sand, pebbles, cobbles, and boulders. It is given that the stream is initially flowing at 50 cm/sec and is slowing to 5 cm/sec. Find the 50 cm/sec value on the axis labeled "Stream Velocity," trace upward to the bold curve, and then trace right to the axis labeled with particle sizes. Note that, at 50 cm/sec, a stream can transport particles as large as pebbles. Note that, at this point, the pebbles are near the lower end of the pebble size range. Find the 5 cm/sec value on the axis labeled "Stream Velocity," trace upward to the bold curve, and then trace right to the axis labeled with particle sizes. Note that, at 5 cm/sec, a stream can transport particles as large as sand. Also note that, at this point, the sand particles are near the upper end of the sand size range. Therefore, most of the particles deposited where the stream velocity decreases from 50 centimeters per second to 5 centimeters per second are small pebbles and large sand.

PART B–2

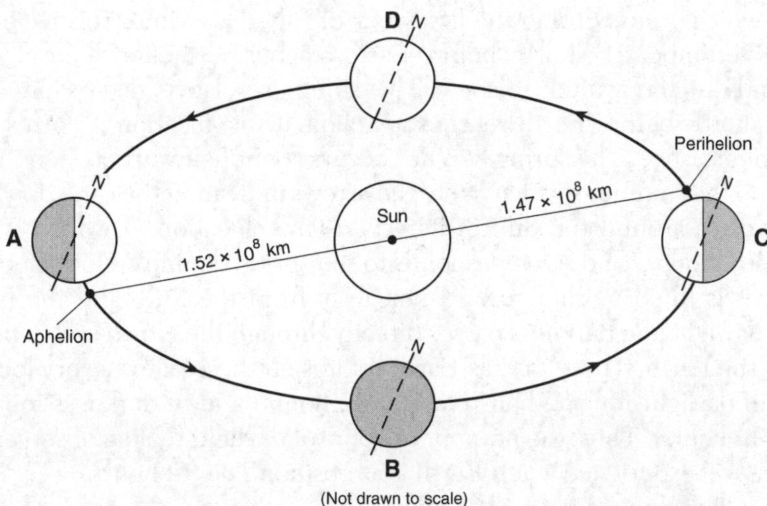

(Not drawn to scale)

51. Earth's spin axis is tilted 23.5° from a line perpendicular to the plane of its orbit. As Earth revolves, its spin axis always points in the same direction in space; that is, the direction of its axis at any given time is parallel to its direction at any other time. Thus, the number of degrees that Earth's axis is tilted from a line perpendicular to the plane of its orbit at each lettered location is 23.5°.

One credit is allowed for any value from **23.4° to 23.5°** or its mathematical equivalent.

Note: Credit is not allowed for a number with a direction, such as 23.5° N, because N indicates a latitude, not an angle.

52. According to the diagram, when Earth is at perihelion, it is closest to location *C*, which represents the first day of one of the four seasons. Note that, at location *C*, Earth's Northern Hemisphere is tilted away from the Sun. Therefore, the Sun's rays will strike Earth's surface less directly in the Northern Hemisphere and will be less intense, resulting in the lower temperatures associated with winter. The shading in the diagram shows that locations surrounding the North Pole are in darkness and will experience 24 hours of darkness as Earth rotates. Thus, location *C* corresponds to the winter solstice, December 21. Thus, when Earth is at perihelion, the season in New York State is winter.

One credit is allowed for **winter**.

53. Earth is closest to the Sun at perihelion (1.47×10^8 km), which occurs during the first week of January. Earth is farthest from the Sun at aphelion (1.52×10^8 km), which occurs during the first week of July. The winter solstice occurs on or about December 21, shortly before Earth reaches perihelion. Thus, location *C* corresponds to the winter solstice. The summer solstice occurs on or about June 21, shortly before Earth reaches aphelion. Thus, location *A* corresponds to the summer solstice. The spring equinox occurs about halfway between the winter solstice and the summer solstice. Note the arrows indicating the direction in which Earth revolves around the Sun in its orbit. Thus, location *D*, which is halfway between locations *C* and *A*, corresponds to the spring equinox. On the equinoxes, Earth's axis is tilted neither toward nor away from the Sun, and the boundary between daylight and darkness passes directly through the North and South Poles, bisecting the Earth. Therefore, as Earth rotates on these dates, every location on Earth is in daylight for one-half rotation (12 hours) and in darkness for one-half rotation (12 hours). Thus, the number of hours of daylight that an observer in New York State will experience when Earth is at position *D* is 12 hours.

One credit is allowed for **12 h.**

54. When you observe the sky, you find that every celestial object (such as stars) changes position over time or is in motion. If you keep track of this motion, you find that, with very few exceptions, all celestial objects appear to move across the sky from east to west along a path that is an arc, or part of a circle. If you measure the rate at which celestial objects are moving, they appear to follow a circular path at a constant rate of 15° per hour, or one complete circle every day (24 h/day × 15°/h = 360°/day). This motion is called "apparent daily motion." In the Northern Hemisphere, all the circles formed by completing the arcs along which celestial objects move are centered very near the star *Polaris*. This is because *Polaris* is located directly above the North Pole and is roughly aligned with Earth's axis of rotation. Because of its position, *Polaris* is often referred to as the North Star. Thus, the name of the star that is aligned with Earth's axis above the North Pole is *Polaris*.

One credit is allowed for ***Polaris*** *or* **North Star.**

55. Find the Geologic History of New York State chart in the *Reference Tables for Physical Setting/Earth Science*. In the column labeled "Era," note that the geologic eras are the Cenozoic, Mesozoic, and Paleozoic. Now, in the column labeled "Period," locate Cambrian. Thus, the Cambrian is not a geologic Era; it is a geologic Period. Therefore, the unit of geologic time that should be substituted for the word "Era" is "Period."

One credit is allowed for **Period.**

56. The passage states that "Then, beginning sixty-five million years ago, an intense earth uplifting occurred that raised the Andes Mountains to their present height." Find the Geologic History of New York State chart in the *Reference Tables for Physical Setting/Earth Science*. On the time scale labeled "Million years ago," locate 65. Trace right to the column labeled "Life on Earth," and note the statement "Mass extinction of dinosaurs, ammonoids, and many land plants." Dinosaurs were animals. In the column labeled "Time Distribution of Fossils," locate the gray bar labeled "ammonoids," and note the circled letter *G*. On the key at the bottom of the chart, locate the drawing of index fossil "*G—Manticoceras*." Note that this drawing shows a shell similar in shape to modern nautiloids and snails, so it is reasonable to infer that ammonoids were animals, not plants.

One credit is allowed for an acceptable answer. Acceptable answers include, but are not limited to:

- **dinosaurs**
- **ammonoids**

Note: Credit is *not* allowed for land plants alone because the question asks for animals.

57. Find the Scheme for Sedimentary Rock Identification table in the *Reference Tables for Physical Setting/Earth Science.* In the column labeled "Rock Name," locate limestone. Trace left to the column labeled "Composition," and note that limestone is composed of calcite. The passage also states that the limestone formed from an "enormous mass of lime and sand that compressed into a limestone and dolomite layer . . . ." According to the Chart of Definitions, lime is "a general term for material containing the element calcium that combined with oxygen." Therefore, the limestone may be mixed with dolomite.

Find the Properties of Common Minerals table in the *Reference Tables for Physical Setting/Earth Science.* In the column labeled "Mineral Name," locate calcite and dolomite. Trace left to the column labeled "Composition," and note that the chemical composition of calcite and dolomite are $CaCO_3$ and the $CaMg(CO_3)_2$, respectively.

One credit is allowed for an acceptable answer. Acceptable answers include, but are not limited to:

- **$CaCO_3$**
- **Ca, C, and O (All three elements must be present for credit.)**
- **$CaMg(CO_3)_2$**

58. It is given in the passage that "As rain ran down from the mountains it formed a great underground water table that slowly began dissolving the limestone." As explained earlier, this limestone may also contain the mineral dolomite. Find the Scheme for Sedimentary Rock Identification in the *Reference Tables for Physical Setting/Earth Science.* In the column labeled "Rock Name," locate limestone. From limestone, trace left to the column labeled "Composition." Note that limestone is composed of calcite. Find the Properties of Common Minerals chart in the *Reference Tables for Physical Setting/Earth Science*. In the column labeled "Mineral Name," locate both calcite and dolomite. From each of these minerals, trace left to the column labeled "Distinguishing Characteristics," and note the statement "bubbles with acid." This indicates that both calcite and dolomite chemically react with acid. Acid can completely dissolve the minerals calcite and dolomite. Carbon dioxide in the atmosphere reacts with water to form carbonic acid. Carbonic acid changes calcite and dolomite into new substances that are soluble in water; that is, they dissolve. Therefore, carbonic acid in rainwater and groundwater that seeps through limestone and dolomite bedrock may almost completely dissolve away the bedrock, leaving large holes or caverns. If the roof of the cavern collapses, a depression called a sinkhole is formed in the ground above it. Thus, the chemical weathering that contributes to the formation of sinkholes is best described as acid in water chemically reacting with the calcite and/or dolomite in the rocks. (**Note:** Calcite and dolomite are also called carbonate minerals because they contain "CO_3," which is called the carbonate anion.)

One credit is allowed for an acceptable answer. Acceptable answers include, but are not limited to:

- **The acid in water chemically reacts with the carbonates in the rocks.**
- **Limestone is chemically altered and changed into new materials.**
- **Slightly acidic groundwater chemically breaks down calcite and/or dolomite.**
- **Water flowing underground dissolves the limestone.**

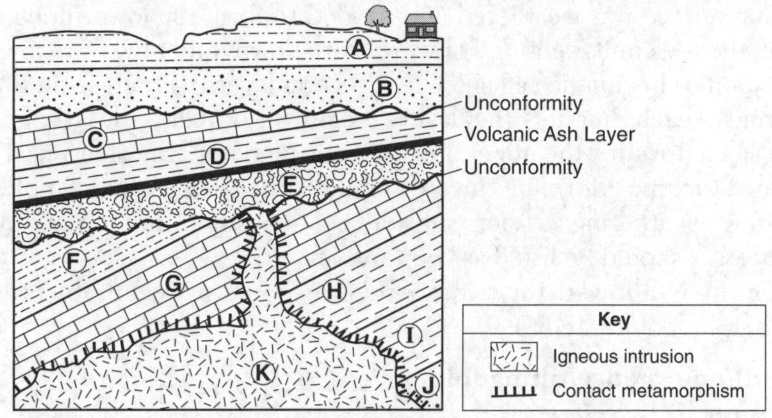

59. According to the key, rock unit *K* is an igneous intrusion. Note that rock unit *K* cuts across rock unit *H* and that the boundary between rock units *K* and *H* is marked with a symbol representing contact metamorphism. An igneous intrusion is younger than any rock layer or structure that it cuts across. In order to be contact metamorphosed, a rock layer must have already existed when the igneous intrusion occurred. Therefore, rock unit *K* is younger than the rock unit *H*. However, rock unit *E* has not undergone contact metamorphism along its boundary with rock unit *K*. Therefore, rock unit E did not yet exist when the igneous intrusion of rock unit *K* occurred. Therefore, rock unit *E* is younger than rock unit *K*. Thus, the correct order of rock unit formation, from oldest to youngest, is *H*, *K*, *E*.

One credit is allowed for the sequence shown below:

Letters: H, K, E

Oldest ⟶ Youngest

60. Layers of rock are generally deposited in an unbroken sequence. However, if forces within Earth cause rocks to be uplifted, deposition ceases. Weathering and erosion may wear away layers of rock before the land surface is low enough for another layer to be deposited. The uplift may cause rock layers to tilt or fold. The result is an unconformity, a break or gap in the sequence of a series of rock layers. Thus, the rocks above an unconformity are quite a bit younger than those below it. Note that the layers beneath the upper unconformity are tilted relative to the unconformity. Note that the layers beneath the lower unconformity are also tilted relative to the unconformity. Thus, it is reasonable to infer that the lowest rock units were uplifted until they emerged from their depositional environment.

Then, their surface was weathered and eroded to form the lower unconformity. After that, the rock units subsided or were worn down until they were low enough for deposition to begin once again. Then, several layers were deposited atop the unconformity, each burying the layers below it. Finally, the whole process occurred again, forming the upper unconformity and uppermost layers. Thus, the formation of the unconformities involved many processes, including uplift, emergence, tilting, weathering, erosion, subsidence, deposition, and burial. Any two of these processes should be listed as your answer.

One credit is allowed for two correct responses. Acceptable responses include, but are not limited to:

- **uplift/emergence/tilting/folding**
- **erosion**
- **submergence/subsidence**
- **weathering**
- **deposition**
- **burial**

61. Contact metamorphism describes changes in rock that result from the extreme heat produced by contact with magma or lava. When magma and lava cool, they form igneous rock. Thus, a zone of contact metamorphism would most likely be found where one of two adjoining bedrock types is igneous. The rock that will undergo contact metamorphism is the one that came in contact with the igneous rock when it was still molten magma or lava. According to the cross section and key, rock unit *K* is igneous rock, and rock unit *G* has undergone contact metamorphism. Find the Scheme for Sedimentary Rock Identification in the *Reference Tables for Physical Setting/Earth Science*. In the column labeled "Map Symbol," locate the symbol corresponding to rock unit *G*. Note that rock unit *G* corresponds to limestone. Now, find the Scheme for Metamorphic Rock Identification in the *Reference Tables for Physical Setting/Earth Science*. In the column labeled "Comments," note that metamorphism of limestone or dolostone forms marble, and metamorphism of various rocks by contact with magma or lava forms hornfels. Therefore, the metamorphic rock that most likely formed within rock unit *G* at the boundary of rock unit *K* is most likely marble or hornfels.

One credit is allowed for **marble** *or* **hornfels**.

Cuba Lake

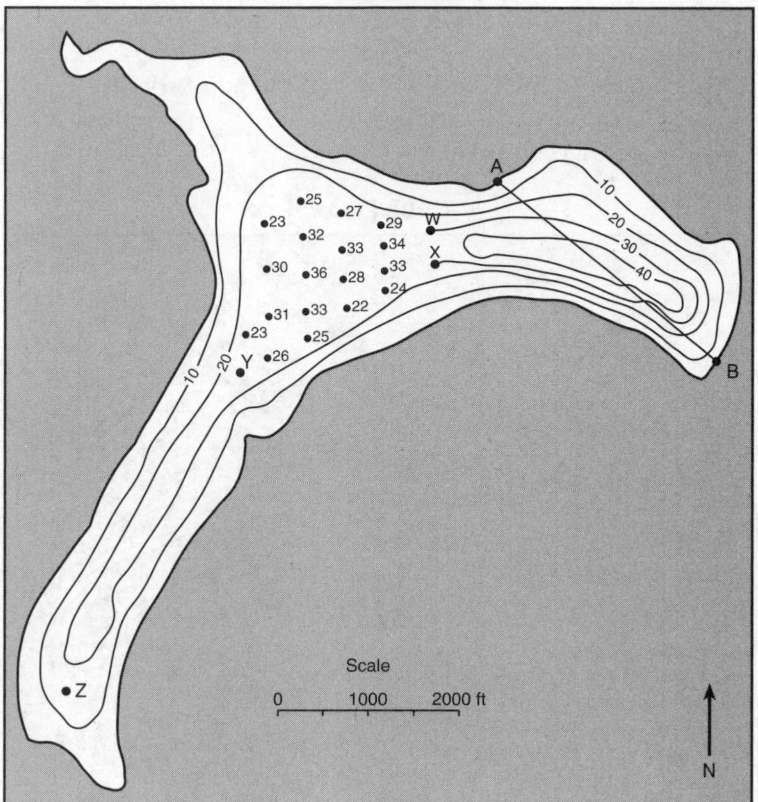

62. An isoline connects points of equal field value, and the field values shown on this map are water depth measured in feet. The 30-foot isoline connects all points that are at a depth of 30 feet. To draw the 30-foot isoline, do the following. Begin at point *W* and note that there are no points marked 30 near it. However, that does not mean that there are no other places on the map with that value. Note that there are values greater than 30 adjacent to values less than 30. For example, note the points labeled "34" and "29" just west of point *W*. It is logical to infer that somewhere between the locations at 34 feet deep and 29 feet deep is a location that is 30 feet deep. Therefore, the 30-foot isoline should pass between these two points. In this way, continue drawing the line west between the points labeled "27" and "33," and then between "25" and "32." Then, bend south between "32" and "23," and connect to the point labeled "30." From there, continue southward between "23" and "31," then turn east and draw the isoline between "35" and "25," then between "28" and "33," and finally between "24" and "33," extending the isoline to point *X*. A correctly drawn 30-foot isoline is shown below.

One credit is allowed if the 30-foot isoline is drawn correctly. If additional isolines are drawn, all must be correct to receive credit.

Note: The isoline must touch points *X*, *W*, and the point for 30.

Example of a 1-credit response:

Cuba Lake

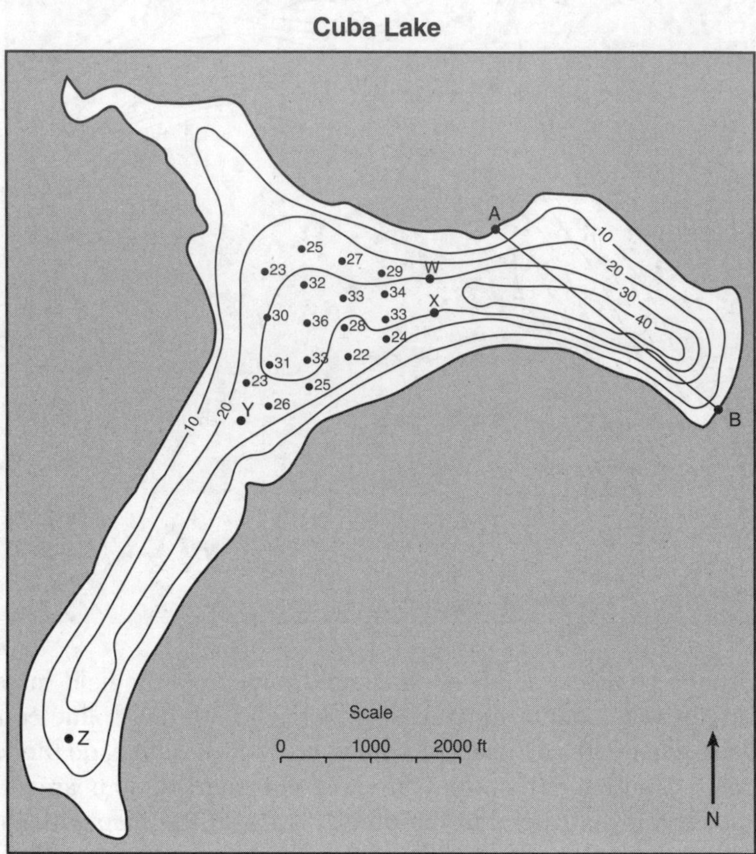

63. To construct a profile of the bottom of Cuba Lake along line *AB*, proceed as follows. Place the straight edge of a piece of scrap paper along the solid line connecting point *A* to point *B*, and mark the edge of the paper at points *A* and *B* and wherever the paper intersects a depth isoline. Wherever the paper intersects a depth isoline, label the mark with the depth of the isoline as shown below. Note that points *A* and *B* are on the shoreline of Cuba Lake. Therefore, the water depth at points *A* and *B* is “0.”

Cuba Lake

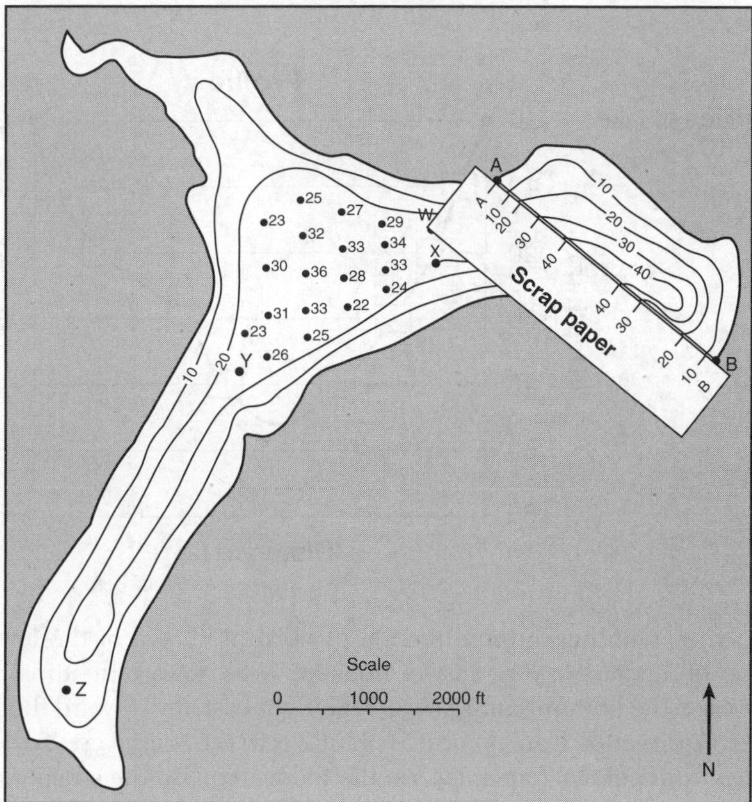

Then, place this paper along the *lower* edge of the grid provided in your answer booklet so that points *A* and *B* on the scrap paper align with points *A* and *B* on the grid. Note that the water depths at points *A* and *B* have already been plotted. At each of the other points where a depth isoline crossed the edge of the scrap paper, draw a plot on the grid at the appropriate depth. Finally, connect all of the plots in a smooth curve to form the finished profile as shown below.

One credit is allowed if the centers of *all eight* plots are within or touch the rectangles shown below and are correctly connected with a line from *A* to *B* that passes within or touches each rectangle. The low point of the line should extend below 40 feet but remain above 50 feet.

Note: Credit is allowed if the line misses a plot but is still within or touches the rectangle.

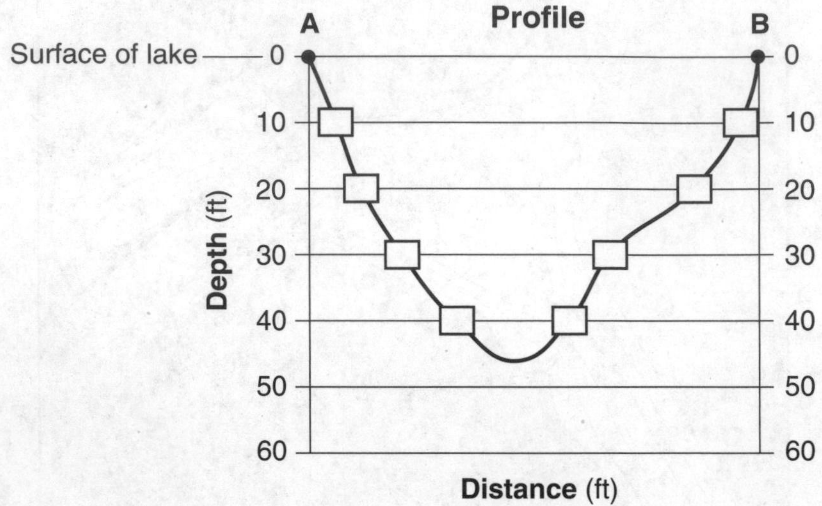

64. An arrow pointing in the direction of north is shown in the lower right-hand corner of the map in your answer booklet. Thus, toward the top of the map is north, toward the bottom south, toward the right east, and toward the left west. Therefore, the direction from point *Y* to point *Z* is to the southwest. To determine the distance from point *Y* to point *Z*, do the following. Use the straight edge of a piece of scrap paper to mark off the distance between points *Y* and *Z*. Compare the distance marked off on the scrap paper to the scale printed beneath the map to determine the distance between points *Y* and *Z*. The distance is about 4000 feet.

One credit is allowed if both responses for compass direction and distance are correct.

Compass Direction:

- **Southwest**
- **SW**
- **SSW**

Distance:

- **Any value from 3800 to 4200 feet**

65. It is given that the coordinates of Cuba Lake are 42°14¢ N, 78°18¢ W. Find the Generalized Bedrock Geology of New York State map in the *Reference Tables for Physical Setting/Earth Science.* Locate the approximate position of 42°14¢ N along the right side of the map, and the approximate position of 78°18¢ W along the bottom of the map. Trace horizontally left from 42°14¢ N and vertically upward from 78°18¢ W until the two intersect. Note that Cuba Lake is located between Jamestown and the Genesee River. Now, find the Generalized Landscape Regions of New York State map in the *Reference Tables for Physical Setting/Earth Science*. Locate the point corresponding to the location of Cuba Lake, and note that it lies within the landscape region labeled "Allegheny Plateau." Note that the Allegheny Plateau in New York State is part of a major geographic province that encompasses several states—the Appalachian Plateau (Uplands).

One credit is allowed for **Allegheny Plateau** *or* **Appalachian Plateau** *or* **Appalachian Uplands**.

PART C

Data for Dwarf Planet Pluto and its Five Moons

Object Name	Classification	Period of Revolution in Earth Days (d)	Eccentricity of Orbit	Diameter (km)
Pluto	Dwarf Planet	90,511.4 (247.8 years)	0.2488	2370
Charon	Moon	6.4	0.0022	1208
Styx	Moon	20.2	0.0058	10 to 15*
Nix	Moon	24.9	0.0020	40
Kerberos	Moon	32.2	0.0033	13 to 34*
Hydra	Moon	38.2	0.0059	33 to 43*

* There is a range in diameters for these moons due to their irregular shapes

66. Find the Solar System Data chart in the *Reference Tables for Physical Setting/Earth Science* and locate the columns labeled "Mean Distance from Sun" and "Period of Revolution." Note that, as distance from the Sun increases, the period of revolution increases. This is generally true of all primaries and their satellites, such as planets and their moons. Thus, the moon of Pluto that has the longest period of revolution is likely the greatest distance from Pluto. According to the data table, Hydra has the longest period of revolution. Therefore, the name of Pluto's moon that most likely has an orbit farthest from Pluto is Hydra, because Hydra has the longest period of revolution.

One credit is allowed for both the moon Hydra and a correct explanation. Acceptable responses include, but are not limited to:

- **Hydra has the longest period of revolution.**
- **As a moon's distance from Pluto increases, the time to make one revolution also increases.**
- **Hydra travels the greatest distance in its orbit because it has the longest period of revolution.**

67. An ellipse is an oval shape obtained by cutting a circular cone with a plane. An ellipse has a major axis and a minor axis, which are lines connecting the two points farthest apart and the two points closest together on the ellipse. It also contains two special points along the major axis, each called a focus (plural is foci), instead of one center. The distance from one focus to any point on the ellipse and back to the other focus is always the same. The closer together the two foci, the more circular the ellipse. The farther apart the foci, the flatter the ellipse. The flatness of an ellipse is called its *eccentricity*. Eccentricity is expressed as the ratio between the distance between the foci and the length of the major axis. A perfect circle would have an eccentricity of 0; a straight line has an eccentricity of 1. According to the data table, neither Pluto nor Pluto's moons have an eccentricity of 0 or 1. Therefore, Pluto and its moons have elliptical, eccentric orbits. Pluto's moons have very small eccentricities of orbit, indicating that their orbits, though elliptical, are nearly circular. Pluto, however, has a much more eccentric orbit—0.2488 compared to 0.0059 or less for its moons. Thus, Pluto's orbit is much more elliptical (less circular) than that of its moons.

One credit is allowed. Acceptable responses include, but are not limited to:

- **elliptical**
- **The orbits are eccentric.**
- **The orbits are nearly, but not perfectly, circular.**
- **oval**
- **Pluto's orbit is elliptical, and the moons of Pluto have a more nearly circular orbit.**

68. Our solar system includes the Sun and all of the objects that orbit it, such as planets, their moons, dwarf planets, asteroids, meteoroids, and numerous comets. All of these objects orbit the Sun because the gravitational attraction of the Sun influences their motion. It is given that Pluto revolves around our Sun and its five moons revolve around Pluto. As Pluto's moons orbit Pluto, they are carried along with Pluto as Pluto revolves around the Sun, so Pluto's moons also revolve around the Sun. Therefore, Pluto and its five moons are considered part of our solar system because they revolve around the Sun.

One credit is allowed. Acceptable responses include, but are not limited to:

- **Pluto and its five moons revolve around the Sun.**
- **All go around/orbit the Sun.**
- **The gravitational attraction of the Sun influences the motion of Pluto and its moons.**
- **Pluto revolves around the Sun, and its moons revolve around Pluto.**
- **Pluto orbits the Sun.**

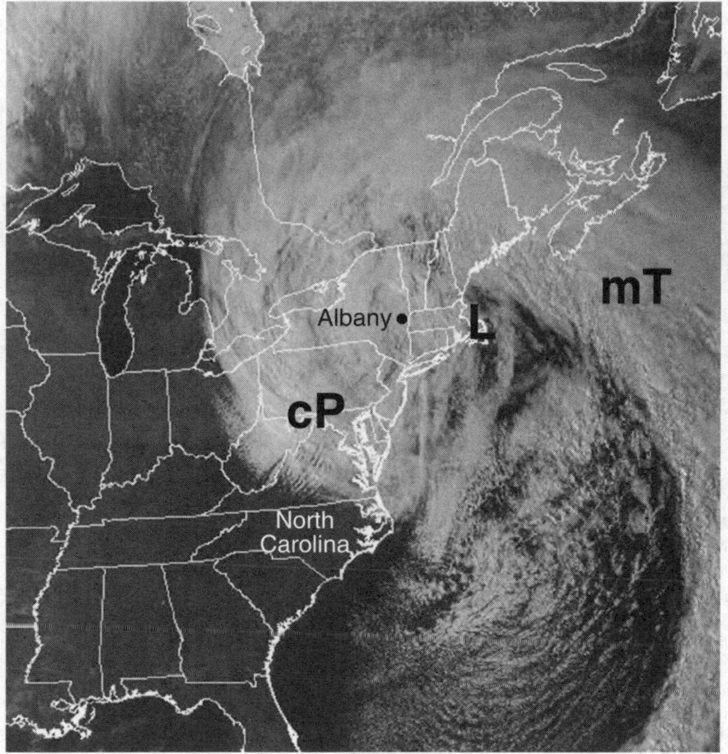

69. The air near the center of a low-pressure system exerts less pressure than the surrounding air because it is less dense than the surrounding air. Therefore, the denser surrounding air moves inward toward the center of the low-pressure system. As the air moves inward toward the low-pressure center, the Coriolis effect causes the moving air to be deflected to the right in the Northern Hemisphere. The combination of these two motions results in winds that blow inward in a counterclockwise spiral. Thus, two characteristics of the circulation pattern of the surface winds around the center of the low-pressure area represented on the map is that it is counterclockwise and inward toward the center.

One credit is allowed for an acceptable response. Acceptable responses include, but are not limited to:

- **counterclockwise and in toward the center**
- **counterclockwise and spiraling toward the center**
- **inward and counterclockwise**
- **counterclockwise and converging**

70. Find the Key to Weather Map Symbols in the *Reference Tables for Physical Setting/Earth Science* and locate the section labeled "Air Masses." Note that cP is the symbol for a continental polar air mass, and mT is the symbol for a maritime tropical air mass. Air masses take on the characteristics of the surface over which they form. Continental air masses form over land and are dry; maritime air masses form over water and are moist. Polar air masses form near the poles and are cold; tropical air masses form near the equator and are warm. Thus, continental polar air masses are cold and dry; maritime tropical air masses are warm and moist. Thus, compared to the cP air mass shown on the map, the mT air mass is warmer and more humid.

One credit is allowed for circling *both* **more humid** and **warmer**.

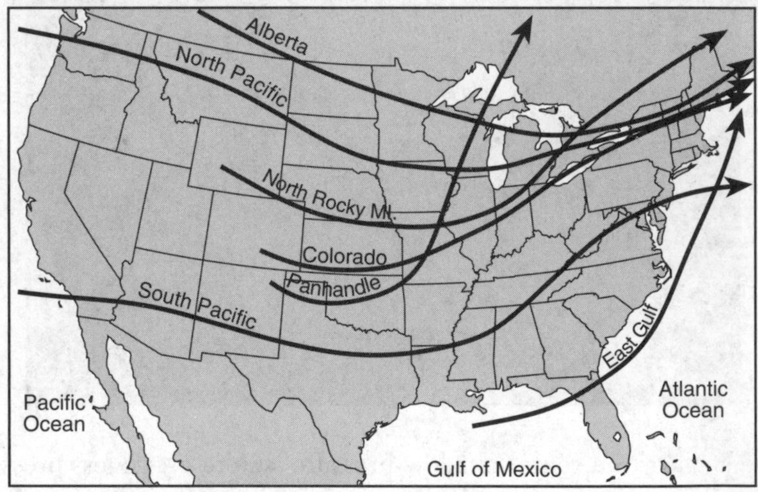

71. According to the passage, "A nor'easter is a large, low-pressure storm system that moves along the east coast of the United States." Note that, of the principal storm tracks shown on the map, the one that moves along the east coast of the United States is labeled "East Gulf."

One credit is allowed for **East Gulf**.

Weather Conditions

Dewpoint	22°F
Barometric Pressure	988.0 mb
Cloud Cover	100%
Present Weather	Snow

72. Find the Key to Weather Map Symbols in the *Reference Tables for Physical Setting/Earth Science*, and locate the section labeled "Station Model Explanation." Note the positions in which the values for temperature, dewpoint, barometric pressure, cloud cover, and present weather are placed on a station model. Note the format in which each of these weather factors is shown on the station model. For example, dewpoint and barometric pressure are shown as values without units, cloud cover is shown by the percent of the inner circle filled in, and present weather is shown as a symbol. Note that the value for barometric pressure is recorded as the last three digits with no decimal point. Therefore, on the station model in your answer booklet, the 22°F dewpoint should be recorded as "22" to the lower left. The 988.0 mb barometric pressure should be recorded as "880" to the upper right. The 100% cloud cover should be represented by completely filling in the central circle. Now, find the section labeled "Present Weather," and note that the symbol representing snow is an asterisk. Thus, the present weather of snow should be shown as an asterisk to the center left. A completed station model is shown on the diagram below.

One credit is allowed if all four weather conditions are correctly indicated, using the proper format.

A 1-credit response is shown below.

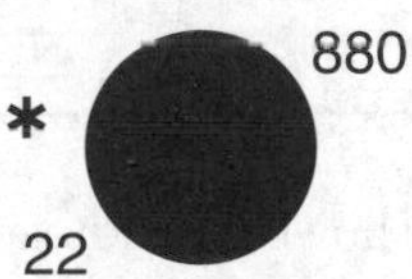

73. Find the Dewpoint Temperatures chart in the *Reference Tables for Physical Setting/Earth Science.* Note that, in order to determine the dewpoint using this chart, the dry-bulb temperature and wet-bulb temperature must be known. The instrument that records both the wet-bulb and dry-bulb temperature is called a psychrometer. A psychrometer consists of two thermometers, one that is dry and one that is kept moist by covering it with a wick moistened with distilled water. Evaporation of water from the wick lowers the temperature (at temperatures above freezing). The temperatures on these two thermometers coincide when the air is fully saturated with water vapor. The drier the air, the greater the difference in their temperatures.

The dewpoint is the temperature to which the air must be cooled to become fully saturated. The dewpoint is related to humidity; the higher the dewpoint, the greater the humidity. Thus, the dewpoint can be determined from the wet-bulb and dry-bulb temperatures using an appropriate chart.

Another instrument used to determine dewpoint is a hygrometer. Modern hygrometers measure the water vapor levels in air by how moisture changes characteristics such as electrical resistance or thermal conductivity. Some older hygrometers use a human hair under tension. Human hair retains moisture, changing length with humidity. Thus, one weather instrument that was most likely used to determine the dewpoint at Albany was a psychrometer *or* a hygrometer.

One credit is allowed for an acceptable response. Acceptable responses include, but are not limited to:

- **psychrometer/sling psychrometer**
- **dry-bulb thermometer/thermometer**
- **wet-bulb**
- **hygrometer**

Data Table

Number of Half-lives	Mass of Carbon-14 (g)
0	64
1	32
2	16
3	8
4	4
5	2
6	1

74. To construct a graph that shows the radioactive decay of carbon-14, use the data in the table to plot the points for the Mass of Carbon-14 (g) after each half-life as follows. For example, the table states that at 0 half-lives, 64 g of carbon-14 remains. To plot these coordinates, locate "0" on the horizontal scale labeled "Number of Half-Lives," trace upward along the "0" line until you intersect the line corresponding to "64" on the vertical scale labeled "Mass of Carbon-14 (g)," and mark a plot at the intersection of the two lines. Repeat this process for each of the remaining coordinates: (1, 32), (2, 16), (3, 8), (4, 4), (5, 2) and (6, 1). Then, connect the plots with a line as shown in the graph below.

One credit is allowed if the centers of *all seven* of the student's plots are within or touch the circles shown, and *all seven* plots are correctly connected with a line that passes within or touches each circle.

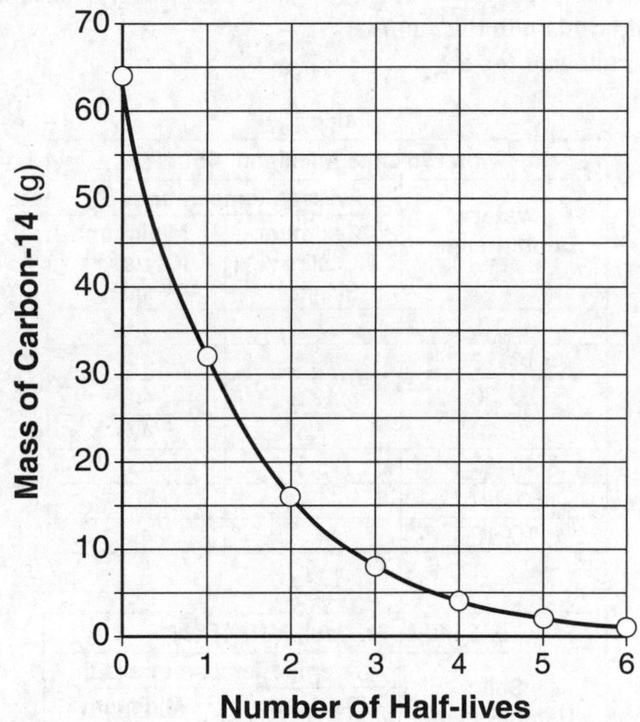

75. When a radioactive isotope decays, its atoms break apart (disintegrate) and form more stable atoms of a different element. Find the Radioactive Decay Data chart in the *Reference Tables for Physical Setting/Earth Science*. In the column labeled "Radioactive Isotope," locate Carbon-14. Trace right to the column labeled "Disintegration," and note that the product of the disintegration of carbon-14 (^{14}C) is nitrogen-14 (^{14}N.)

One credit is allowed for an acceptable response. Acceptable responses include, but are not limited to:

- **^{14}N**
- **nitrogen-14**
- **N-14**

Note: Credit is *not* allowed for nitrogen alone because nitrogen has more than one isotope, and the stable disintegration product is needed.

76. Find the Radioactive Decay Data chart in the *Reference Tables for Physical Setting/Earth Science*. In the column labeled "Radioactive Isotope," locate Carbon-14. Trace right to the column labeled "Half-Life (years)," and note that carbon-14 has a half-life of 5.7×10^3 (5,700) years. Thus, in 17,100 years, three half-lives would have occurred. According to the data table, after 3 half-lives, 8 grams of carbon-14 would remain in this sample.

One credit is allowed for **8 g**.

Table 1

Water in Lake Michigan, 45° N		
Water Depth Range (m)	**Average Temperature (°C)**	
	Maximum (Occurs in Summer)	**Minimum** (Occurs in Winter)
0–10	24.0	4.0
10–20	19.5	4.0
20–30	12.5	4.0
30–40	7.5	4.0
40–60	5.5	4.0
60–110	4.5	4.0
110–150	4.0	4.0

Table 2

Land in St. Paul, MN, 45° N		
Soil Depth Range (m)	**Average Temperature (°C)**	
	Maximum (Occurs in Summer)	**Minimum** (Occurs in Winter)
0–0.1	25.5	–6.0
0.1–1	23.0	–1.5
1–2	19.5	2.0
2–3	16.0	4.5
3–4	13.5	7.0
4–5	12.5	8.5
5–6	11.0	10.0

77. On Table 1, in the column labeled "Average Temperature (°C)," locate the sub-column labeled "Maximum (occurs in summer)." In this sub-column, locate "12.5." Trace left to the column labeled "Water Depth (m)," and note that the maximum temperature of 12.5°C occurs at a water depth of 20–30 m. On Table 2, in the column labeled "Average Temperature (°C)," locate the sub-column labeled

"Maximum (occurs in summer)." In this sub-column, locate "12.5." Trace left to the column labeled "Soil Depth Range (m)," and note that the maximum temperature of 12.5°C occurs at a soil depth of 4–5 m. Thus, the water depth range and soil depth range, in meters, that have the same average maximum temperature of 12.5°C are water depth range of 20–30 m and soil depth range of 4–5 m.

One credit is allowed for a **water depth range of 20–30 m** and a **soil depth range of 4–5 m.**

78. On Table 1, the average temperature in the summer is recorded in the sub-column labeled "Maximum (occurs in summer)." Note that in summer, as water depth range increases, the average water temperature decreases. On Table 2, the average temperature in the summer is also recorded in the sub-column labeled "Maximum (occurs in summer)." Note that in summer, as soil depth range increases, the average temperature also decreases. Thus, the general relationship between depth and average temperature for both water and soil in the summer is an inverse relationship; that is, as depth increases, temperature decreases.

One credit is allowed for an acceptable response. Acceptable responses include, but are not limited to:

- **As depth increases, temperature decreases.**
- **It gets colder, the greater the depth.**
- **inverse relationship/negative relationship**
- **The shallower the depth, the greater the average temperature for soil and water.**

79. According to the data tables, Lake Michigan and St. Paul, MN are at the same latitude. Therefore, both places receive roughly the same intensity of insolation. However, land surfaces have a different composition from water surfaces. Find the Specific Heats of Common Materials chart in the *Reference Tables for Physical Setting/Earth Science*. Note that liquid water has a specific heat of 4.18 Joules/gram • C°. Note, too, that basalt and granite (common materials from which land is derived) have specific heats of 0.84 and 0.79 Joules/gram • C°, respectively. This means that it takes more energy to change the temperature of water than it takes to change the temperature of land. For example, if 1 gram of water loses 4.18 Joules of heat, the temperature of the water will decrease by 1°C (−4.18 J/g ÷ 4.18 J/g • °C = −1°C). However, if 1 gram of basalt loses 4.18 Joules of heat, the temperature of the basalt will decrease by about 5°C (−4.18 J/g ÷ 0.84 J/g • °C = −4.97°C). Thus, when they lose the same amount of heat, a land surface will become cooler than a water surface. Therefore, the winter water surface is warmer than the winter land surface because they are composed of different substances, which have different specific heat values. Water has a higher specific heat than land and changes temperature more slowly. Therefore, water

will not cool off as quickly as land in the winter, and the winter water surface will be warmer than the winter land surface.

One credit is allowed for an acceptable response. Acceptable responses include, but are not limited to:

- **Water has a higher specific heat.**
- **Water changes temperature more slowly.**
- **Land has a lower specific heat.**
- **It takes more energy to change the temperature of water than land.**
- **Land cools off more quickly than water.**

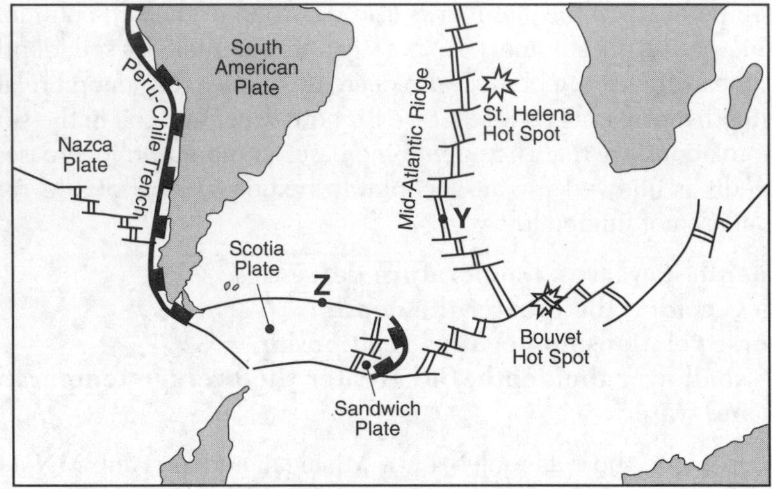

80. Note that location Z is along the northern boundary of the Scotia Plate. Find the Tectonic Plates map in the *Reference Tables for Physical Setting/Earth Science.* Locate the Scotia Plate, and note the symbol along its northern border—a solid black line with arrows pointing in different directions north and south of the boundary. According to the key at the bottom of the map, this represents a Transform plate boundary (transform fault). Transform faults are a special category of strike-slip faults, a type of fault in which the two sides of the fault slide past one another laterally instead of vertically.

One credit is allowed for an acceptable response. Acceptable responses include, but are not limited to:

- **transform boundary**
- **transform fault**
- **strike-slip fault**

81. Note that the tectonic plate to the south of the Bouvet Hot Spot extends across the entire lower edge of the map. Note, too, that the tectonic plate to the north of the Bouvet Hot Spot encompasses the southern tip of Africa. Find the Tectonic Plates map in the *Reference Tables for Physical Setting/Earth Science*. Locate the Bouvet Hot Spot near the lower right-hand corner. Note that the tectonic plate stretching across the bottom of the map that is south of the Bouvet Hot Spot is the Antarctic Plate. Note that the tectonic plate north of the Bouvet Hot Spot that encompasses the southern tip of Africa is the African Plate. Thus, the tectonic plates on each side of the Bouvet Hot Spot are the Antarctic Plate and the African Plate.

One credit is allowed for **African Plate and Antarctic Plate**.

82. Note that point *Y* is located directly on the Mid-Atlantic Ridge. Find the Inferred Properties of Earth's Interior chart in the *Reference Tables for Physical Setting/Earth Science*. In the cross section, locate the region labeled "Mid-Atlantic Ridge." Note that beneath the Mid-Atlantic Ridge the arrows showing motion in the asthenosphere indicate rising currents that diverge beneath the lithosphere.

Thus, the arrowheads should be drawn on each of the four bold lines to indicate rising currents and a divergent movement in the asthenosphere as shown in the example below.

One credit is allowed for *four* arrowheads/arrows that indicate rising currents and a divergent movement in the asthenosphere.

Example of a 1-credit response:

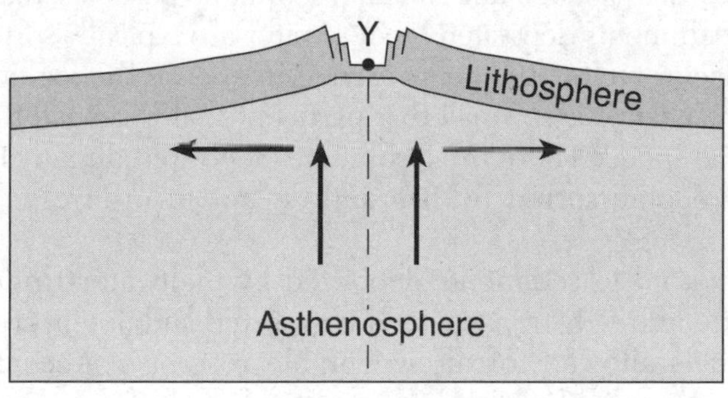

(Not drawn to scale)

83. According to the passage, "As the glacier thickens, more pressure is created and the glacier moves faster." Thus, the relationship between the thickness of a glacier and its rate of downhill motion is a direct relationship—as the thickness of a glacier increases, the rate of movement also increases.

One credit is allowed for an acceptable response. Acceptable responses include, but are not limited to:

- **As the thickness of a glacier increases, the rate of movement increases.**
- **Thicker glaciers move faster.**
- **direct relationship**

84. Glaciers move down the slopes of mountains from higher elevations to lower elevations along the easiest paths. Those paths are usually stream valleys that were eroded before the glacier formed. When a glacier moves down a stream valley, the glacier alters the valley's shape. The ice does not meander back and forth as readily as water. So, the ice often breaks through obstacles, forming a straighter path. The ice also cuts back the walls of the valley, making them steeper. The result is a wider, straighter, steeper-sided, U-shaped valley. Thus, the most likely shape of the cross section of a valley formed as a result of erosion by a mountain glacier is U-shaped.

One credit is allowed for an acceptable response. Acceptable responses include, but are not limited to:

- **U-shaped**
- **wide valley with steep sides**

85. When a glacier melts, numerous streams of meltwater spread out over the land and deposit glacial sediments in a wide sheet. Sediments deposited by running water are typically sorted. Horizontal sorting occurs when a stream gradually slows and deposits smaller and smaller sediment particles that settle through water. Thus, sediments deposited by meltwater are typically sorted and layered.

When a glacier melts, sediments carried in and on the ice fall directly to the ground. Air resistance is so small that particles of all sizes fall through air at virtually the same speed. Therefore, sediments deposited directly by a melting glacier do not become sorted by size and do not form layers due to different settling rates.

Thus, compared to sediments deposited by meltwater from a glacier, sediments deposited directly by a glacier are unsorted and not layered.

One credit is allowed for an acceptable response. Acceptable responses include, but are not limited to:

- **unsorted**
- **mixed piles**
- **not layered**

Topic	Question Numbers (Total)	Wrong Answers (x)	Grade
Standards 1, 2, 6, and 7: Skills and Application			
Skills Standard 1 Analysis, Inquiry, and Design	2, 3, 5, 7, 8, 10, 11, 15, 17–21, 23–26, 28, 30, 33, 34, 35, 37, 39, 42–47, 50, 55-63, 65, 66, 68, 72, 74–85		$\frac{100(56-x)}{56} = \%$
Standard 2 Information Systems			
Standard 6 Interconnectedness, Common Themes	1, 4, 6, 8, 13, 14, 21, 24, 27–29, 31–38, 40–43, 45, 47–49, 51–54, 59–65, 67–74, 78, 80–82, 85		$\frac{100(51-x)}{51} = \%$
Standard 7 Interdisciplinary Problem Solving	12		$\frac{100(1-x)}{1} = \%$
Standard 4: The Physical Setting/Earth Science			
Astronomy The Solar System (MU 1.1a, b; 1.2d)	39, 40, 41, 67, 68		$\frac{100(5-x)}{5} = \%$
Earth Motions and Their Effects (MU 1.1c, d, e, f, g, h, i)	1, 4, 5, 6, 29, 30, 51, 52, 53, 54		$\frac{100(10-x)}{10} = \%$
Stellar Astronomy (MU 1.2b)	45, 46, 47		$\frac{100(3-x)}{3} = \%$
Origin of Earth's Atmosphere, Hydrosphere and Lithosphere, (MU 1.2e, f, h)			
Theories of the Origin of the Universe and Solar System (MU 1.2a,c)	2, 3, 66		$\frac{100(3-x)}{3} = \%$

Topic	Question Numbers (Total)	Wrong Answers (x)	Grade
Meteorology Energy Sources for Earth Systems (MU 2.1a, b)			
Weather (MU 2.1c, d, e, f, g, h)	10, 31, 32, 36-38, 69–73		$\frac{100(11-x)}{11} = \%$
Insolation and Seasonal Changes (MU 2.1i; 2.2a,b)	7, 11, 14, 16, 34, 77–79		$\frac{100(8-x)}{8} = \%$
The Water Cycle and Climates (MU 1.2g; 2.2c,d)	8, 9, 12, 13, 15		$\frac{100(5-x)}{5} = \%$
Geology Minerals and Rocks (MU 3.1a, b, c	23, 25, 26, 28, 35, 57, 61		$\frac{100(7-x)}{7} = \%$
Weathering, Erosion, and Deposition (MU 2.1s, t, u, v, w)	22, 27, 48, 49, 50, 58, 60, 83–85		$\frac{100(10-x)}{10} = \%$
Plate Tectonics and Earth's Interior(MU 2.1j, k, l, m, n, o)	20, 42–44, 80-82		$\frac{100(7-x)}{7} = \%$
Geologic History (MU 1.2 i, j)	17–19, 21, 55, 56, 59, 74–76		$\frac{100(10-x)}{10} = \%$
Topographic Maps and Landscapes (MU 2.1p, q, r)	24, 33, 62, 63, 64, 65		$\frac{100(6-x)}{6} = \%$
ESRT *2011 Edition Reference Tables for Physical Setting/Earth Science*	2, 3, 7, 10, 11, 15, 17–21, 23–26, 28, 30, 34, 35, 37, 42, 44, 50, 55–57, 61, 65, 72, 75, 76, 80–82		$\frac{100(34-x)}{34} = \%$

To further pinpoint your weak areas, use the Topic Outline in the front of the book. MU = Major Understanding (see Topic Outline)

Examination June 2019

Physical Setting/Earth Science

PART A

Answer all questions in this part.

Directions (1–35): For *each* statement or question, choose the word or expression that, of those given, best completes the statement or answers the question. Some questions may require the use of the *2011 Edition Reference Tables for Physical Setting/Earth Science*. Record your answers in the space provided.

1 The map below shows four major time zones of the United States. The locations of Boston and San Diego are shown.

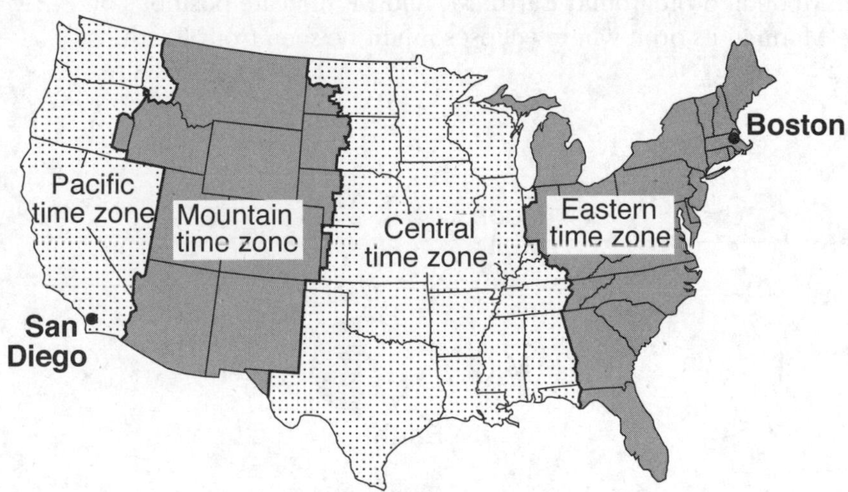

What is the time in Boston when it is 11 A.M. in San Diego?

(1) 8 a.m.
(2) 2 p.m.
(3) 3 p.m.
(4) noon

1 ____

2 The diagram below represents the spectral lines from the light emitted from a mixture of two gaseous elements in a laboratory on Earth.

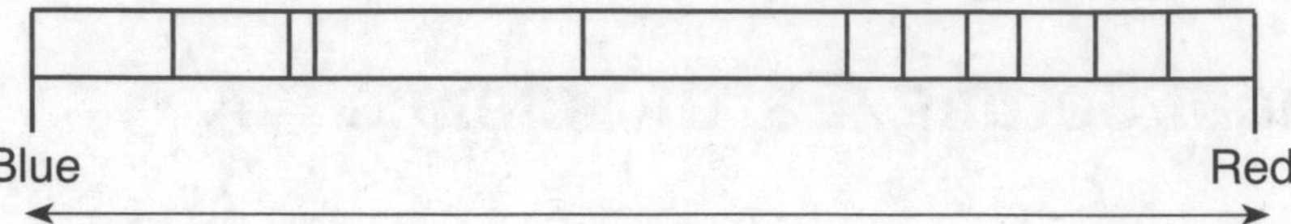

If the same two elements were detected in a distant star that was moving away from Earth, how would the spectral lines appear?

(1) The entire set of spectral lines would shift toward the red end.
(2) The entire set of spectral lines would shift toward the blue end.
(3) The spectral lines of the shorter wavelengths would move closer together.
(4) The spectral lines of the longer wavelengths would move closer together. 2 ____

3 The diagram below shows Earth in orbit around the Sun, and the Moon in orbit around Earth. M_1 and M_2 indicate positions of the Moon in its orbit where eclipses might be seen from Earth.

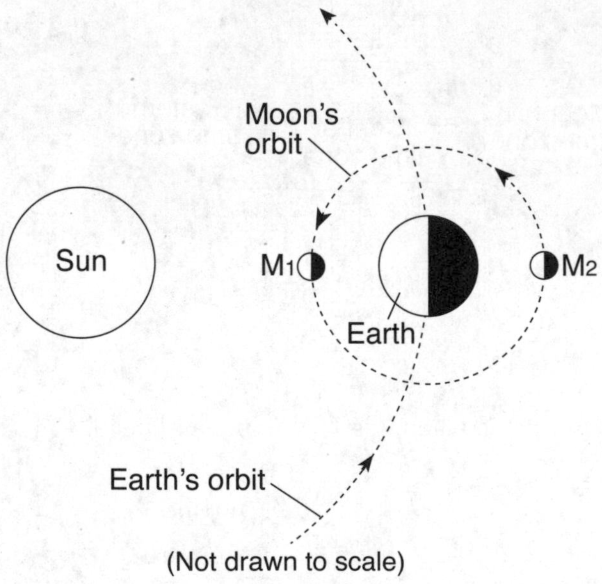

Which table correctly matches each type of eclipse with the orbital position of the Moon and the cause of each eclipse?

Type of Eclipse	Moon's Position	Cause of Eclipse
Solar	M_1	Earth's shadowfalls on Moon
Lunar	M_2	Moon's shadow falls on Earth

(1)

Type of Eclipse	Moon's Position	Cause of Eclipse
Solar	M_1	Moon's shadow falls on Earth
Lunar	M_2	Earth's shadowfalls on Moon

(2)

Type of Eclipse	Moon's Position	Cause of Eclipse
Solar	M_2	Earth's shadowfalls on Moon
Lunar	M_1	Moon's shadow falls on Earth

(3)

Type of Eclipse	Moon's Position	Cause of Eclipse
Solar	M_2	Moon's shadow falls on Earth
Lunar	M_1	Earth's shadowfalls on Moon

(4)

3 _____

4 The timeline below represents the entire geologic history of Earth. The lettered dots on the timeline represent events in Earth's history.

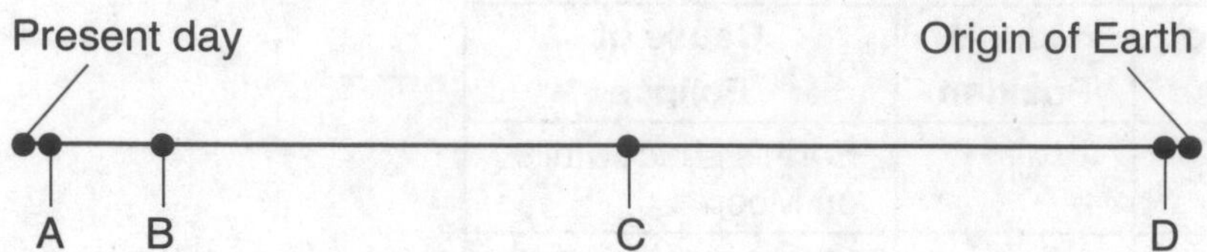

Which lettered dot best indicates the geologic time when humans first appeared on Earth?

(1) *A* (3) *C*
(2) *B* (4) *D* 4 _____

5 Compared to our solar system, the universe is

(1) younger, smaller, and contains fewer stars
(2) younger, larger, and contains more stars
(3) older, smaller, and contains fewer stars
(4) older, larger, and contains more stars 5 _____

6 Which motion allows an observer on Earth to view different constellations throughout the year?

(1) Earth orbiting the Sun
(2) constellations orbiting Earth
(3) Earth orbiting the constellations
(4) constellations orbiting the Sun 6 _____

7 What is the approximate location of the Tasman Hot Spot in the Pacific Ocean?

(1) 36° N 160° W (3) 160° N 36° W
(2) 36° S 160° E (4) 160° S 36° E 7 _____

8 Most ozone is found in a region of Earth's atmosphere between 10 and 20 miles above Earth's surface. This temperature zone of the atmosphere is known as the

(1) thermosphere (3) stratosphere
(2) mesosphere (4) troposphere 8 _____

9 The Coriolis effect, which causes the curving of planetary winds, is a direct result of the

(1) distance between Earth and the Sun
(2) inclination of Earth's axis
(3) orbiting of Earth around the Sun
(4) spinning of Earth on its axis 9 _____

10 What is the dewpoint if the dry-bulb temperature is 18°C and the relative humidity is 64%?

(1) 14°C (3) 9°C
(2) 11°C (4) 4°C 10 _____

11 When one gram of liquid water at its boiling point is changed into water vapor

(1) 334 J/g is gained from the surrounding environment
(2) 334 J/g is released into the surrounding environment
(3) 2260 J/g is gained from the surrounding environment
(4) 2260 J/g is released into the surrounding environment 11 _____

12 The cross section below represents some processes of the water cycle. Arrows represent the infiltration of water. The dashed line labeled *X* represents the uppermost level of Earth material that is saturated by groundwater.

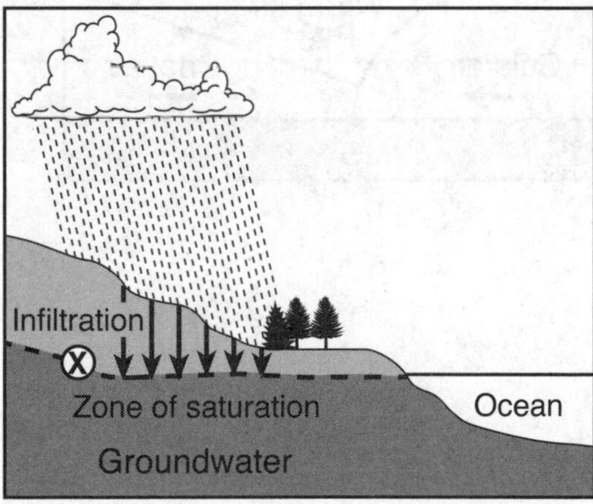

What is indicated by the dashed line labeled *X*?

(1) watershed
(2) water table
(3) impermeable bedrock
(4) impermeable soil 12 ____

13 Which index fossil in sedimentary surface bedrock most likely indicates that a marine environment once existed in the region where the sediments were deposited?

(1) Mastodont
(2) Condor
(3) Eospirifer
(4) Coelophysis 13 ____

14 The three diagrams below represent three frontal boundaries with the surface locations of the fronts labeled *A*, *B*, and *C*. Arrows indicate direction of air movement.

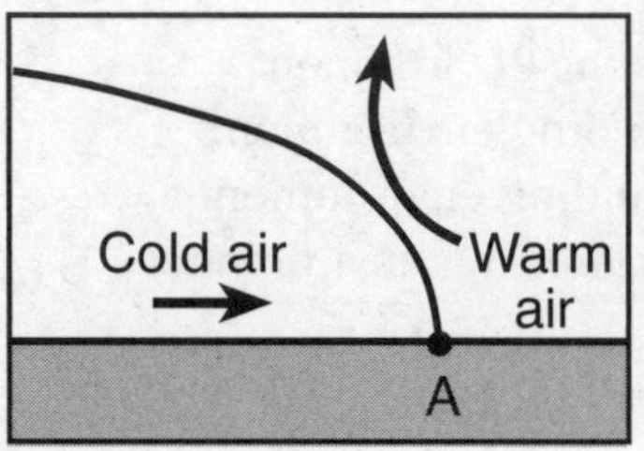

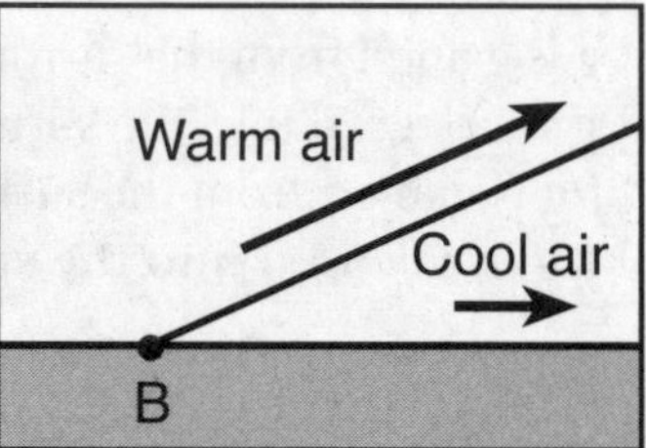

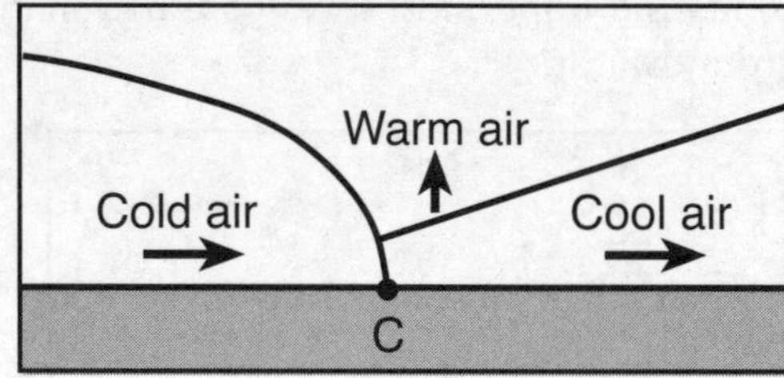

Which table correctly matches each letter with the type of frontal boundary it represents?

Letter	Type of Frontal Boundary
A	Cold Front
B	Warm Front
C	Occluded Front

(1)

Letter	Type of Frontal Boundary
A	Warm Front
B	Cold Front
C	Occluded Front

(2)

Letter	Type of Frontal Boundary
A	Cold Front
B	Warm Front
C	Stationary Front

(3)

Letter	Type of Frontal Boundary
A	Warm Front
B	Cold Front
C	Stationary Front

(4)

14 ____

15 The cross section below indicates the geologic ages of the bedrock beneath the state of Michigan. These rocks formed from sediments deposited in an ancient depositional basin. This region is called the Michigan Basin. Glacial deposits cover most of the surface.

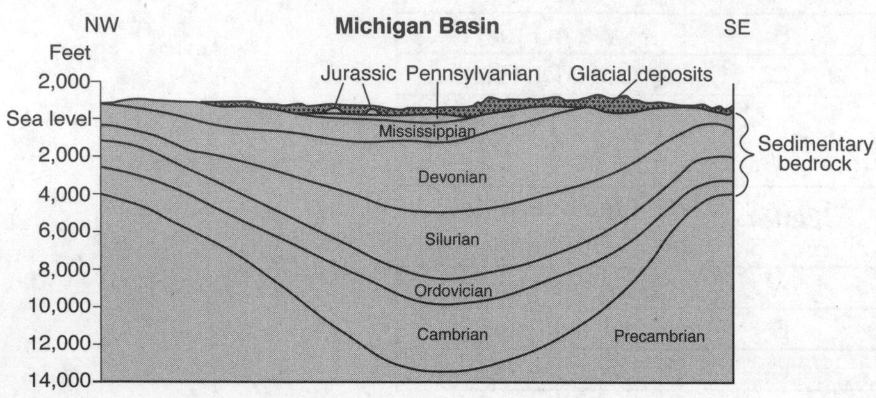

Which process most likely caused the formation of the Michigan Basin?

(1) uplift
(2) faulting
(3) metamorphism
(4) downwarping

15 ____

16 The cross section below represents prevailing winds, shown by arrows, moving over a coastal mountain range. Letters *A* through *D* represent locations on Earth's surface.

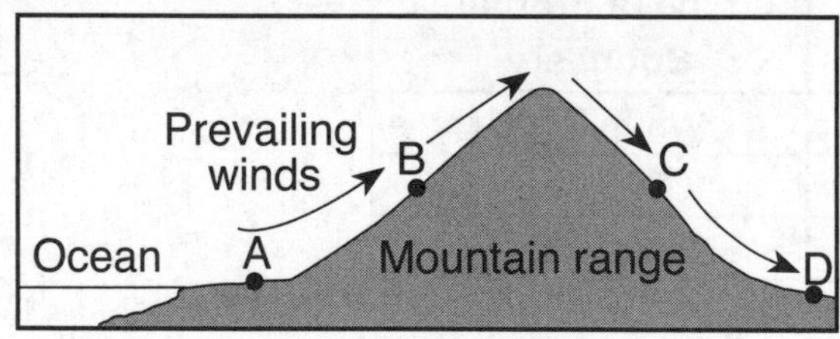

Which location would most likely have cloud cover and precipitation?

(1) *A*
(2) *B*
(3) *C*
(4) *D*

16 ____

17 Which Earth surface would most likely absorb the greatest amount of insolation on a sunny day if all of these surfaces have equal areas?

(1) blacktop parking lot
(2) white sand beach
(3) surface of a calm lake
(4) snow covered mountain slope

17 ____

18 Most of the infrared radiation given off by Earth's surface is absorbed in Earth's atmosphere by greenhouse gases such as water vapor, carbon dioxide, and

(1) hydrogen
(2) nitrogen
(3) oxygen
(4) methane

18 ____

19 The *X* on the map below indicates the region where the state of Washington is located on the present-day North American continent.

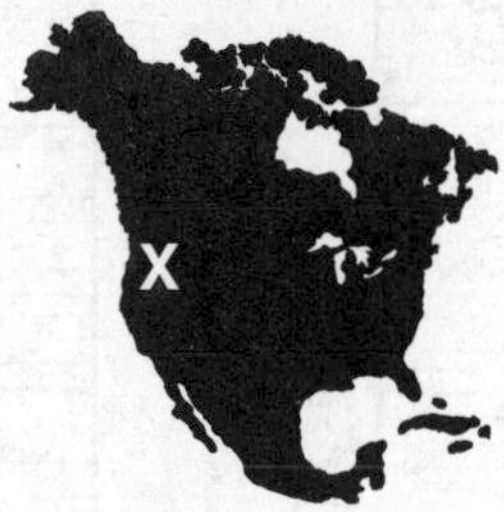

X

During which geologic period was the region of Washington State closest to the equator?

(1) Cretaceous
(2) Triassic
(3) Mississippian
(4) Ordovician

19 ____

20 A climate event that occurs when surface water in the eastern equatorial area of the Pacific Ocean becomes warmer than normal and may cause a warm, dry winter in New York State is

(1) an air mass formation
(2) the Doppler effect
(3) an El Niño
(4) a monsoon

20 ____

21 In the New York State rock record, there is *no* bedrock from the Permian Period. Which two other entire geologic periods have no sediment or bedrock in the New York State rock record?

(1) Cretaceous and Quaternary Periods
(2) Paleogene and Neogene Periods
(3) Triassic and Jurassic Periods
(4) Mississippian and Pennsylvanian Periods 21 ____

22 The cross sections below represent two widely separated bedrock outcrops, 1 and 2. Letters *A*, *B*, *C*, and *D* identify some rock layers. Line *XY* represents a fault. The rock layers have *not* been overturned.

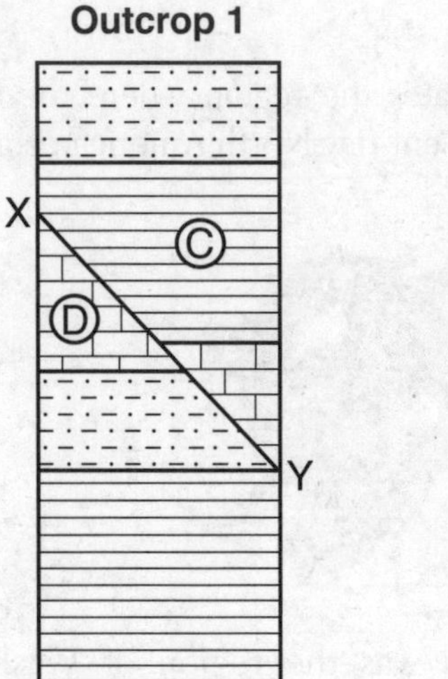

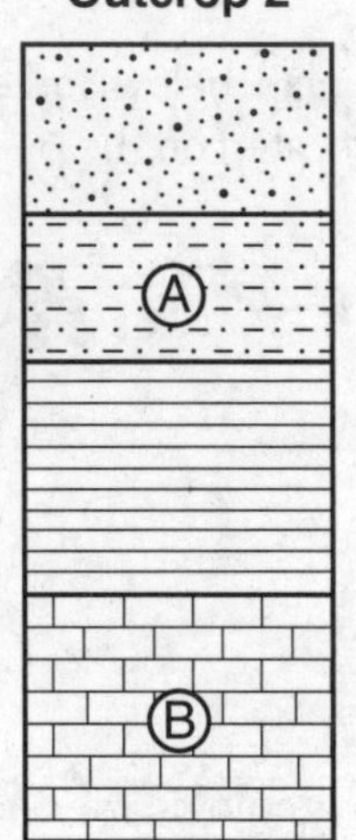

Which lettered rock layer is the youngest?

(1) *A*
(2) *B*
(3) *C*
(4) *D* 22 ____

23 The block diagram below represents the surface features in a landscape region.

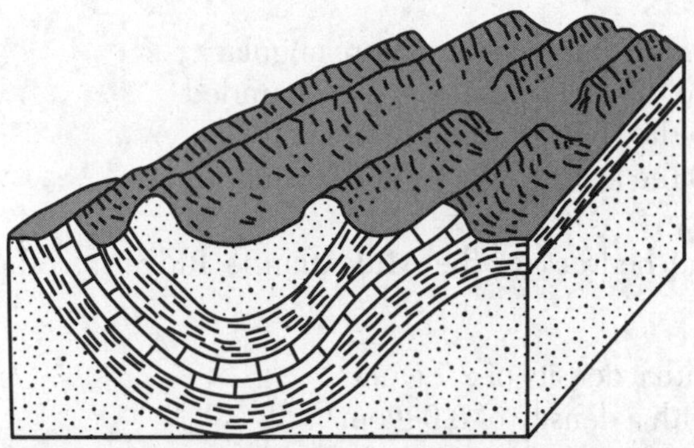

Which diagram best represents the general stream drainage pattern of this entire region?

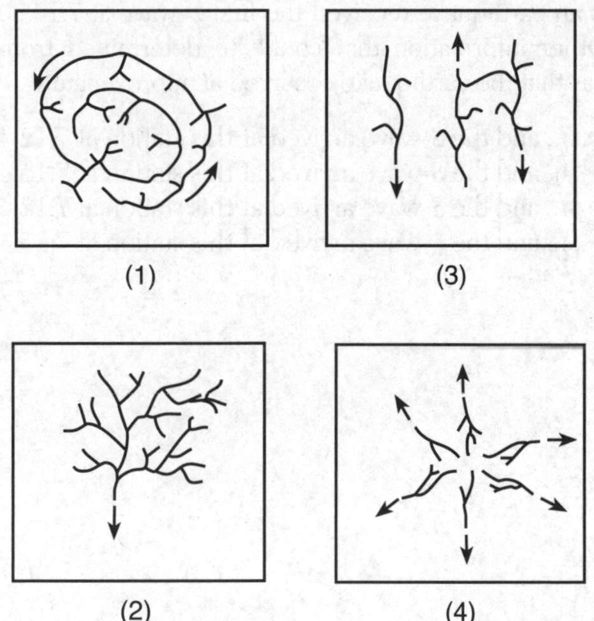

23 _____

24 Compared to a landscape that develops in a cool, dry climate, a landscape that develops in a warm, rainy climate will most likely weather and erode

(1) slower, so the landforms are more angular
(2) slower, so the landforms are more rounded
(3) faster, so the landforms are more angular
(4) faster, so the landforms are more rounded 24 ____

25 Earth's crustal bedrock at the Mid-Atlantic Ridge is composed mostly of

(1) basalt, with a density of 2.7 g/cm^3
(2) basalt, with a density of 3.0 g/cm^3
(3) granite, with a density of 2.7 g/cm^3
(4) granite, with a density of 3.0 g/cm^3 25 ____

26 A seismic recording station located 4000 kilometers from the epicenter of an earthquake received the first *P*-wave at 7:10:00 P.M. (h:min:s). Other information that could be determined from this recording was that the earthquake occurred at approximately

1 7:03:00 P.M., and the *S*-wave arrived at this station at 7:12:40 P.M.
2 7:03:00 P.M., and the *S*-wave arrived at this station at 7:15:40 P.M.
3 7:17:00 P.M., and the *S*-wave arrived at this station at 7:12:40 P.M.
4 7:17:00 P.M., and the *S*-wave arrived at this station at 7:15:40 P.M. 26 ____

27 Which graph best represents the percentage by mass of elements of Earth's crust?

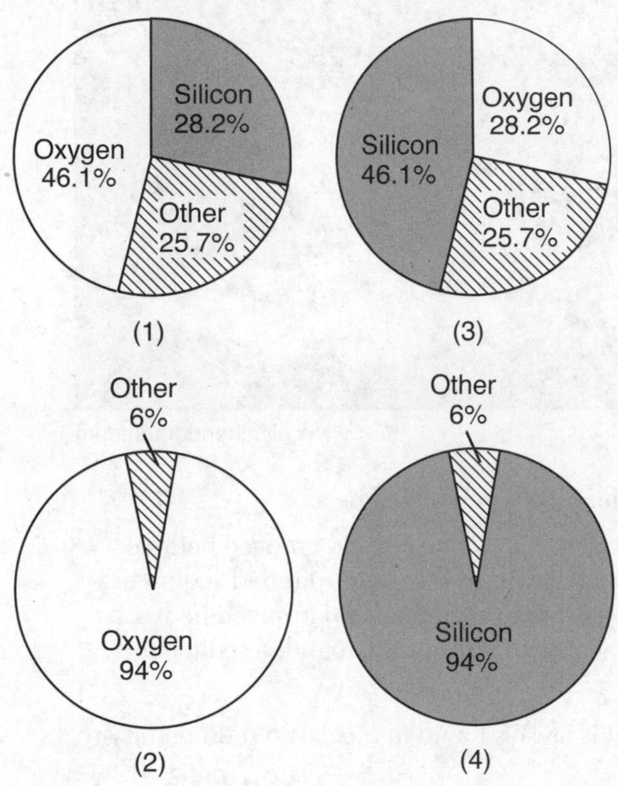

27 _____

28 The aerial photograph below shows an elongated hill near the Finger Lakes formed by glacial deposition in New York State.

This elongated hill is best identified as a

(1) kettle
(2) moraine
(3) sand dune
(4) drumlin

28 _____

29 The photograph below shows a valley glacier altering the landscape.

www.alicehenderson.com

The result of this glacial action will be a

(1) U-shaped valley with scratched and grooved bedrock
(2) U-shaped valley with well-sorted, rounded sediment
(3) V-shaped valley with scratched and grooved bedrock
(4) V-shaped valley with well-sorted, rounded sediment 29 ____

30 Which element is always found in bioclastic sedimentary rocks?

(1) iron
(2) sodium
(3) carbon
(4) sulfur 30 ____

31 What are two major factors that control the development of soil in a given location?

(1) vegetative cover and slope
(2) tectonic activity and elevation
(3) erosion and transpiration
(4) bedrock composition and climate 31 ____

32 The cross section below represents rock units within Earth's crust. The geologic ages of two of the layers are shown.

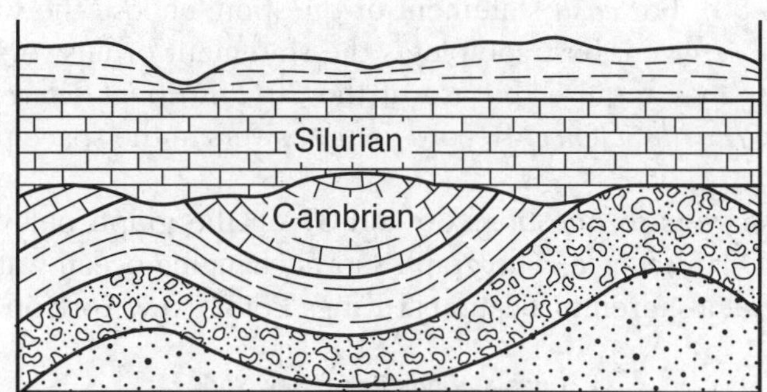

Which sequence of geologic events occurred after the formation of the Cambrian limestone and before the formation of the Silurian limestone?

(1) uplift → weathering → erosion → subsidence
(2) uplift → subsidence → erosion → weathering
(3) subsidence → weathering → erosion → uplift
(4) subsidence → erosion → uplift → weathering 32 _____

33 Which two rocks contain the mineral quartz?

(1) gabbro and schist
(2) dunite and sandstone
(3) granite and gneiss
(4) pumice and scoria 33 _____

34 Which chemical formula represents the compo-sition of a mineral that usually exhibits fracture, *not* cleavage?

(1) $Mg_3Si_4O_{10}(OH)_2$
(2) NaCl
(3) $CaCO_3$
(4) $(Fe,Mg)_2SiO_4$ 34 _____

35 Which mineral is commonly used as the "lead" in pencils?

(1) pyrite
(2) graphite
(3) galena
(4) fluorite 35 _____

PART B–1

Answer all questions in this part.

Directions (36–50): For *each* statement or question, choose the word or expression that, of those given, best completes the statement or answers the question. Some questions may require the use of the *2011 Edition Reference Tables for Physical Setting/Earth Science*. Record your answers in the space provided.

Base your answers to questions 36 and 37 on the graph below and on your knowledge of Earth science. The graph shows changing ocean water levels, over a 3-day period, at a shoreline location at Kings Point, New York on Long Island.

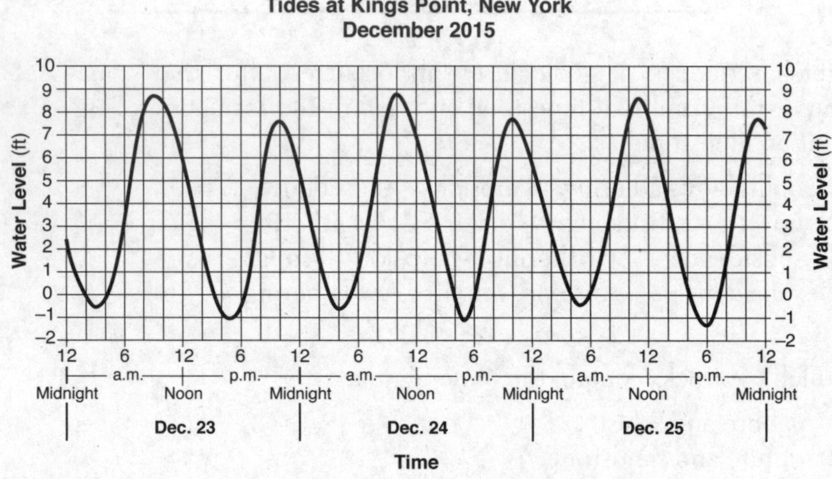

36 Based on the graph, the first low tide on December 26 occurred at approximately

(1) 6 A.M.
(2) 11 A.M.
(3) 6 P.M.
(4) 11 P.M.

36 _____

37 These Long Island tides show a pattern that is

(1) cyclic and predictable
(2) cyclic and unpredictable
(3) noncyclic and predictable
(4) noncyclic and unpredictable

37 _____

Base your answers to questions 38 through 41 on the diagram below and on your knowledge of Earth science. The diagram represents Earth orbiting the Sun. Four positions of Earth in its orbit are labeled *A*, *B*, *C*, and *D*. Letter N represents the North Pole. Distances are indicated for aphelion (Earth's farthest position from the Sun around July 4) and perihelion (Earth's closest position to the Sun around January 3). Arrows indicate directions of movement.

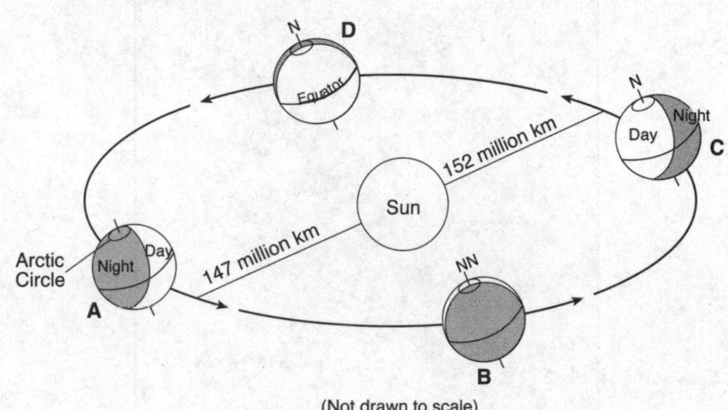

(Not drawn to scale)

38 During which season in the Northern Hemisphere is Earth at aphelion?

(1) winter
(2) spring
(3) summer
(4) fall

38 _____

39 Between which pair of lettered positions is the Sun's vertical ray moving from the equator southward to the Tropic of Capricorn?

(1) *A* and *B*
(2) *B* and *C*
(3) *C* and *D*
(4) *D* and *A*

39 _____

40 Approximately how many times does Earth rotate as it moves in its orbit from position *A* back to position *A*?

(1) 1 time
(2) 15 times
(3) 24 times
(4) 365 times

40 _____

41 What is the tilt of Earth's rotational axis relative to a line perpendicular to the plane of Earth's orbit?

(1) 15°
(2) 23.5°
(3) 66.5°
(4) 90°

41 _____

Base your answers to questions 42 through 44 on the map below and on your knowledge of Earth science. The map shows a composite of Doppler radar images. Darker shadings indicate the precipitation pattern of a large storm system over the eastern United States.

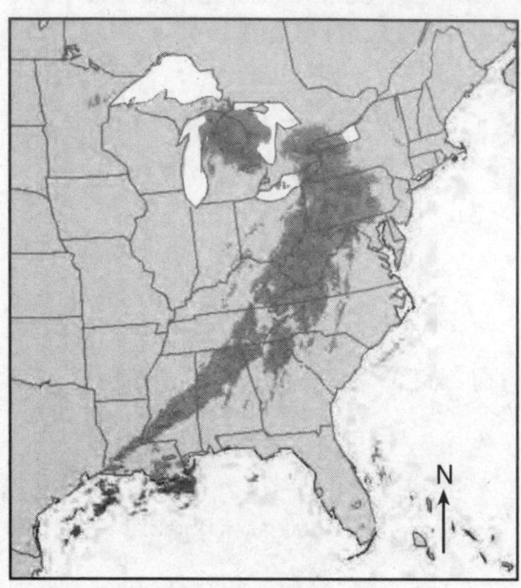

42 The surface wind circulation pattern around the center of this storm system is

(1) inward and clockwise
(2) inward and counterclockwise
(3) outward and clockwise
(4) outward and counterclockwise

42 _____

43 The best evidence on a weather map to indicate high-speed winds near the center of this storm system would most likely be

(1) cloud cover of 100%
(2) type of precipitation
(3) temperature and dewpoint values
(4) isobars drawn close together

43 _____

44 As this storm system follows a normal storm track, it will most likely move toward the

(1) southeast
(2) southwest
(3) northeast
(4) northwest

44 _____

Base your answers to questions 45 through 48 on the passage and map below and on your knowledge of Earth science. The map shows the location of the epicenter (✳) of a major earthquake that occurred about 1700 years ago. Point A represents a location on a tectonic plate boundary. Plates *X* and *Y* represent major tectonic plates. The island of Crete; the Anatolian Plate, which is a minor tectonic plate; and the Hellenic Trench have been labeled. Arrows indicate the relative directions of plate motion.

Crete Earthquake

Scientists have located the geological fault, off the coast of Crete in the Mediterranean Sea, that likely shifted, causing a huge earthquake in the year 365 that devastated life and property on Crete. The southwestern coastal region of Crete was uplifted, as evidenced by remains of corals and other sea life now found on land 10 meters above sea level. Scientists measured the age of these corals to verify when this event occurred. This earthquake caused a tsunami that devastated the southern and eastern coasts of the Mediterranean Sea. It is estimated that earthquakes along the fault, associated with the Hellenic Trench, may occur about every 800 years.

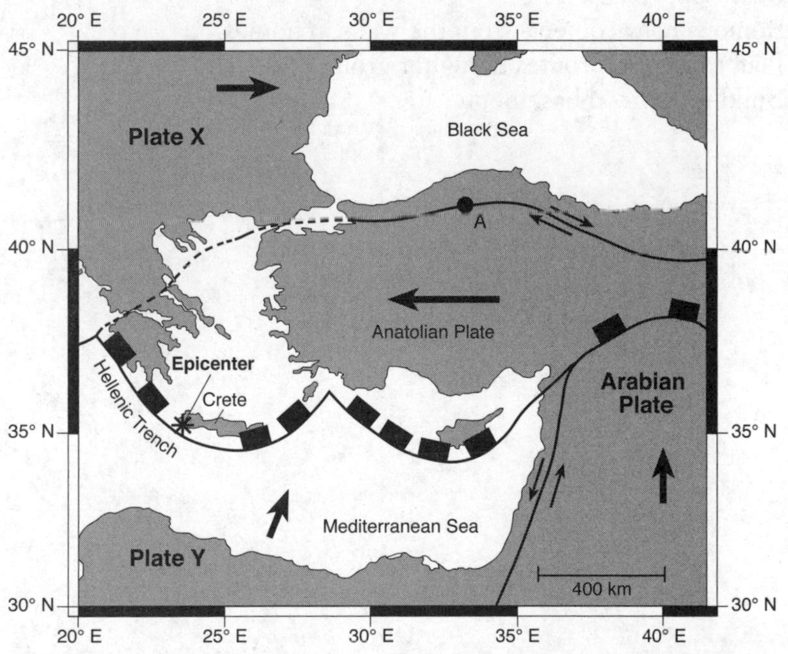

45 Which type of plate boundary is represented at point *A*?

(1) divergent
(2) convergent
(3) transform
(4) complex

45 ____

46 What are the names of the major tectonic plates *X* and *Y*?

(1) *X* = Eurasian Plate; *Y* = African Plate
(2) *X* = Eurasian Plate; *Y* = Arabian Plate
(3) *X* = Indian-Australian Plate; *Y* = African Plate
(4) *X* = Indian-Australian Plate; *Y* = Arabian Plate

46 ____

47 Which two New York State index fossils are most closely related to the corals that were radioactively dated in this study?

(1) *Eucalyptocrinus* and *Ctenocrinus*
(2) *Elliptocephala* and *Phacops*
(3) *Maclurites* and *Platyceras*
(4) *Lichenaria* and *Pleurodictyum*

47 ____

48 Which activity could best prepare residents along the Mediterranean coast to reduce the loss of human life during a future tsunami?

(1) Board up windows.
(2) Remove heavy objects from the walls in homes.
(3) Plan evacuation routes to higher ground.
(4) Build reinforced basements.

48 ____

Base your answers to questions 49 and 50 on the photograph below and on your knowledge of Earth science. The photograph shows a sandstone erosional feature that formed near the Grand Canyon, in southwestern United States.

www.nationalgeographic.com

49 Which erosional agent is most likely sandblasting this rock formation?

(1) wind
(2) waves
(3) running water
(4) moving ice

49 ____

50 What is the range of grain sizes that are most commonly found in rock making up this feature?

(1) 0.0004 cm–0.006 cm
(2) 0.006 cm–0.2 cm
(3) 0.2 cm–6.4 cm
(4) 6.4 cm–25.6 cm

50 ____

PART B–2

Answer all questions in this part.

Directions (51–65): Record your answers in the spaces provided. Some questions may require the use of the *2011 Edition Reference Tables for Physical Setting/Earth Science*.

Base your answers to questions 51 through 53 on the data table below, on the graph below, and on your knowledge of Earth science. The data table shows the projected percentages of radioactive isotope *X* remaining and its disintegration product *Z* forming over 6.5 billion years. The graph shows the disintegration of radioactive isotope *X*.

Radioactive Isotope X (%)	**Disintegration Product Z** (%)	**Time** (billion years)
100	0	0
50	50	1.3
25	75	2.6
12.5	87.5	3.9
6.25	93.75	5.2
3.125	96.875	6.5

51 On the grid *below*, construct a line graph by plotting the percentages of disintegration product *Z* forming over 6.5 billion years. Connect *all six plots* with a line. The percentages of radioactive isotope *X* have already been plotted. [1]

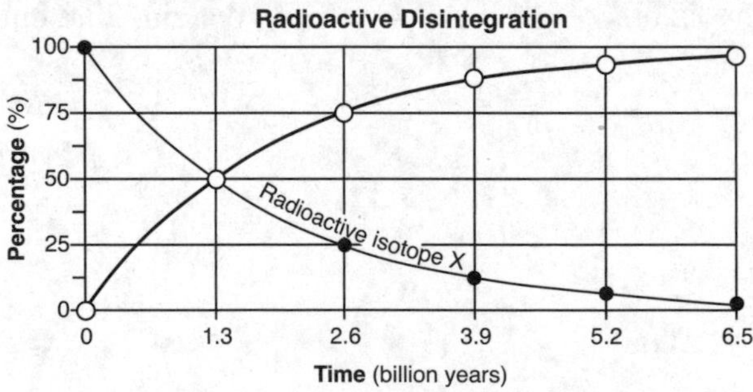

52 Identify radioactive isotope *X*. [1]

____

53 Calculate the amount, in grams, of an original 300-gram sample of radioactive isotope *X* remaining after 3.9 billion years. [1]

______________ g

Base your answers to questions 54 through 57 on the weather map below and on your knowledge of Earth science. The map shows the location of a low-pressure system over New York State during summer. Isobar values are recorded in millibars. The darker shading indicates areas of precipitation. Some New York State locations are indicated.

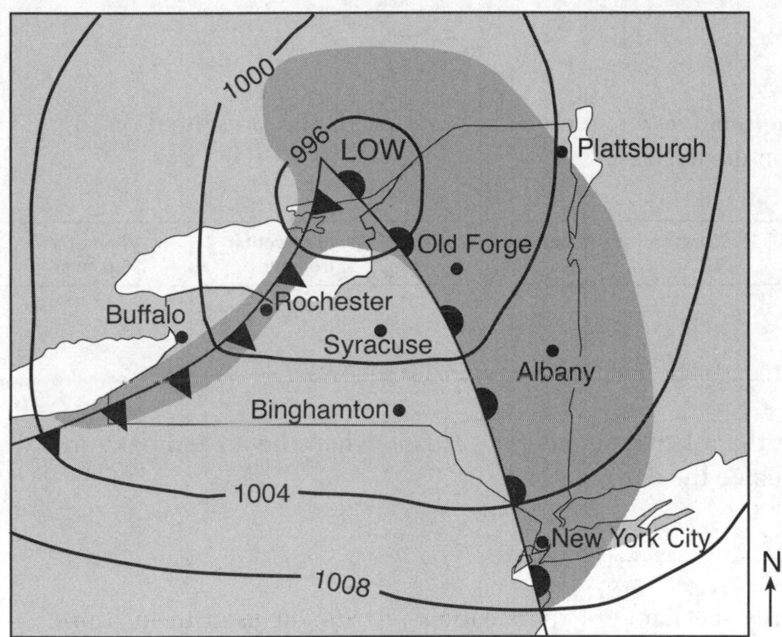

54 Describe the change in air pressure that will most likely occur at Rochester by the time that the cold front has reached Syracuse. Then describe what will most likely happen to the amount of cloud cover in Rochester with this change in air pressure and location of the cold front. [1]

Change in air pressure: ________________

Amount of cloud cover: ________________

55 The station model below represents the weather conditions at Buffalo, New York, at the time that this map was prepared.

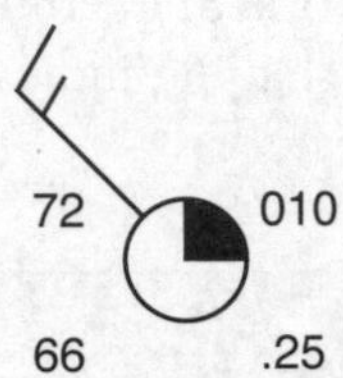

In the table *below*, record the weather data represented on this station model. [1]

Air Temperature (°F)	Barometric Pressure (mb)	Wind Direction from the	Wind Speed (knots)

56 State the relative humidity at Albany when the air temperature is equal to the dewpoint. [1]

________________ %

57 Identify the name of the weather instrument used to measure the wind speed at Plattsburgh. [1]

____

Base your answers to questions 58 through 61 on the diagram and passage below and on your knowledge of Earth science. The diagram represents the orbits of Earth, Comet Tempel-Tuttle, and planet *X*, another planet in our solar system. Arrows on each orbit represent the direction of movement.

Orbit of Comet Tempel-Tuttle

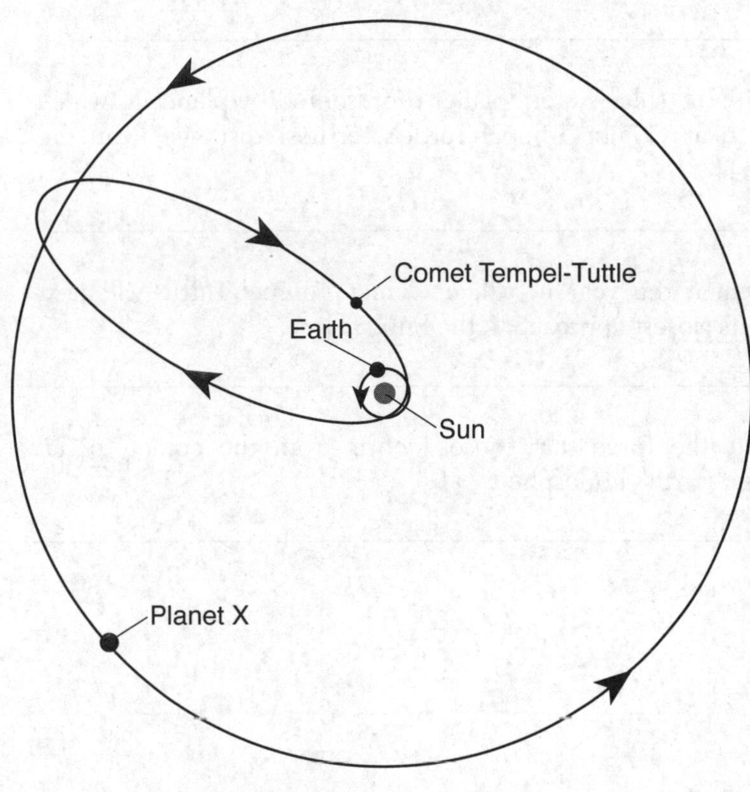

(Not drawn to scale)

Comet Tempel-Tuttle

Comet Tempel-Tuttle orbits our Sun and is responsible for the Leonid meteor shower event observed from Earth. This meteor shower occurs every year in November and is visible in the night sky as Earth passes through the debris left in space by this comet. The debris from the comet produces meteors that are smaller than a grain of sand, which enter Earth's atmosphere and burn up in the mesosphere temperature zone. Comet Tempel-Tuttle's orbital distance from the Sun ranges from about 145 million kilometers at its closest approach to 2900 million kilometers at its farthest distance. Its two most recent closest approaches to the Sun occurred in 1965 and one revolution later in 1998.

58 Identify the name of the object located at one of the foci of the elliptical orbit of Comet Tempel-Tuttle. [1]

____

59 Identify the solar system planet represented by planet X, which orbits near Comet Tempel-Tuttle's farthest distance from the Sun. [1]

____

60 Determine the year in which Comet Tempel-Tuttle will next make its closest approach to the Sun. [1]

____

61 Identify the force that causes debris from the comet to fall through Earth's atmosphere. [1]

____

Base your answers to questions 62 through 65 on the diagram below and on your knowledge of Earth science. The diagram represents the apparent path of the Sun across the sky as seen by an observer on Earth's surface on June 21. Points *A*, *B*, *C*, and *D* represent positions of the Sun at different times of the day. The angle of *Polaris* above the horizon as seen in the nighttime sky is indicated.

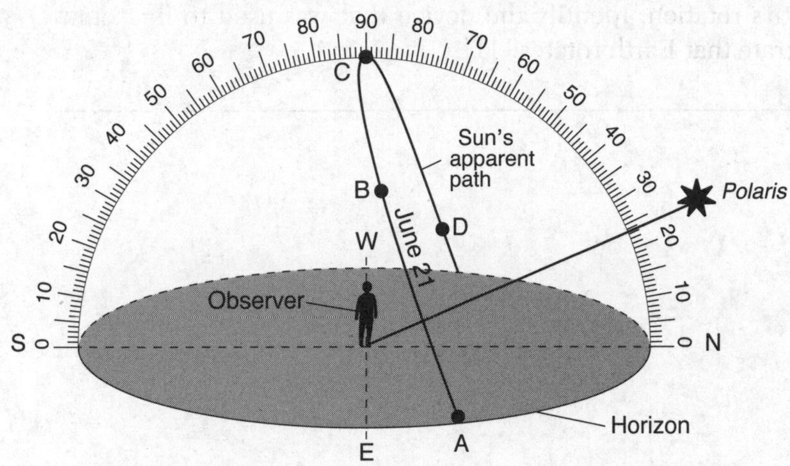

62 Describe the changes in the length of the observer's shadow as the Sun appears to move from position *A* to position *D*. [1]

____

____

63 Describe *one* piece of evidence from the diagram that supports the inference that the observer is located at the Tropic of Cancer. [1]

____

____

64 State the number of daylight hours at this location on September 23. [1]

______________ h

65 The Sun's apparent movement through the sky is caused by Earth's rotation. Identify the device that was used to first demonstrate that Earth rotates. [1]

____

PART C

Answer all questions in this part.

Directions (66–85): Record your answers in the spaces provided. Some questions may require the use of the *2011 Edition Reference Tables for Physical Setting/ Earth Science*.

Base your answers to questions 66 through 68 on the models below and on your knowledge of Earth science. The models represent cutaway views of four planets in our solar system, showing their inferred interior struc-tures. Each planet is shown in relation to the size of Earth.

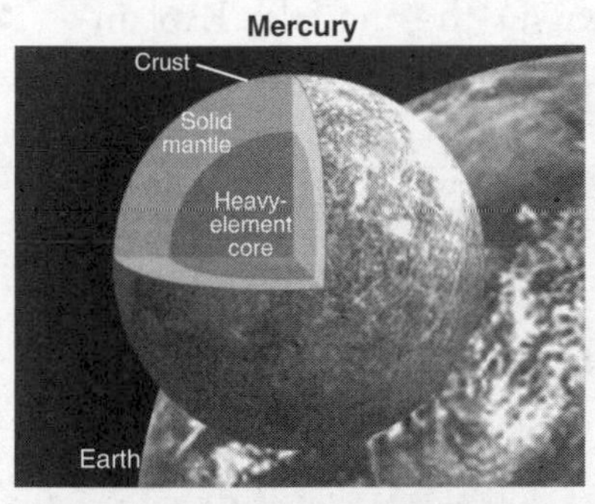

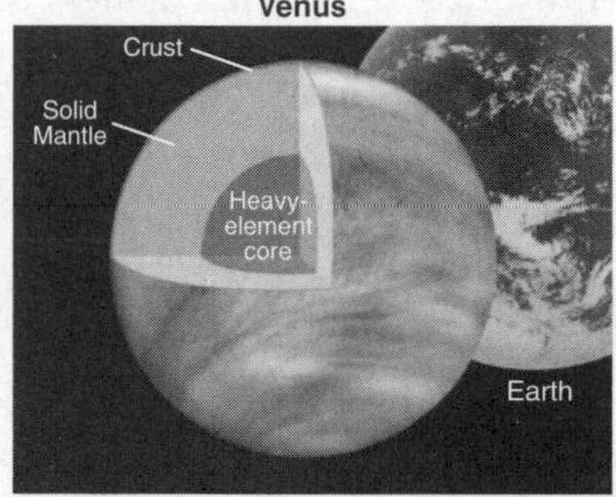

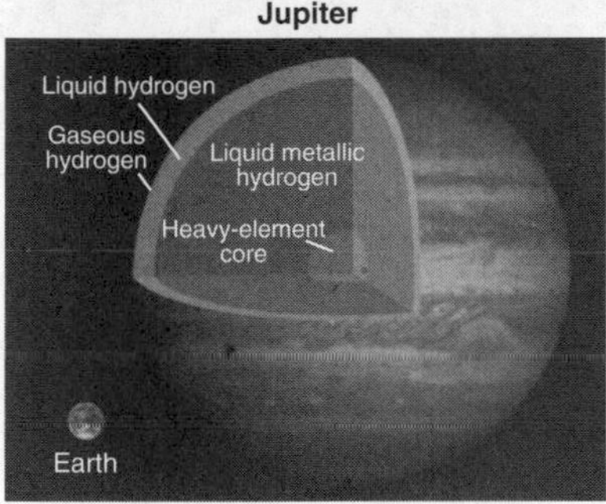

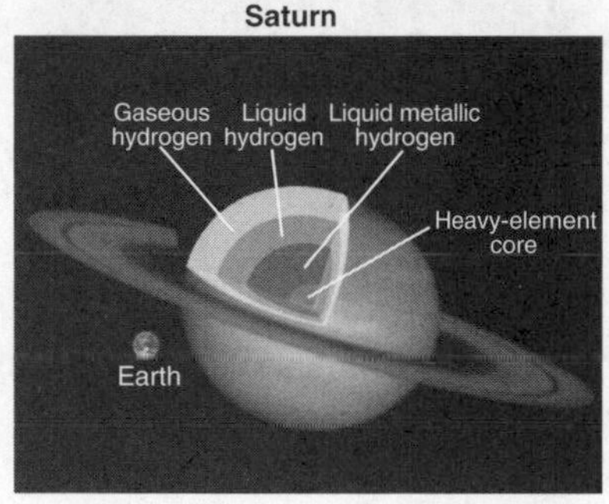

Seeds, Michael and Backman, Dana. 2011. *The Solar System.*

66 Determine how many times larger Jupiter's equatorial diameter is, compared to Earth's equatorial diameter. [1]

______________________________ **times larger**

67 Explain why Jupiter appears brighter in the night sky than Mercury, despite Jupiter's greater distance from Earth. [1]

____

____

68 Identify *two* terrestrial planets shown in the models. Explain why they are considered terrestrial planets. [1]

Terrestrial planets: ______________________________________ and

____

Explanation: ____

____

Base your answers to questions 69 through 72 on the diagram below and on your knowledge of Earth science. The diagram represents several streams converging and eventually flowing into a lake. Points *X* and *Y* indicate locations on either side of a meander in the stream. Points *A* and *B* indicate locations in the streams where the stream discharge was measured in cubic meters per second. The circled region labeled *C* represents a depositional feature.

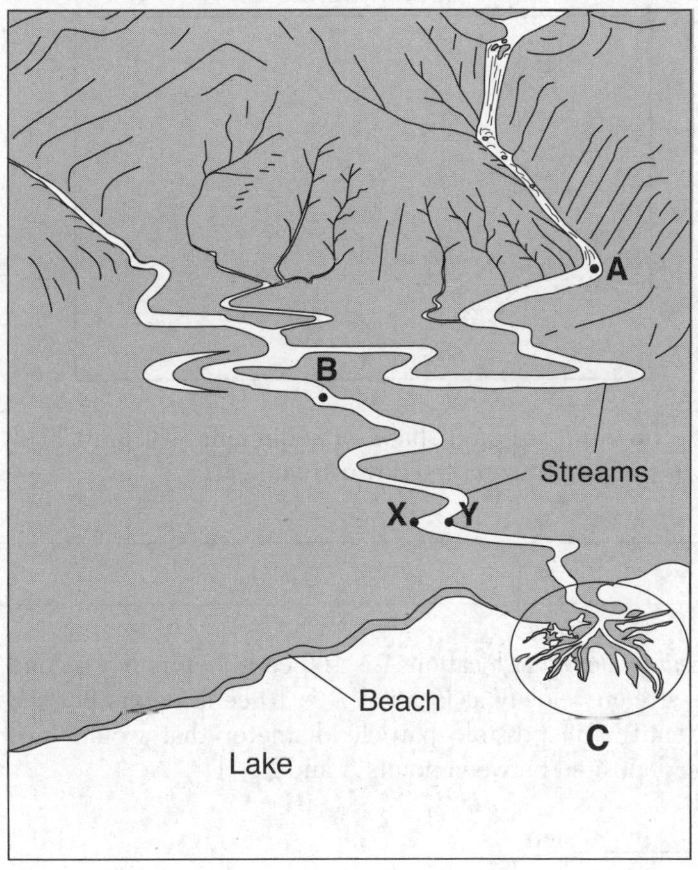

69 Identify the name of the depositional feature labeled *C*. [1]

____

70 In the box *below*, draw a cross-sectional view of the general shape of the stream bottom from point *X* to point *Y*. [1]

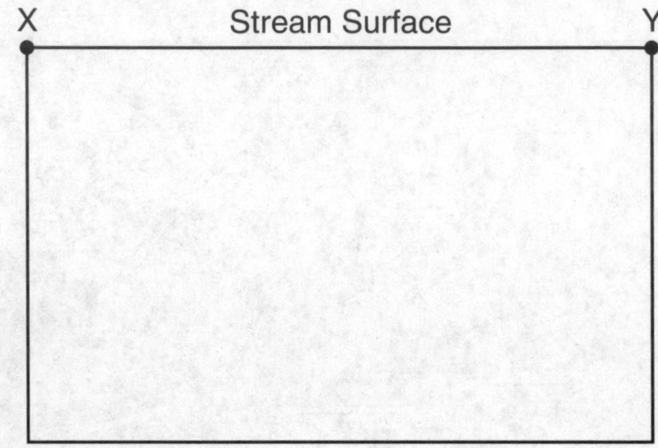

71 Describe how the size and shape of sediments will most likely change as they are transported downstream. [1]

Size: ______________________________________

Shape: _____________________________________

72 The stream velocity at location *A* is 100 centimeters per second, and the stream velocity at location *B* is 10 centimeters per second. Identify *one* possible particle diameter that would most likely be deposited between points *A* and *B*. [1]

_______________ **cm**

Base your answers to questions 73 through 76 on the map and graphs below and on your knowledge of Earth science. The map shows the locations of two cities, Hastings, Nebraska, and Riverhead, New York. The graphs show average monthly air temperatures for Hastings and Riverhead.

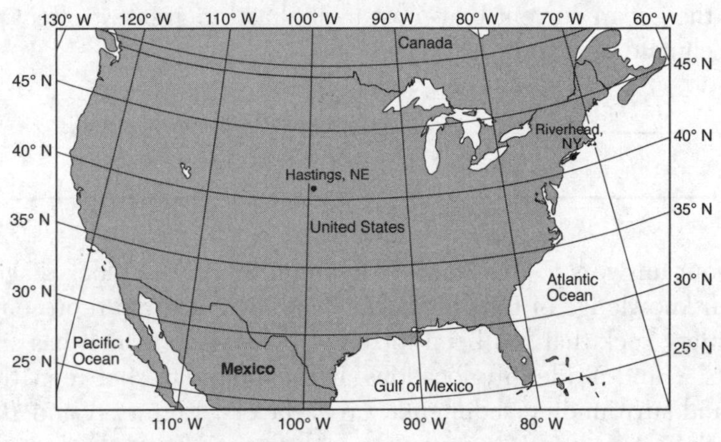

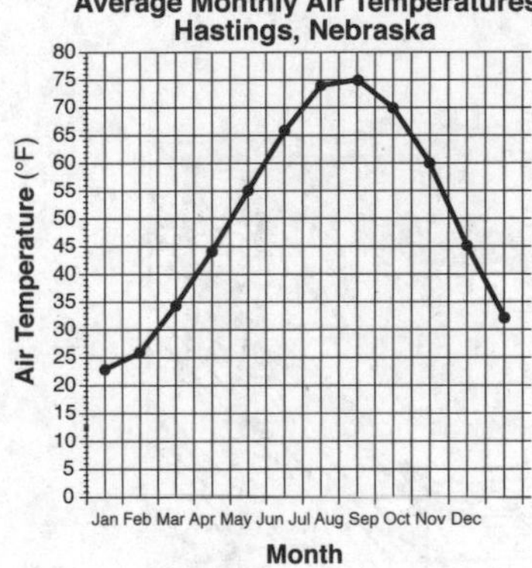

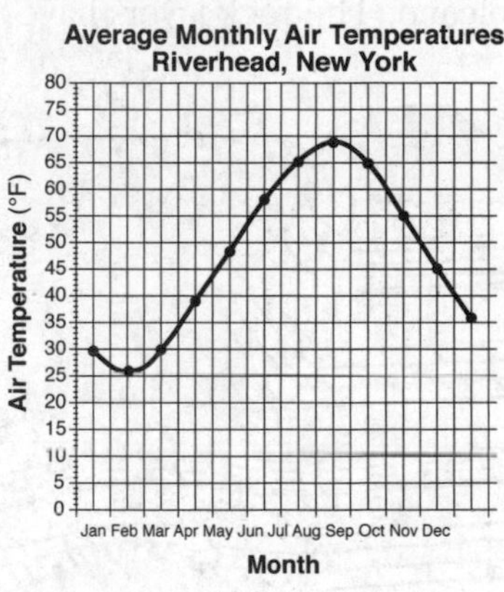

73 Explain why Hastings has a greater annual temperature range than Riverhead. [1]

____

____

74 Explain why the angle of insolation at both locations is approximately the same at solar noon on any given day. [1]

____

____

75 Identify the planetary wind belt that primarily influences the climates of *both* Hastings and Riverhead. [1]

____

76 Name the ocean current that most likely has the greatest effect on the climate of Riverhead. [1]

______________________________ **Current**

Base your answers to questions 77 through 80 on the block diagram below and on your knowledge of Earth science. The block diagram represents a region of sedimentary rock that has been intruded by magma, which has since solidified. Points *X* and *Y* identify locations at the boundary between the igneous intrusion and surrounding sedimentary rock layers. Letters *A* and *B* represent specific rock units. Letter *C* represents rock formed from the lava flow of the nearby volcano. The rock layers have *not* been overturned.

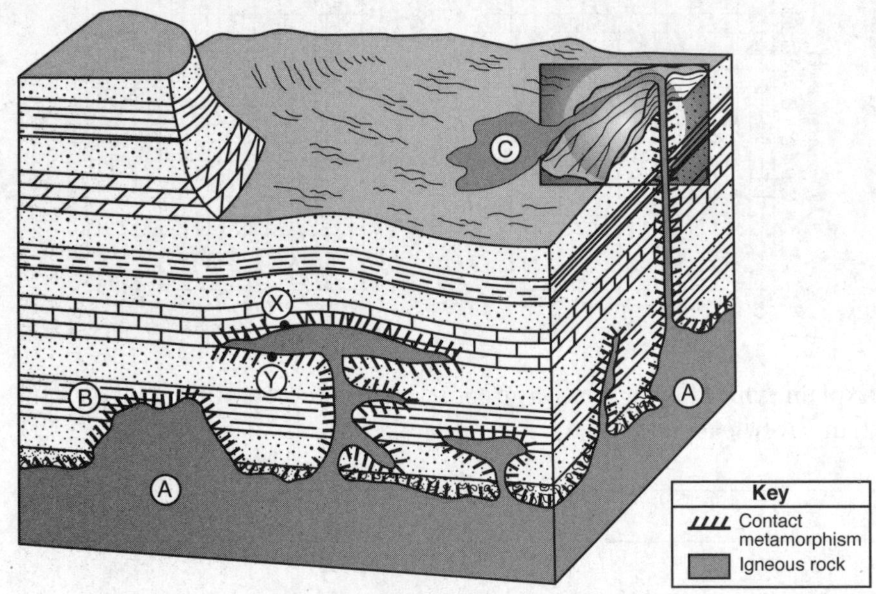

Adapted from www.brocku.ca/earthsciences

77 Identify *two* processes that formed the sedimentary rock layers represented in the diagram. [1]

1: ____________________ 2: ____________________

78 Describe *one* piece of evidence shown in the diagram that indicates that rock unit *A* is younger than rock unit *B*. [1]

____

____

79 Explain why the igneous rock that formed at location *C* is composed of crystals less than 1 millimeter in size. [1]

____

____

80 State the names of *two different* metamorphic rocks that are most likely found in the zone of contact metamorphism at locations *X* and *Y*. [1]

Location *X*: ____________________________________

Location *Y*: ____________________________________

Base your answers to questions 81 through 85 on the passage and map below, the field map below, and on your knowledge of Earth science. The map shows the location of Crater Lake in Oregon in the western United States. The field map shows lake depths and some isolines in Crater Lake recorded in meters. Line *AB* and line *CD* are reference lines. Letter *X* represents a location on the lake bottom.

Crater Lake

Crater Lake is the deepest lake in the United States. The lake formed in the crater at the top of volcanic Mount Mazama after it exploded in a violent eruption approximately 7700 years ago. The rim of the crater is approximately 2300 meters (7500 feet) above sea level and is mostly composed of the rock andesite. The average yearly air temperature at the lake is 38°F, and snowfall often occurs from October through June. Hydrothermal activity (heating of the water) is ongoing under the lake, indicating that this region is still volcanically active.

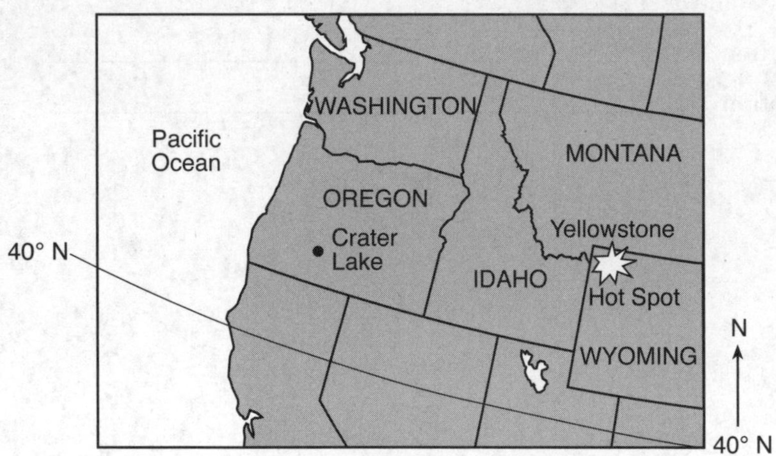

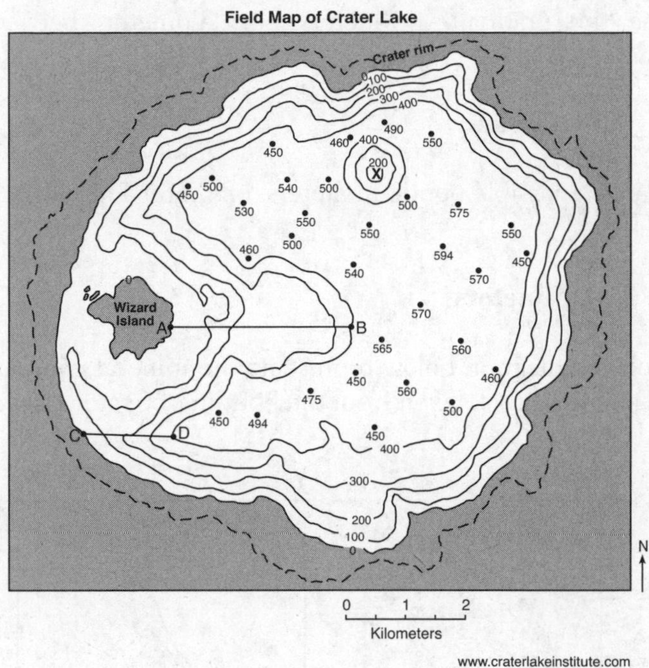

81 On the field map *above*, draw the 500-meter depth isoline. [1]

82 On the grid *below*, construct a profile along line *AB* by plotting the lake depth of each isoline that crosses line *AB*. Connect *all* plots with a line to complete the profile. [1]

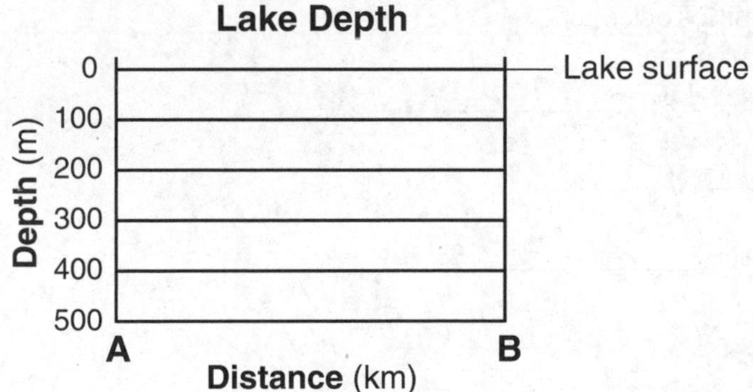

83 Determine the gradient, in meters per kilometer, between points *C* and *D*. [1]

______________ **m/km**

84 Determine *one* possible depth, in meters, of Crater Lake at location *X*. [1]

______________ **meters**

85 Line *YZ* on the diagram below represents the mineral composition of an andesitic rock taken from the bottom of Crater Lake.

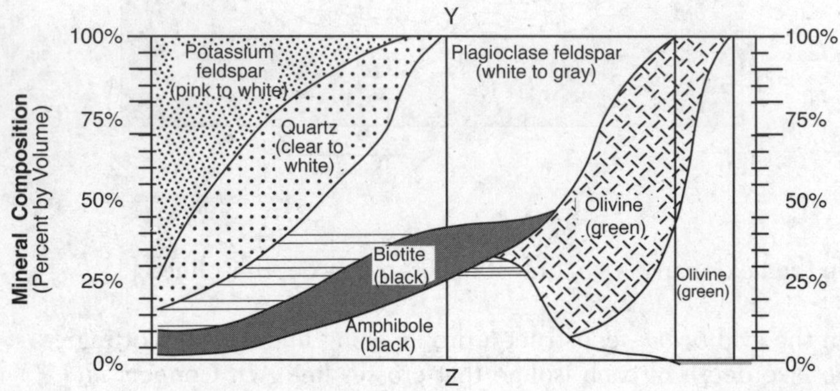

Identify the percent by volume of each of the three minerals in this andesitic rock. [1]

Plagioclase feldspar: ______________ %

Biotite: ______________ %

Amphibole: ______________ %

Answers June 2019

Physical Setting/Earth Science

Answer Key

PART A

1. 2	**8.** 3	**15.** 4	**22.** 1	**29.** 1
2. 1	**9.** 4	**16.** 2	**23.** 3	**30.** 3
3. 2	**10.** 2	**17.** 1	**24.** 4	**31.** 4
4. 1	**11.** 3	**18.** 4	**25.** 2	**32.** 1
5. 4	**12.** 2	**19.** 4	**26.** 2	**33.** 3
6. 1	**13.** 3	**20.** 3	**27.** 1	**34.** 4
7. 2	**14.** 1	**21.** 2	**28.** 4	**35.** 2

PART B-1

36. 1	**39.** 4	**42.** 2	**45.** 3	**48.** 3
37. 1	**40.** 4	**43.** 4	**46.** 1	**49.** 1
38. 3	**41.** 2	**44.** 3	**47.** 4	**50.** 2

PART B–2 and **PART C**. *See* **Answers Explained**.

Answers Explained

PART A

1. **2**

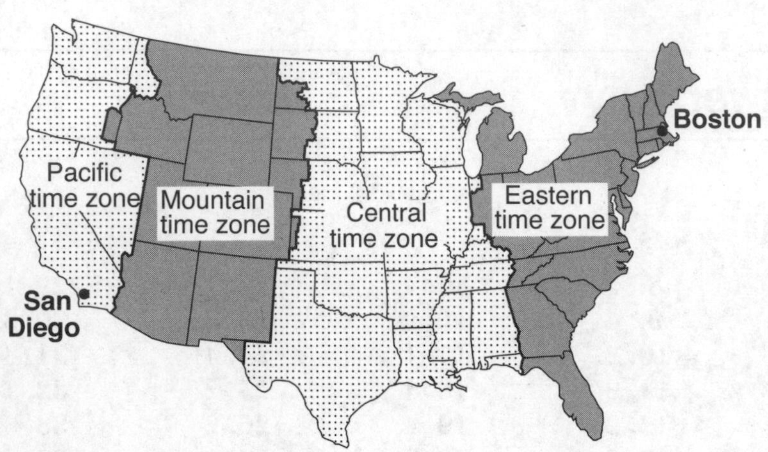

Earth rotates 15° per hour. So time zones were set up around a series of meridians spaced at 15° intervals starting at the prime meridian, which passes through Greenwich, England. Earth rotates from west to east. So for every time zone you travel west of the prime meridian, the clocks are set one hour earlier. According to the map, San Diego is located within the shaded region labeled "Pacific time zone" and Boston is located within the shaded region labeled "Eastern time zone." Note that the Pacific time zone is three time zones to the west of the Eastern Time Zone. Therefore, clocks in the Pacific time zone are set 3 hours earlier than those in the Eastern Time Zone. Thus, when it is 11 a.m. in San Diego, it is 2 p.m. in Boston.

2. **1** Find the Electromagnetic Spectrum chart in the *Reference Tables for Physical Setting/Earth Science.* Note that visible light at the red end of the spectrum has a longer wavelength than visible light at the blue end of the spectrum. Thus, a shift toward the longer wavelength end of the spectrum is a shift toward the red end of the spectrum. If a source of electromagnetic waves is moving away from an observer at the same time as the source is emitting light of a particular wavelength, fewer wave crests will reach the eye of the observer each second. The eye will interpret this as meaning that the light has a longer wavelength than it actually has. In other words, the light will appear shifted toward the red end of

the spectrum. It is given that the two elements were detected in a distant star that was moving away from Earth. Thus, all of the elements in the distant star are moving away from Earth. Therefore, the entire set of spectral lines would shift toward the red end of the spectrum.

3. **2**

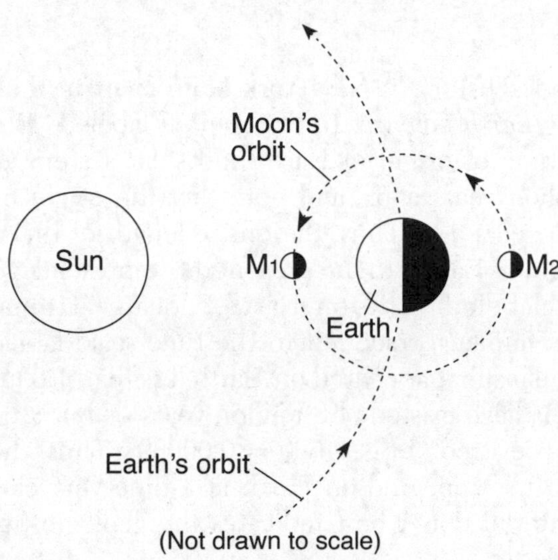

(Not drawn to scale)

A solar eclipse occurs when the Moon passes directly between the Sun and Earth, causing the Moon to cast a shadow on Earth. Note that when the Moon is in position M_1, it is directly between the Sun and Earth, the Moon's shadow falls on Earth and a solar eclipse occurs.

A lunar eclipse occurs when the Moon moves into Earth's shadow. When Earth's shadow falls on the Moon, Earth blocks sunlight from reaching the Moon. Earth's shadow is located on the side of Earth opposite the Sun. Therefore, in order for a lunar eclipse to occur, the Moon must be on the side of Earth directly opposite the Sun. Note that when the Moon is in position M_2, it is on the side of Earth directly opposite the Sun, causing Earth's shadow to fall on the Moon and a lunar eclipse to occur.

Only table (2) correctly matches each type of eclipse with the orbital position of the Moon and the cause of each eclipse.

4. **1**

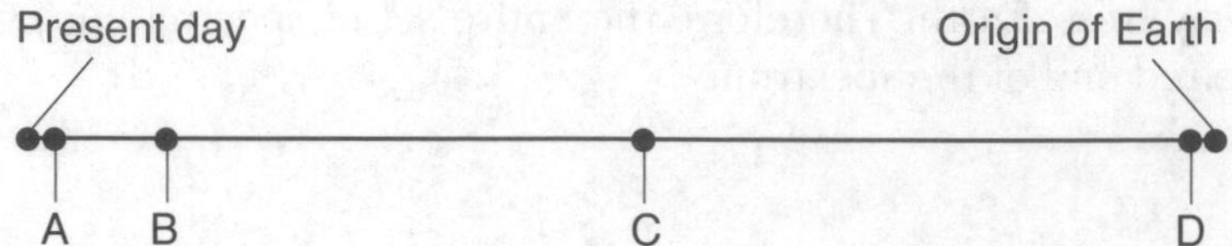

Find the Geologic History of New York State chart in the *Reference Tables for Physical Setting/Earth Science*. In the column labeled "Era," find the statement "Estimated time of origin of Earth and solar system." Trace left to the scale labeled "Million years ago," and note that the origin of Earth occurred about 4600 million years ago. Thus, the entire length of the timeline from the point labeled "Origin of Earth" to the present day represents 4600 million years.

In the column labeled "Life on Earth," locate "Humans, mastodonts, mammoths." From humans, trace left to the time scale labeled "Million years ago." Note that humans first appeared on Earth less than 1.8 million years ago.

The time humans have existed (1.8 million years) is but a tiny fraction of the total time Earth has existed ($1.8 \div 4600 = 0.00039$). Thus, the timeline should show that humans first appeared on Earth at a time very close to the present day. Therefore, lettered dot *A* best indicates the geologic time when humans first appeared on Earth.

5. **4** Observations indicate that the universe is expanding. If the motion of galaxies is traced back in time, there is a point about 13–14 billion years ago at which the galaxies were very close to each other. Thus, we have a model in which the universe started out with all its matter and energy in a very small volume and then expanded outward in all directions. This motion is similar to an explosion, hence the name for this model—the Big Bang theory. According to this theory, the universe was initially smaller, denser, much hotter, and filled with a uniform glowing fog of hydrogen plasma. As the universe exploded outward in all directions from its origin point, both the plasma and the radiation filling it spread out and grew cooler. As the universe cooled, some of the matter condensed to form stars and their solar systems. Thus, the universe is older than our solar system.

The universe is everything that we can perceive. It includes all of the stars, gas clouds, galaxies, and other objects that we can detect by the radiation they emit. The universe contains billions of galaxies, each of which consists of hundreds of billions of stars. Our solar system consists of just a single star and all of the objects (such as planets) that orbit that star. Thus, the universe is larger than our solar system and contains more stars than the universe.

Thus compared to the solar system, the universe is older, larger, and contains more stars.

6. **1** As Earth revolves around the Sun, the side of Earth facing the Sun experiences day and the side facing away from the Sun experiences night. Since the stars are only visible at night, the portion of the universe whose stars (and the constellations they form) are visible to an observer on Earth varies cyclically as Earth revolves around the Sun. Thus, the motion that allows an observer on Earth to view different constellations throughout the year is Earth orbiting the Sun.

WRONG CHOICES EXPLAINED

(2) The idea that the stars in constellations orbit around Earth is part of the geocentric theory, which has been proven to be incorrect. Furthermore, Earth is too small and too far from the stars in constellations for Earth to exert enough gravitational attraction to hold the stars in orbit around Earth. Therefore, this motion does not occur and cannot allow an observer on Earth to view different constellations throughout the year.

(3) Earth orbits the Sun, not the stars in constellations. The stars in constellations are too far away to exert enough gravitational attraction on Earth to hold it in orbit. Therefore, this motion does not occur and cannot allow an observer on Earth to view different constellations throughout the year.

(4) Objects orbit a body that is both massive enough and close enough to exert a gravitational attraction strong enough to hold the objects in orbit. The Sun has too little mass and is too far from the stars in constellations to exert a strong enough gravitational attraction to hold them in orbit. For example, *Sirius*, in the constellation Canis Major, is located about 8.6 light-years from the Sun and is about twice as massive as the Sun. Therefore, this motion does not occur and cannot allow an observer on Earth to view different constellations throughout the year.

7. **2** Find the Tectonic Plates map in the *Reference Tables for Physical Setting/Earth Science*. In the key, note the symbol for a "Mantle hot spot." On the Tectonic Plates map, locate the Tasman Hot Spot off the southeast coast of Australia. Draw a horizontal line through the center of the Tasman Hot Spot, and extend it to intersect with the latitude scales along the left and right edges of the map. Note the value at these intersections—36°. Recall that 0° on this scale represents the equator. Latitudes above the 0° mark are north of the equator, and those below the 0° mark are south of the equator. Therefore, the Tasman Hot spot is located at about 36° S latitude. Now draw a vertical line through the Tasman Hot Spot, and extend it to intersect with the longitude scales along the top and bottom edges of the map. Note the value at these intersections—160°. Recall that 0° on the longitude scale represents the prime meridian. Longitudes to the left of the prime meridian from 0° to 180° are west longitudes; longitudes to the right of the prime meridian from 0° to 180° are east longitudes. Note that this particular map

is not centered on the prime meridian! The east longitudes to the right of the prime meridian wrap around and continue from the left edge of the map. Therefore, although the Tasman Hot is located to the left of the prime meridian on the map, it is beyond the 180° longitude line. In fact, the Tasman Hot Spot lies between 0° and 180° to the right of the prime meridian. Therefore, the Tasman Hot Spot is located at 160° E longitude. When taken together, the approximate location of the Tasman Hot Spot is 36° S 160° E.

8. **3** Find the Selected Properties of Earth's Atmosphere chart in the *Reference Tables for Physical Setting/Earth Science*. Locate 10–20 miles on the "Altitude" scale. Note that miles (mi) are listed on the right-hand side of the scale. Trace right to the section labeled "Temperature Zones," and note that this altitude range corresponds to the atmospheric temperature zone labeled "Stratosphere." Thus, most ozone is located in the temperature zone of the atmosphere known as the stratosphere.

9. **4** Planetary winds tend to blow in a straight line from regions of high pressure toward regions of low pressure. As these winds are blowing, Earth is spinning on its axis. This rotation causes the winds to appear to be turning toward the right in the Northern Hemisphere and toward the left in the Southern Hemisphere. This phenomenon is called the Coriolis effect. Thus, the Coriolis effect occurs as a result of the spinning of Earth on its axis.

WRONG CHOICES EXPLAINED

(1) The Coriolis effect does not change even though Earth's distance from the Sun varies. Therefore, it is unlikely that the Coriolis effect is the direct result of Earth's distance from the Sun.

(2) The inclination of Earth's axis causes unequal heating of the planet's surface. This causes differences in air temperature and therefore differences in air pressure. However, these differences in air pressure cause winds to blow directly from regions of high pressure to regions of low pressure, not cause them to be deflected.

(3) Earth's orbiting around the Sun causes seasonal changes in temperature and can lead to planetary regions of high and low pressure. However if Earth did not rotate, the planetary winds would blow in a straight line from regions of high pressure to regions of low pressure regions without being deflected.

10. **2** Find the Relative Humidity (%) chart in the *Reference Tables for Physical Setting/Earth Science*. Find the "Dry-Bulb Temperature (°C)" scale along the left side of the chart, and locate the row labeled "18." Trace this row to the right to the cell containing the value "64." From this cell, trace vertically upward to the

"Difference Between Wet-Bulb and Dry-Bulb Temperatures (°C)" scale along the top of the chart. Note the value listed there is "4." Thus, when the dry-bulb temperature is 18°C and the relative humidity is 64%, the difference between the wet-bulb and dry-bulb temperatures is 4°C.

Now find the Dewpoint (°C) chart in the *Reference Tables for Physical Setting/Earth Science*. Locate the column headed "4" in the "Difference Between Wet-Bulb and Dry-Bulb Temperatures (°C)" scale along the top of the chart. Locate the row corresponding to 18°C on the "Dry-Bulb Temperature (°C)" scale along the left side of the chart. Trace to the right from the row labeled "18" until you intersect the column headed "4." At this intersection, note the value "11." Thus, when the dry bulb temperature is 18°C and the relative humidity is 64%, the dewpoint is 11°C.

11. **3** The process by which liquid water changes into water vapor is called evaporation, or vaporization. Find the Properties of Water chart in the *Reference Tables for Physical Setting/Earth Science.* Note that the heat gained during vaporization is 2260 J/g. Thus, in order for one gram of liquid water at its boiling point to change into water vapor, 2260 Joules of heat is gained from the surrounding environment.

12. **2**

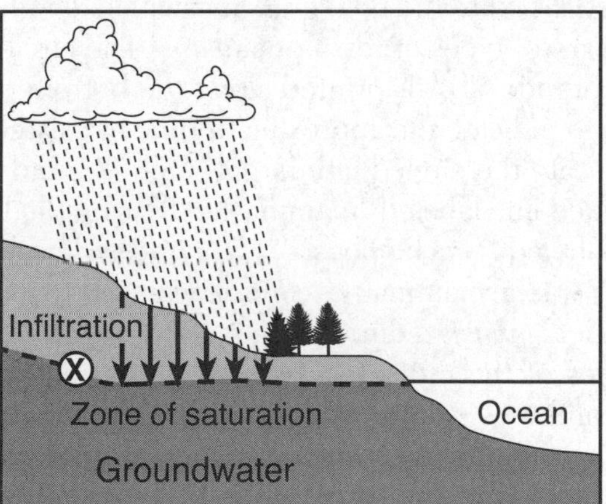

By definition, the uppermost level of Earth material that is saturated by groundwater is known as the water table. The water table marks the boundary between the zone of saturation (dark gray shaded area) in which water fills all the pore spaces in the Earth material and the zone of aeration (light gray shaded area) in which air fills all the pore spaces. Thus, the dashed line labeled *X*, which marks the upper boundary of the zone of saturation, is the water table.

WRONG CHOICES EXPLAINED

(1) A watershed is the area drained by a stream and its tributaries, not the upper level of Earth material that is saturated with water.

(3) and (4) Note the arrows labeled "Infiltration" passing through the light gray shaded area of the ground above the dashed line labeled *X*. Infiltration is the downward movement of liquid water into and through interconnected openings among soil particles due to gravity. An Earth material through which water can infiltrate is said to be permeable. It is given that dashed line *X* represents the upper level of Earth material that is saturated by groundwater. In order for the Earth material to become saturated, its pore spaces must be interconnected so that the water can pass through the Earth material to fill empty pores. A substance through which water can pass is said to be permeable. Thus, both the Earth material above and below dashed line *X* are permeable. Therefore, it is unlikely that dashed line *X* indicates either impermeable bedrock or impermeable soil.

13. **3** Find the Geologic History of New York State chart in the *Reference Tables for Physical Setting/Earth Science.* Locate Mastodont, Condor, *Eospirifer, and Coelophysis* among the drawings of index fossils along the bottom of the chart. Note that mastodont corresponds to the circled letter "O," condor to "S," *Eospirifer* to "Y," and *Coelophysis* to "L." Now find the column labeled "Time Distribution of Fossils." Note the bold gray lines labeled with names and circled letters. The names are of types of fossil organisms. The circled letters are keyed to the illustrations of index fossils printed along the bottom of the chart and are placed on the lines to indicate the approximate time of existence of each specific important fossil. Locate the circled letters "O," "S," "Y," and "L." Note that "O" lies on the gray shaded line labeled "mammals." "S" is on the line labeled "birds." "Y" is on the line labeled "brachiopods." "L" is on the line labeled "dinosaurs." Thus, a mastodont was a mammal, a condor was a bird, an *Eospirifer* was a brachiopod, and a *Coelophysis* a dinosaur.

From the pictures of the index fossils, note that a mastodont is shown as a hairy elephant, a condor as a large winged bird, *Coelophysis* as a dinosaur that walked upright on its hind legs, and an *Eospirifer* as a shelled organism. Present-day elephants live on land, so it is reasonable to infer that an elephant-like mammal such as mastodont also lived on land. Therefore, mastodont fossils would not indicate a marine environment. Condors still exist today and live in mountainous regions, so it is reasonable to infer that condors in the past also lived in such regions. Therefore, condor fossils would not indicate a marine environment. Marine dinosaurs had flippers, not legs. Thus, *Coelophysis* was most likely a dinosaur that lived on land and *Coelophysis* fossils would not indicate a marine environment. However, present-day brachiopods are shelled organisms similar in appearance to clams or scallops and live in shallow marine environments. It is reasonable to infer that in the past, brachiopods also lived in such environments. Thus, the presence of the index fossil *Eospirifer* in sedimentary surface bedrock would most likely indicate that a marine environment once existed in the region where the sediments were deposited.

14. **1**

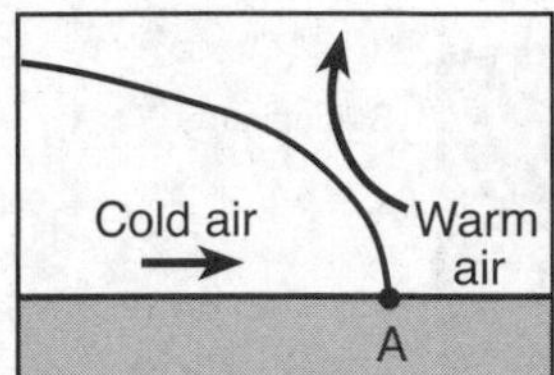

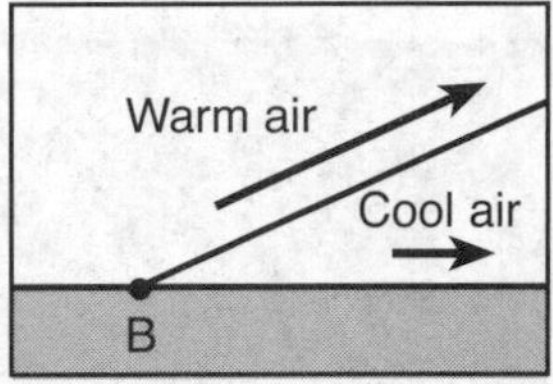

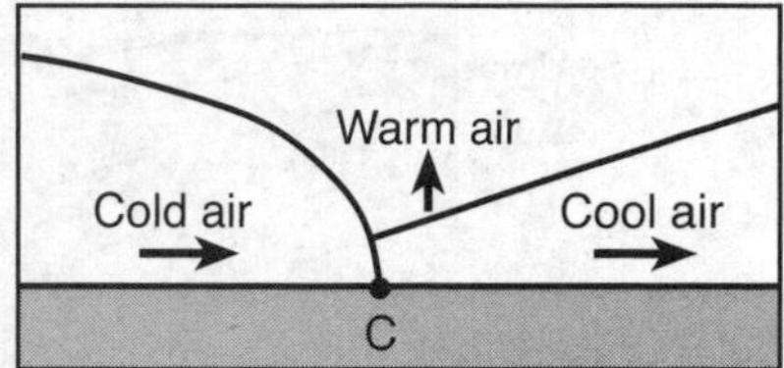

Cold fronts marks the boundary along which a cold air mass is advancing against a warmer air mass. The cold air mass is denser than the warmer air ahead of it. So the cold air mass pushes against and under the warm air mass like a wedge, forcing the warmer air upward. The result is a boundary that curves sharply upward in the direction of the cold air. Note that in diagram *A*, cold air is advancing against warm air and that the warm air is being forced upward. Note, too, that the boundary between the cold and warm air curves sharply upward toward the cold air. Thus, diagram *A* represents a cold front.

A warm front marks the boundary along which a warm air mass is advancing against a cold air mass. The warmer air mass is less dense than the colder air ahead of it. So as the warmer air pushes against the cooler air ahead of it, the warmer air is forced up and over the cooler air, forming a gently sloping boundary. Note that in diagram *B*, warm air is advancing against cool air and moving up and over the cool air, forming a gently sloping boundary. Thus, diagram *B* represents a warm front.

When a fast-moving cold front overtakes a slow-moving warm front, the cold front pushes completely under the warm air until it collides with the cool air ahead of the warm air and lifts the warm air entirely aloft, forming an occluded front. Note that in diagram *C*, cold air has pushed completely under the warm air and has reached the cool air ahead, lifting the warm air entirely aloft. Thus, diagram *C* most likely represents an occluded front.

A stationary front forms along the boundary between a warm air mass and a cold air mass when neither air mass moves appreciably in either direction. A stationary front slowly takes on the shape of a warm front as the denser cold air slowly slides beneath the less dense, warmer air. Note that none of the diagrams show a frontal boundary in which neither warm nor cold air is advancing. Thus, none of the diagrams shows a stationary front.

Therefore, the table that best matches each letter with the type of frontal boundary it represents is shown in Choice 1—*A* is a cold front, *B* is a warm front, and *C* is an occluded front.

15. **4**

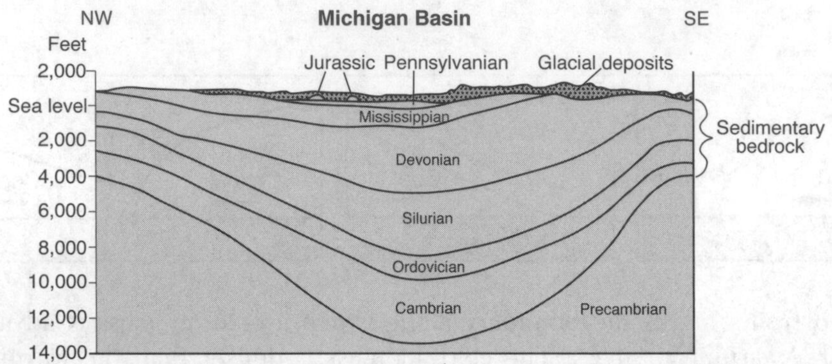

It is given that the sedimentary bedrock formed from sediments deposited in an ancient depositional basin. Basins are circular structures that arch downward, forming a bowl-like depression. Basins are formed by gentle downward bending, or downwarping, of rock due to compression forces acting on the crust. Note that the ancient Precambrian bedrock forms such a depression. As sediments were washed into this low-lying area, they accumulated in layers as shown in the cross section. Thus, the process that most likely caused the formation of the Michigan Basin is downwarping.

WRONG CHOICES EXPLAINED

(1) Uplift would cause the surface to be raised, not lowered. The uplifted surface would become a site of erosion, not a depositional environment. Therefore, it is unlikely that uplift caused the formation of the Michigan Basin

(2) The cross section shows no fault lines and no evidence of rock layers being displaced in either the sedimentary bedrock or the Precambrian bedrock. Therefore, there is no evidence that faulting caused the formation of the Michigan Basin.

(3) The glacial deposits at the surface are the result of erosion and deposition, not metamorphism. Note the rock layers labeled "Sedimentary bedrock" in the cross section. Sedimentary rocks form by the compaction and cementation of sediments, not metamorphism. There is no evidence in the cross section that the Precambrian bedrock has undergone metamorphism. Therefore, there is no evidence that metamorphism caused the formation of the Michigan Basin.

16. **2**

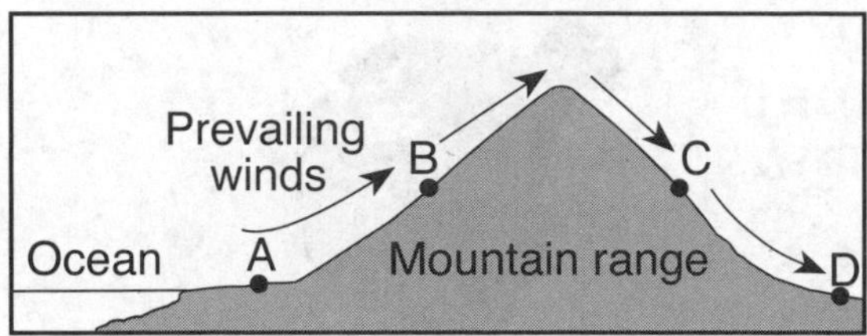

The cross section shows prevailing winds forcing air to rise along the windward side of the mountain range. As the air rises to higher elevations, air pressure decreases, causing the air to expand and cool. When the air cools to its dewpoint temperature, water vapor begins to condense, forming clouds. If enough water vapor condenses and if large water droplets form, precipitation will occur on the windward side. However, precipitation on the windward side is much more likely at location *B* than at location *A*, because at location *A*, the air has not yet been forced to rise much above sea level and has not yet been cooled to its dewpoint.

As the drier air sinks on the leeward side of the mountains, air pressure increases, causing the air to be compressed and warm. The air is warmed above its dewpoint, and any available moisture evaporates. Therefore, clouds and precipitation are much less likely to form on the leeward side of the mountain (locations *C* and *D*).

Thus, the location that would most likely have cloud cover and precipitation is *B*.

17. **1** All other factors being equal, dark-colored materials absorb more insolation than light-colored materials. Due to the high specific heat of water, land surfaces increase in temperature more than water surfaces when insolation strikes. Thus of the choices given, the darkest-colored land surface is the blacktop parking lot. Therefore, the blacktop parking lot will absorb the most insolation.

18. **4** Greenhouse gases allow sunlight to enter the atmosphere and strike Earth's surface but absorb the infrared radiation reradiated by Earth's surface, trapping the heat in the atmosphere and resulting in global warming. Some of these greenhouse gases occur in nature, such as water vapor, carbon dioxide, methane, ozone, and nitrous oxide. Others are exclusively human-made, such as the chlorofluorocarbons used in aerosol sprays. Thus in addition to water vapor and carbon dioxide, only choice (4)—methane—contains another major greenhouse gases found in Earth's atmosphere.

19. **4**

Note the relative position of the state of Washington on the North American continent. Find the Geologic History of New York State chart in the *Reference Tables for Physical Setting/Earth Science*. In the column labeled "Inferred Positions of Earth's Landmasses," locate the diagram in which the position of the state of Washington on the North American continent is closest to the equator. Note that this diagram is labeled "458 million years ago." On the time scale labeled "Million years ago" to the right of the column labeled "Epoch," locate the point corresponding to "458." From this point, trace left to the column labeled "Period," and note that 458 million years ago corresponds to the Ordovician Period. Therefore, the geologic period during which the region of Washington State was closest to the equator is the Ordovician.

20. **3** Along the eastern sides of many ocean basins, the global trade winds blow parallel to the coast and warm surface waters are carried offshore, producing strong coastal upwelling. The nutrient-rich water brought to the surface by this upwelling makes these coastal areas some of the ocean's richest fishing grounds. El Niño originally referred to the changes in surface currents along the coasts of Peru and Chile. Every year, usually in December, the trade winds subside, the surface ocean currents flowing offshore slow, upwelling slacks off, and the water gets warmer. Local fishermen have known about this for centuries since it marks the end of the peak fishing season. Since the change in currents comes around Christmas, they called it El Niño—"The Christ Child." Every few years, however, the change is greater than normal. The surface water gets much warmer, upwelling ceases altogether, and the surface ocean current reverses to inshore. The warming of the surface waters during an El Niño causes a large region of low pressure to form over the southeastern Pacific Ocean. This, in turn, causes changes in the jet stream winds, which profoundly affect global weather patterns. On average, during the last ten El Niño climate events, changes to the jet stream winds have resulted in warmer, drier winters in New York State. Thus, the climate event that occurs when surface waters in the eastern equatorial area of the Pacific Ocean become warmer than normal and may cause a warm, dry winter in New York State is an El Niño.

WRONG CHOICES EXPLAINED

(1) Air mass formation occurs all the time, not just when the eastern equatorial Pacific surface waters are warmer than usual.

(2) The Doppler effect refers to the change in the perceived wavelength of sound or light emitted by a source that is moving toward or away from an observer, not a climate event.

(4) A monsoon is a seasonally reversing system of surface winds caused by temperature differences between land and ocean, not the warmer than normal surface water in the eastern equatorial Pacific Ocean.

21. **2** Find the Geologic History of New York State chart in the *Reference Tables for Physical Setting/Earth Science*, and locate the column labeled "NY Rock Record." The thick black or shaded lines indicate that rocks or sediments from that time period exist in New York State. If there is no line, there is no rock or sediment of that age in New York State. Trace the areas where there is no line in the "NY Rock Record" column to the left to the column labeled "Period." Note that New York State contains no rocks corresponding to any part of the Neogene, Paleogene, and Permian Periods. Thus in addition to the Permian, two other entire geologic periods for which there is no sediment or bedrock in the New York State rock record are the Paleogene and Neogene Periods.

22. **1**

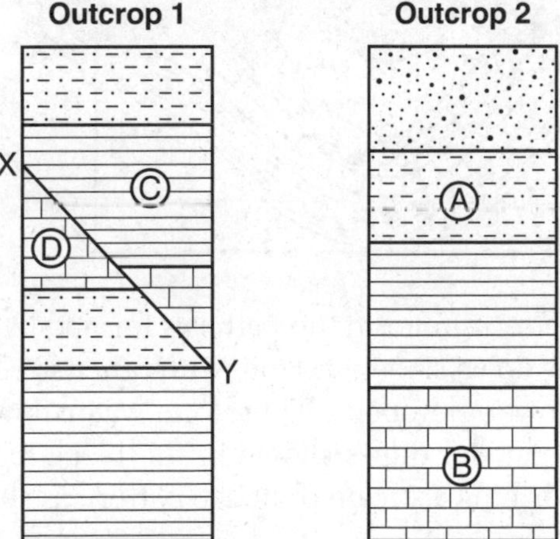

Find the Scheme for Sedimentary Rock Identification in the *Reference Tables for Physical Setting/Earth Science*. In the column labeled "Map Symbol," note the symbols that correspond to those shown in the outcrops. From each map symbol used in the outcrops, trace left to the column labeled "Rock Name." Note the name of the rock associated with each symbol.

Rock layers can sometimes be correlated on the basis of distinct similarities in physical characteristics, such as composition, color, thickness, and fossil remains. Based on composition, thickness, and sequence, the limestone, shale, siltstone sequence in the two outcrops can be correlated. Thus, limestone rock layer *D* correlates with limestone layer *B*; shale rock layer *C* correlates with the shale layer above *B* in Outcrop 2; and the siltstone rock layer above rock layer *C* correlates with siltstone layer *A*.

Note that all of the rock layers in the two outcrops are horizontal and do not appear to have been overturned. In fact, the question specifically states that the rock layers have *not* been overturned. In such cases, the principle of superposition states that the uppermost layer is the youngest. Therefore in outcrop 1, rock layer *C* is younger than rock layer *D*. Rock layers *D* and *B* correlate, so they are the same age. In outcrop 2, rock layer *A* is younger than rock layer *B*. Rock layer *A* is also younger than the shale layer beneath it that correlates with rock layer *C*. Therefore, rock layer *A* is younger than rock layer *C*. Thus, *A* is the youngest lettered rock layer.

23. **3**

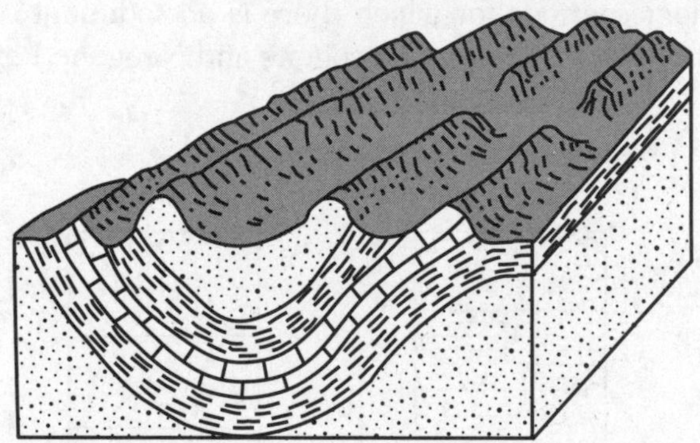

Stream drainage patterns refer to the patterns formed by the system of streams in an area as they flow down slopes and join with other streams. Water flows from higher elevations to lower elevations. Therefore, water flows along either side of the tops of the long, parallel ridges shown in the block diagram into the valleys between the ridges, forming a stream drainage pattern as shown in map (3).

WRONG CHOICES EXPLAINED

(1) Drainage diagram (1) shows a series of concentric, circular streams. This type of drainage pattern forms where there is series of concentric circular ridges and valleys. The block diagram in the question shows a series of long, parallel ridges and valleys, not concentric circular ridges and valleys.

(2) On a flat plain, the slope is gentle. Streams flow downhill slowly and typically form a dendritic stream drainage pattern. In a dendritic stream drainage pattern, there are numerous small streams that join together to form larger streams in a treelike pattern as shown in drainage pattern (2). The block diagram in the question shows a series of long, parallel ridges and valleys, not a flat plain.

(4) Drainage diagram (4) shows a series of streams flowing outward from a central area. Water flows from higher elevations to lower elevations. Therefore, the central area in this drainage diagram must be at a higher elevation than the surrounding areas. This is typically the case on rounded surfaces, such as conical hills or domes. The block diagram in the question shows a series of long, parallel ridges and valleys, not a conical hill or dome.

24. **4** Water is important to many weathering processes. Many chemical weathering processes require water. Water is a reactant in hydration and carbonation. Additionally, water provides a medium in which acid reactions can occur. Chemical reactions also occur faster at warm temperatures than at cool temperatures. Warm, rainy climates also support increased biological activity, ranging from plant action to production of humic acid as plant matter decomposes. Therefore, a warm, rainy climate will result in more pronounced weathering than will a cool, dry climate. An increase in rainfall will also result in an increase in erosion. Abundant rainfall and warm temperatures also promote the growth of a protective layer of vegetation that protects soil from rapid runoff and erosion, leading to rounded slopes. Thus compared to a landscape that develops in a cool, dry climate, a landscape that develops in a warm, rainy climate will most likely weather and erode faster, so the landforms are more rounded.

25. **2** Find the Inferred Properties of Earth's Interior chart in the *Reference Tables for Physical Setting/Earth Science.* In the cross section at the top of the chart, locate "Mid-Atlantic Ridge." Note that the Mid-Atlantic Ridge lies within the section labeled "Atlantic Ocean." Therefore, the crustal bedrock at the Mid-Atlantic Ridge is considered oceanic crust. Locate the "Density (g/cm^3)" scale along the right side of the cross section. Note that oceanic crust is listed as basaltic and has a density of 3.0 g/cm^3. Thus, Earth's crustal bedrock at the Mid-Atlantic Ridge is composed mostly of basalt, with a density of 3.0 g/cm^3.

26. **2** When an earthquake occurs, *P*-waves and *S*-waves start moving outward from the focus at the same time. However since these waves travel at different speeds, they do not arrive at a seismic station at the same time. It is given that the first *P*-wave traveled 4000 kilometers from the epicenter of the earthquake to the seismic station. Find the Earthquake *P*-Wave and *S*-Wave Travel Time graph in

the *Reference Tables for Physical Setting/Earth Science*. Locate 4000 kilometers (4.0×10^3 km) along the horizontal "Epicenter Distance ($\times 10^3$ km)" axis at the bottom of the graph. From this point, trace vertically until you intersect the bold line labeled "P." From this intersection, trace horizontally to the left to the "Travel Time (min)" axis. Read the *P*-wave travel time—7 minutes. If the first *P*-wave takes 7 minutes to travel the 4000 kilometers from the epicenter to the seismic station and arrives at 7:10:00 p.m., it must have left the epicenter 7 minutes before 7:10:00 p.m., or 7:03:00 p.m. Thus, the earthquake occurred at 7:03:00 p.m.

Now that you know the origin time of the earthquake, you can use this procedure to determine the travel time of the *S*-wave and its arrival time. From 4000 kilometers on the horizontal "Epicenter Distance ($\times 10^3$ km)" axis, trace vertically until you intersect the bold line labeled "S." From this intersection, trace horizontally to the left to the "Travel Time (min)" axis. Read the *S*-wave travel time—12 minutes 40 seconds. If the first *S*-wave takes 12 minutes 40 seconds to travel the 4000 kilometers from the epicenter to the seismic station and the earthquake started out at 7:03:00 p.m., it would arrive at the seismic station at 7:15:40 p.m.

Therefore, other information that could be determined from this recording was that the earthquake occurred at approximately 7:03.00 p.m., and the *S*-wave arrived at 7:15:40 p.m.

27. **1** Find the Average Chemical Composition of Earth's Crust, Hydrosphere, and Troposphere chart in the *Reference Tables for Physical Setting/Earth Science*. In the section labeled "Crust," locate the column headed "Percent by mass." According to the chart, the two elements that make up the largest percent by mass of Earth's crust are oxygen (46.10%) and silicon (28.20%). These percentages are best represented by the graph in Choice (1).

28. **4**

Note that the elongated hill in the aerial photograph has a teardrop shape and slopes downward from the wider end toward the narrower end. It is given that the hill formed by glacial deposition. An elongated, teardrop-shaped hill formed by glacial deposition is called a drumlin. Drumlins are formed when the ice of a glacier slides over previously deposited piles of sediment, causing them to become elongated into a teardrop shape. Thus, the elongated hill in the aerial photograph is best identified as a drumlin.

WRONG CHOICES EXPLAINED

(1) Kettles are bowl-shaped depressions formed when buried blocks of glacial ice melt. They are not elongated, teardrop-shaped hills.

(2) Moraine is material transported by a glacier and then deposited, typically as ridges at the glacier's edges or end. A drumlin may have started out as a pile of sediment deposited by a glacier (moraine), but then it was reshaped as the glacier advanced over it once again.

(3) A sand dune is a mound of sand deposited by wind. It is not a feature formed by glacial deposition.

29. **1**

www.alicehenderson.com

When a glacier moves down a valley, it alters the valley's shape. Since ice does not meander back and forth as readily as water does, the glacier often breaks through obstacles to form a straighter path and a flat bottom. The ice also cuts back the walls of the valley, making them steeper. The result is a wide, straight, U-shaped valley with a flat bottom and steep sides. As the glacier moves along, loose rock may freeze into the ice at the bottom of the glacier. As the glacier moves over bedrock, these rocks scrape parallel grooves or scratches, called striations, into the bedrock beneath the glacier. Therefore, the result of the glacial action of the valley glacier shown in the photograph will be a U-shaped valley with scratched and grooved bedrock.

30. **3** Find the Scheme for Sedimentary Rock Identification in the *Reference Tables for Physical Setting/Earth Science.* In the column labeled "Texture," locate the term "bioclastic." Note that "bioclastic" occurs in two rows. Trace these rows to the right to the column labeled "Composition." Note that bioclastic sedimentary rocks may be composed of calcite or carbon. Find the Properties of Common Minerals table in the *Reference Tables for Physical Setting/Earth Science.* In the column labeled "Mineral Name," locate calcite. Trace this row to the left to the column labeled "Composition," and note the formula for calcite – $CaCO_3$. Calcite is a compound consisting of elements represented by the symbols Ca, C, and O. In the "Chemical symbols" key at the bottom of the chart, note that "C" is the symbol for the element carbon. Therefore, the element always found in bioclastic rocks is carbon.

31. **4** Soil is the accumulation of loose, weathered material that covers much of Earth's land surface. The main component is weathered rock. However, a true soil also contains water, air, bacteria, and decayed plant and animal material (humus).

Since weathered rock is the main component of soil, one of the most important factors controlling the development of soil is the type of rock that was weathered. Therefore, bedrock composition is a major factor controlling the development of soil in a given location. Another important factor controlling the development of soil is the weathering process that converts the bedrock to soil. The single most important factor affecting weathering is climate. Climate is typically expressed in terms of two factors: temperature and precipitation. Both of these factors influence the type and rate of weathering. Warm climates favor chemical weathering; cold climates favor physical weathering. In both cases, the more moisture that is present, the more pronounced the weathering. Therefore, two major factors that control the development of soil in a given location are bedrock composition and climate.

32. **1**

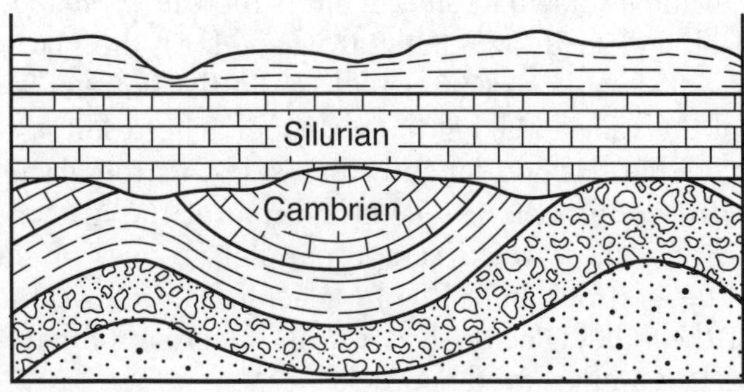

Layers of rock are generally deposited in an unbroken sequence. However if forces within Earth uplift rocks, deposition ceases. Erosion may wear away layers of rock before the land surface is low enough for another layer to be deposited. The uplift may cause rock layers to tilt or fold. The result is an unconformity, which is a break or gap in the sequence of a series of rock layers. Thus, the rocks above an unconformity are quite a bit younger than those below it. The erosional surface marking the gap between the older and younger rocks is generally drawn as a wavy line.

On the cross section, note that the Cambrian and Silurian limestones are separated by a wavy line. This wavy line represents an unconformity. Note, too, that below the wavy line the rock layers are folded and that above the wavy line the layers are flat.

Find the Geologic History of New York State chart in the *Reference Tables for Physical Setting/Earth Science*. In the column labeled "Period," locate the Cambrian and the Silurian. Trace right to the time scale labeled "Million years ago." Note that the Cambrian Period ended 488 million years ago and that the Silurian Period began 444 million years ago. So the Silurian limestone above the wavy line is more that 40 million years younger than the Cambrian limestone below the wavy line. In other words, there is a gap of more than 40 million years between the two limestone layers.

During those more than 40 million years, a number of things most likely happened. First, it can be inferred that uplift occurred because sediments stopped being deposited in an unbroken sequence. When rocks are uplifted, deposition stops because the surface is no longer at a low point where sediments settle to the ground. The folding in the lower rock layers indicates that forces acting on the crust were great enough to bend and fold rock. The folding pushed rock upward like a piece of paper on a desk when you push the two ends toward each other. Once uplifted, the rock would be exposed to weathering and be broken down into sediments. Erosion would then carry away the sediments, and the land surface would be worn down until it resembled the wavy line of the erosional surface.

Find the Scheme for Sedimentary Rock Identification in the *Reference Tables for Physical Setting/Earth Science.* In the column labeled "Rock Name," locate "Limestone." Trace left to the column labeled "Comments," and note that limestone is made up of "Precipitates of biologic origin or cemented shell fragments." This indicates that limestone forms in water, typically in shallow seas. In order for the Silurian limestone to be deposited, the land surface had to subside and become a seafloor. Thus, the last geologic event that occurred before the Silurian limestone was deposited was subsidence.

When taken together, the sequence of geologic events that occurred after the formation of the Cambrian limestone and before the formation of the Silurian limestone was uplift, weathering, erosion, and subsidence.

33. **3** Find the Scheme for Igneous Rock Identification in the *Reference Tables for Physical Setting/Earth Science*. In the chart at the top of the scheme, locate the "Granite." Then trace vertically down to the graph labeled "Mineral Composition (relative by volume)." Note that this column intersects the region labeled "Quartz (clear to white)." Thus, granite contains quartz.

Find the Scheme for Metamorphic Rock Identification in the *Reference Tables for Physical Setting/Earth Science*. In the column labeled "Rock Name," locate "Gneiss." Trace the row to the left to the column labeled "Composition." Note that gneiss contains quartz. Therefore, the two rocks that contain quartz are granite and gneiss.

WRONG CHOICES EXPLAINED

(1) Find the Scheme for Igneous Rock Identification in the *Reference Tables for Physical Setting/Earth Science*. In the chart at the top of the scheme, locate "Gabbro." Then trace vertically down to the graph labeled "Mineral Composition (relative by volume)." Note that this column does not intersect the region labeled "Quartz (clear to white)." Thus, gabbro does not contain quartz.

(2) Find the Scheme for Igneous Rock Identification in the *Reference Tables for Physical Setting/Earth Science*. In the chart at the top of the scheme, locate "Dunite." Then trace vertically down to the graph labeled "Mineral Composition (relative by volume)." Note that this column does not intersect the region labeled "Quartz (clear to white)." Thus, dunite does not contain quartz.

(4) Find the Scheme for Igneous Rock Identification in the *Reference Tables for Physical Setting/Earth Science*. In the chart at the top of the scheme, locate "Scoria." Then trace vertically down to the graph labeled "Mineral Composition (relative by volume)." Note that this column does not intersect the region labeled "Quartz (clear to white)." Thus, scoria does not contain quartz.

34. **4** Find the Properties of Common Minerals table in the *Reference Tables for Physical Setting/Earth Science*. In the column labeled "Fracture," note that only seven of the minerals on the chart display fracture. From each of these minerals, trace right to the column labeled "Composition," and check if the formula listed is one of the answer choices. Note that only Choice (4), $(Fe,Mg)_2SiO_4$, is a mineral that usually exhibits fracture, not cleavage.

35. **2** Find the Properties of Common Minerals table in the *Reference Tables for Physical Setting/Earth Science*. In the column labeled "Uses," locate "pencil lead, lubricants." From this phrase, trace right to the column labeled "Mineral Name." Note that the name of the mineral used to make pencil lead is graphite.

PART B-1

36. **1**

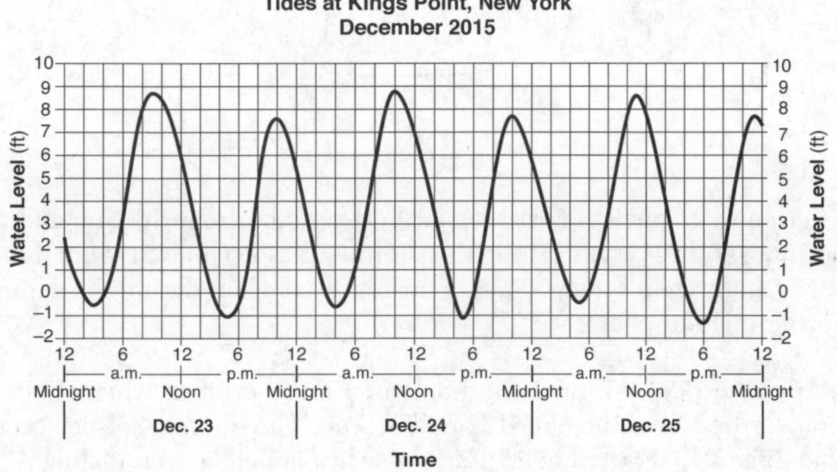

Locate the first low tide on Dec. 23. Trace vertically downward to the horizontal axis, and note that it occurred at 3:00 a.m. Now locate the first low tide on Dec. 24, and note that it occurred at 4:00 a.m., about 25 hours later. Next, locate the first low tide on Dec. 25, and note that it occurred at 5:00 a.m., about 25 hours after the first low tide on Dec. 24. Thus, the pattern is that the first low tide for each day shown on the graph occurs about 25 hours after the first low tide the day before. Therefore, if the pattern continues, the first low tide on Dec. 26 occurred at 6:00 a.m., 25 hours after the first low tide on Dec. 25.

37. **1** Note that the line representing water levels at Kings Park rises and falls in a repeating pattern. In other words, the changes in water level are cyclic. Note, too, that all of these changes repeat in a specific period of time. Therefore, they are predictable. For example, the first low tide on Dec. 23 occurs at 3:00 a.m., the first low tide on Dec. 24 occurs at 4:00 a.m., and the first low tide on Dec. 25 occurs at 5:00 a.m. Each of the first low tides are 25 hours apart. Thus, the changing water levels at Kings Park show a pattern that is cyclic and predictable.

38. **3**

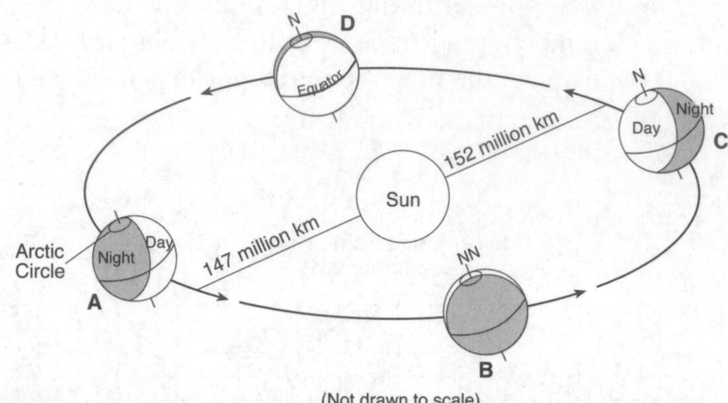

It is given that aphelion is Earth's farthest position from the Sun and occurs around July 4. July 4 occurs during the summer season (June 21–September 21) in the Northern Hemisphere. Thus, Earth is at aphelion during the summer in the Northern Hemisphere.

39. **4** On the diagram, note the point in Earth's orbit at which Earth is at aphelion, farthest from the Sun (152 million km). The summer solstice occurs on or about June 21, shortly before Earth reaches aphelion around July 4. Thus, location *C* corresponds to the summer solstice. The winter solstice occurs on about December 21, shortly before Earth reaches perihelion, closest to the Sun (147 million km). Thus, location *A* corresponds to the winter solstice. The fall equinox occurs about halfway between the summer solstice and the winter solstice. Note the arrows indicating the direction in which Earth revolves around the Sun in its orbit. Thus, location *D*, which is halfway between locations *C* and *A*, corresponds to the fall equinox. Therefore, location *B* corresponds to the spring equinox.

The location at which the Sun's vertical rays strike Earth's surface varies in an annual cycle with the seasons. For example, on June 21, Earth's Northern Hemisphere is tilted farthest toward the Sun and the Sun's vertical rays strike Earth's surface at 23.5° N latitude (Tropic of Cancer). At the equinoxes in March and September, Earth is tilted neither toward nor away from the Sun, and the Sun's vertical rays strike Earth's surface at the equator. On December 21, Earth's Southern Hemisphere is tilted farthest toward the Sun and the Sun's vertical rays strike Earth's surface at 23.5° S latitude (Tropic of Capricorn). Therefore, the Sun's vertical ray is moving from the equator southward to the Tropic of Capricorn between September 21 (fall equinox) and December 21 (winter solstice), that is, between positions *D* and *A*.

40. **4** As explained in the answer to question 39, position *A* corresponds to the winter solstice in the Northern Hemisphere. In moving from position *A* back to position *A*, Earth goes from one winter solstice to the next winter solstice, a period of one year. One year is roughly 365 days. Earth completes one rotation in 24 hours, or one day. Thus, in one year Earth, rotates about 365 times. Therefore, as Earth moves in its orbit from position *A* back to position *A*, it rotates 365 times.

41. **2** Earth's rotational axis is tilted at an angle of 23.5° from a perpendicular to Earth's orbital plane.

42. **2**

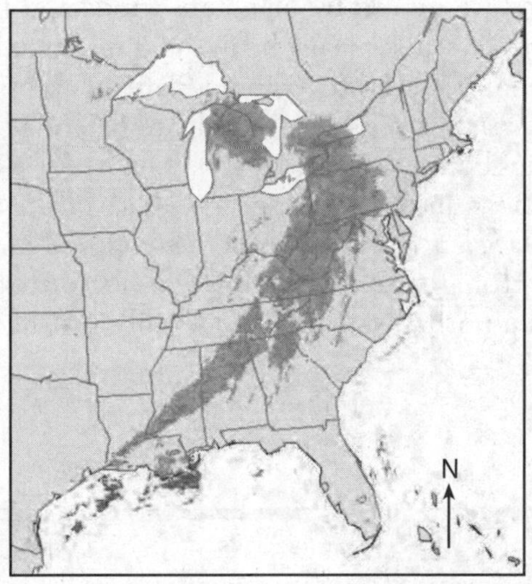

Most storm systems are associated with low-pressure centers. Air moves from regions of higher pressure toward regions of lower pressure. In a low-pressure system, the pressure gradient causes the air to move inward toward the low-pressure center. At the same time, the Coriolis Effect causes the moving air to be deflected to the right in the Northern Hemisphere. The combination of these two motions results in winds that blow inward in a counterclockwise pattern.

43. **4** Isobars are lines connecting points of equal air pressure. The spacing of isobars indicates the nature of the pressure gradient. The more closely spaced the isobars are on a weather map, the greater the pressure gradient and the higher the wind speed. Thus, the best evidence on a weather map to indicate high-speed winds near the center of the storm system would most likely be isobars drawn close together.

WRONG CHOICES EXPLAINED

(1) Clouds may cover the sky at many different wind speeds or even if there is no wind at all. Therefore, cloud cover of 100% would provide no information about wind speed.

(2) The type of precipitation is simply the form of moisture falling to the ground. Indicating the type of precipitation would provide no information about wind speed.

(3) High winds can occur at a wide range of temperature and de point values. Therefore, indicating temperature and dewpoint values would provide no information about wind speed.

44. **3** Storm systems are moved by planetary winds and by the polar front jet stream. So the storm track will typically be in the direction of the planetary winds and polar front jet stream. Find the Tectonic Plates map in the *Reference Tables for Physical Setting/Earth Science*. Note that the region of North America covered by this storm system lies between 30° N and 60° N latitude. Now find the Planetary Wind and Moisture Belts in the Troposphere chart in the *Reference Tables for Physical Setting/Earth Science*. Note that between 30° N and 60° N latitude, the planetary winds blow from the southwest toward the northeast. Therefore as this storm system follows a normal storm track, it will most likely move toward the northeast.

45. **3**

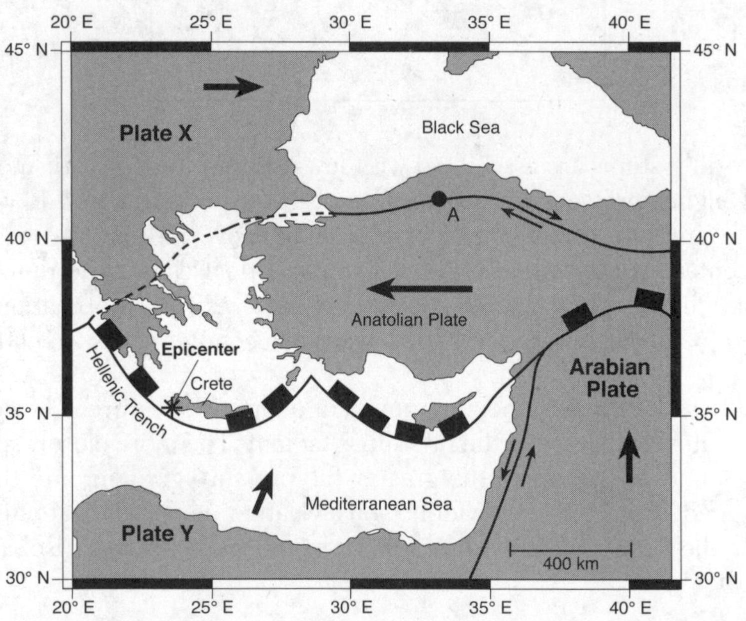

Note that the plate boundary at point *A* is shown as a solid line with arrows on either side of the line pointing in opposite directions parallel to the line. Find the Tectonic Plates map in the *Reference Tables for Physical Setting/Earth Science.* In the key at the bottom of the map, note that this symbol corresponds to a transform boundary. Therefore, the plate boundary represented at point *A* is a transform plate boundary.

WRONG CHOICES EXPLAINED

(1) Find the Tectonic Plates map in the *Reference Tables for Physical Setting/Earth Science.* In the key at the bottom of the map, locate the symbol representing a divergent plate boundary. Note that the line and arrows at point *A* do not correspond to this symbol.

(2) Find the Tectonic Plates map in the *Reference Tables for Physical Setting/Earth Science.* In the key at the bottom of the map, locate the symbol representing a convergent plate boundary. Note that the solid line and arrows at point *A* do not correspond to this symbol.

(4) Find the Tectonic Plates map in the *Reference Tables for Physical Setting/Earth Science.* In the key at the bottom of the map, Note that symbol for a complex plate boundary is a dashed line, not the solid line and arrows shown at point *A*.

46. **1** The segment of plate *X* shown on the map lies just north of about 40° N latitude between 20° E and 40° E longitude. Find the Tectonic Plates map in the *Reference Tables for Physical Setting/Earth Science.* Locate the region just north of 40° N latitude, between 20° E and 40° E longitude. Note that this region corresponds to the Eurasian Plate.

The segment of plate *Y* shown on the map lies north of about 30° N latitude between 20° E and 35° E longitude. Find the Tectonic Plates map in the *Reference Tables for Physical Setting/Earth Science.* Locate 40° N latitude, between 20° E and 35° E longitude, just west of the Arabian Plate. Note that this region corresponds to the African Plate. (Note that Africa is split in this map projection; one-half is near the right edge and one-half is near the left edge of the map. The Arabian Plate is shown next to the part of Africa near the left edge of the map.) Therefore, plate *X* is the Eurasian Plate, and plate *Y* is the African Plate.

47. **4** The two New York State index fossils most closely related to the corals that were radioactively dated in this study are themselves corals. Find the Geologic History of New York State chart in the *Reference Tables for Physical Setting/Earth Science.* In the column labeled "Time Distribution of Fossils (including important fossils of New York)," locate the vertical bar labeled "Corals." Note that

the circled letters representing index fossils on this bar are "T," "U," and "V." Thus, the index fossils at the bottom of the chart corresponding to the circled letters "T," "U," and "V" are corals. Locate the drawings of "T," "U," and "V" at the bottom of the chart, and note their names: "T"—*Lichenaria*, "U"—*Cystiphyllum*, and "V"—*Pleurodictyum*. Thus, the two New York State index fossils most closely related to the corals that were radioactively dated in this study are choice (4)—*Lichenaria* and *Pleurodictyum*.

48. **3** Tsunamis may be only a few meters high, but they have very long wavelengths and travel much more rapidly than do ordinary ocean waves. Tsunamis have been clocked moving faster than 500 km/hr with wavelengths as great as 200 kilometers. When tsunamis reach shallow water, they are slowed by friction with the ocean bottom and begin to pile up as large waves. In bays and narrow channels, their high speed and long wavelength may be funneled into huge breaking waves more than 20 m high. The force exerted by such a huge, fast-moving mass of water can do extensive damage and cause widespread coastal flooding. The best thing to do if a tsunami approaches is to evacuate to higher ground. Therefore, the activity that could best prepare residents along the Mediterranean coast to reduce loss of human life during a future tsunami is to plan evacuation routes to higher ground.

WRONG CHOICES EXPLAINED

(1) Boarding up windows will do little to reduce the loss of human life because buildings are often completely inundated or washed away during a tsunami.

(2) Removing heavy objects from walls will do little to reduce the loss of human life because this will do nothing to prevent water from entering the room and drowning the inhabitants.

(4) A basement is likely the worst place to be during a tsunami because water will flood into the basement and drown the inhabitants.

49. **1**

www.nationalgeographic.com

Note that the feature shown in the photograph is located above the surrounding land, and that there is no vegetation covering the nearby land surface. The main agent of erosion in dry climates is wind. Sandblasting is the process by which sand carried by the wind collides with and wears away solid objects. When the sand particles collide with the solid object, the impact causes small pieces of the object to break off. This can cause the object being hit with particles to become smooth or worn. Thus, the agent of erosion that is most likely sandblasting this rock formation is wind.

50. **2** It is given that the photograph shows a sandstone erosional feature. Therefore, the rock making up this feature is sandstone. Find the Scheme for Sedimentary Rock Identification in the *Reference Tables for Physical Setting/ Earth Science.* In the column labeled "Rock Name," locate "Sandstone." Then trace left to the column labeled "Grain Size." Note that sandstone is made up of sand grains that range from 0.006 cm–0.02 cm.

PART B-2

51.

Radioactive Isotope X (%)	Disintegration Product Z (%)	Time (billion years)
100	0	0
50	50	1.3
25	75	2.6
12.5	87.5	3.9
6.25	93.75	5.2
3.125	96.875	6.5

To construct a graph that shows the percentages of disintegration product Z forming over 6.5 billion years, use the data in the table to plot the points for disintegration product Z and time in billion years as follows. The table states that at 0 billion years, there is 0% of Z. To plot these coordinates, locate "0" on the horizontal scale labeled "Time (billion years)" and locate "0" on the vertical scale labeled "Percentage (%)." Plot the point of intersection. Next, the table states that at 1.3 billion years, there is 50% of Z. Locate "1.3" on the horizontal scale labeled "Time (billion years)." Trace upward along the "1.3" line until you intersect the line corresponding to "50" on the vertical scale labeled "Percentage (%)." Plot the point of intersection. Repeat this process for each of the remaining coordinates: (2.6, 75), (3.9, 87.5), (5.2, 93.75), and (6.5, 96.875). Then connect the plotted intersection points with a line as shown in the graph below.

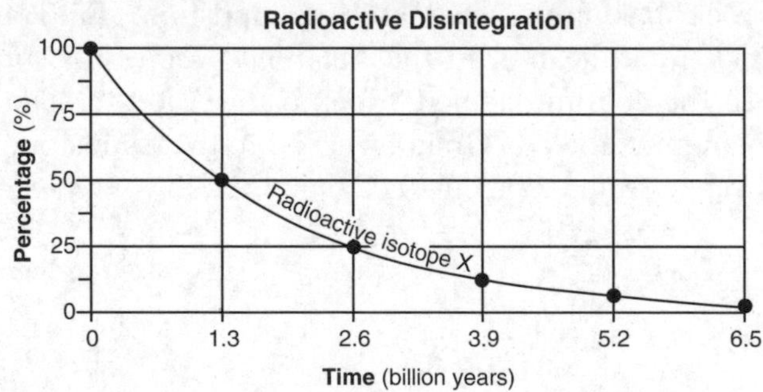

One credit is allowed **if the centers of *all six* plots are within or touch the circles shown and are correctly connected with a line that passes within or touches each circle.**

Note: Credit is allowed if the student-drawn line does *not* pass through the student plots but is still within or touches the circles.

52. Half-life is the time required for one-half (50%) of the atoms of a radioactive isotope to decay into a stable disintegration product. According to the table, every 1.3 billion years, half of the remaining radioactive isotope *X* decays into disintegration product Z. Therefore, the half-life of radioactive isotope *X* is 1.3 billion years. Find the Radioactive Decay Data chart in the *Reference Tables for Physical Setting/Earth Science*. In the column labeled "Half-Life (years)," locate 1.3 billion (1.3×10^9). Trace left to the column labeled "Radioactive Isotope," and note that potassium-40 has a half-life of 1.3×10^9 (1.3 billion) years. Thus, radioactive isotope *X* is potassium-40.

One credit is allowed for an acceptable response. Acceptable responses include but are not limited to:

- **potassium-40**
- **^{40}K**
- **K-40**

53. On the table, locate "3.9" in the column labeled "Time (billion years)." Trace left to the column labeled "Radioactive Isotope *X* (%)." Note that after 3.9 billion years, 12.5% of radioactive isotope *X* will remain. Therefore, after 3.9 billion years, 12.5% of the original 300-gram sample of radioactive isotope *X*, or 37.5 grams, will remain ($300 \times 0.125 = 37.5$).

One credit is allowed for **37.5 g.**

54.

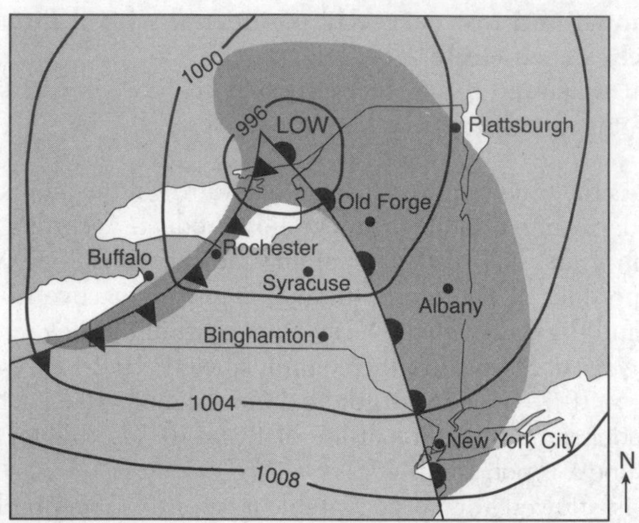

Find the Key to Weather Map Symbols in the *Reference Tables for Physical Setting/Earth Science.* In the section labeled "Fronts," note that the symbol for a cold front is a bold line with triangles. Locate the symbol corresponding to a cold front on the map, and note that it is located very near Rochester. A cold front marks the boundary where cold air is advancing against warmer air. The side of the front on which the symbols are drawn indicates the direction in which the cold air is advancing. Thus, the air mass west of the cold front is a cold air mass and the air mass east of the cold front is a warm air mass. Therefore, when the cold front reaches Syracuse, it will be well past Rochester and Rochester will be located in a cold air mass, not in a warm air mass. Cold air is denser than warm air, so cold air exerts a higher air pressure. Therefore by the time the cold front has reached Syracuse, the air pressure at Rochester will have increased.

Note the dark shaded area along the cold front indicating areas of precipitation. By the time the cold front reaches Syracuse, this area will be well past Rochester. Cold air masses contain less moisture than do warm air masses, so clouds are less likely to form with cold air masses. Therefore, the cloud cover in Rochester will most likely decrease.

One credit is allowed if both responses are correct. Acceptable responses include but are not limited to:

Change in air pressure:

- **slight decrease, then a steady increase**
- **generally increasing/rising**
- **lower to higher**
- **greater**

Amount of cloud cover:

- **It decreased.**
- **lower percent**
- **clear/0%**
- **There are fewer clouds.**
- **little cloud cover**

55. Find the Key to Weather Map Symbols in the *Reference Tables for Physical Setting/Earth Science.* In the section labeled "Station Model Explanation," note how the values for air temperature, barometric pressure, wind direction, and wind speed are represented on a station model. Note the station model for Buffalo, New York. The value corresponding to air temperature is "72," which is 72°F. The value corresponding to barometric pressure is "010," which is 1001.0 mb. (Air pressure rarely rises above 1050.0 mb or drops below 950.0 mb. Therefore, the general rule is to place a 9 in front of the digits on the station model if they are greater than 500 and to place a 10 in front of the digits if they are less than 500 and then insert a decimal between the last two digits.) The line representing wind direction extends to the northwest, corresponding to a NW wind direction. There is one whole feather and one half feather on the wind direction line, corresponding to a wind speed of 15 knots.

One credit is allowed if ***all four*** **weather variables are correct as shown below.**

Air Temperature (°F)	Barometric Pressure (mb)	Wind Direction from the	Wind Speed (knots)

Note: Credit is allowed if the student places the correct units in the boxes as part of their response. Do *not* allow credit for a barometric pressure of 010, as this is the format used on a station model, not the actual barometric pressure.

56. When air temperature cools to the dewpoint, water vapor in the air begins to condense and the relative humidity is 100%. Therefore, the relative humidity at Albany when the air temperature is equal to the dew point is 100%.

One credit is allowed for **100%.**

57. One of the most common instruments used to measure wind speed is an anemometer. A simple anemometer has cups mounted on a shaft that cause the shaft to spin when wind blows against them. The faster the wind blows, the faster the cups spin the shaft and the higher the number that registers on the anemometer's scale. Small handheld anemometers, called wind gauges, contain small fan blades that spin in the wind. There are also a variety of other instruments that can be used to measure wind speed. A wind sock is a cone-shaped tube made of fabric that resembles a giant sock but is open at both ends. As wind speed increases, the sock rises like a flag in a breeze. The higher the wind speed, the closer the sock flies to horizontal. Some handheld digital wind speed meters contain a metal wire that is heated by an electric current. The wire is cooled by air flow, and the meter instantly calculates wind speed. Doppler radar can also be used to determine wind speed. For Doppler radar, atmospheric objects moving inbound (toward the radar) produce a positive shift in frequency of the radar signal. Objects moving away from the radar (outbound) produce a negative shift in frequency. This change in frequency can then be analyzed to determine wind speed.

One credit is allowed. Acceptable responses include but are not limited to:

- **anemometer**
- **wind sock**
- **wind speed meter**
- **wind gauge**
- **Doppler radar/radar**

58.

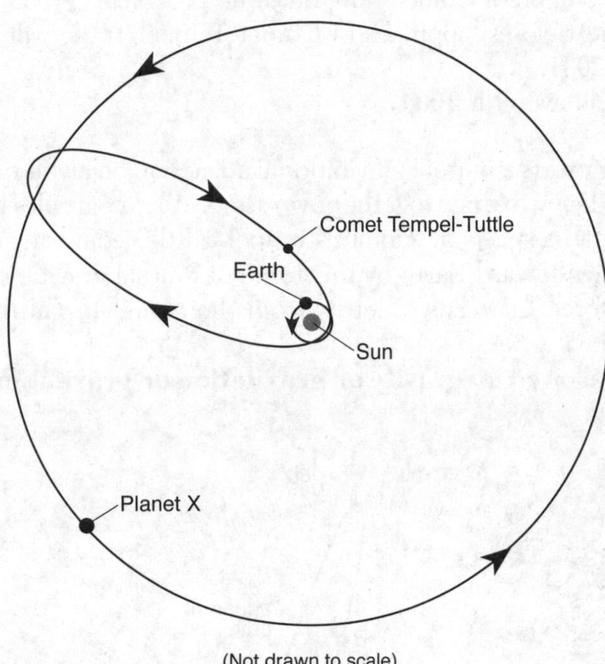

A comet in an elliptical orbit is orbiting the foci of that ellipse. The reading passage states that Comet Tempel-Tuttle orbits our Sun. Therefore, our Sun is located at one of the foci of the elliptical orbit of Comet Tempel-Tuttle.

One credit is allowed for the **Sun.**

59. The diagram shows that the orbit of planet *X* is slightly less than the farthest distance of Comet Tempel-Tuttle from the Sun. The reading passage states that Comet Tempel-Tuttle's orbital distance from the Sun is "2900 million kilometers at its farthest distance." Find the Solar System Data chart in the *Reference Tables for Physical Setting/Earth Science.* In the column labeled "Mean Distance from Sun (million km)," locate the value closest to but less than 2900 million kilometers, which is 2871.0. From this value, trace left to the column labeled "Celestial Object" Note that Uranus is located 2871.0 million kilometers from the Sun. Therefore, the solar system planet represented by planet *X* is Uranus.

One credit is allowed for **Uranus.**

60. The reading passage states that Comet Tempel-Tuttle's "two most recent closest approaches to the Sun occurred in 1965 and one revolution later in 1998." Thus, the period of revolution of Comet Tempel-Tuttle is 33 years (1998 − 1965 = 33). Therefore, the next closest approach of Comet Tempel-Tuttle will occur 33 years after 1998, or in 2031.

One credit is allowed for **2031.**

61. All matter exerts a mutual gravitational attraction on all other matter. Thus, matter has a tendency to draw together over time. When particles of matter, such as the debris left in space by Comet Tempel-Tuttle, come near Earth, those particles are drawn toward Earth by this force of mutual gravitational attraction. Therefore, the force that causes debris from the comet to fall through Earth's atmosphere is gravity.

One credit is allowed for **gravity *or* gravitation *or* gravitational pull.**

62.

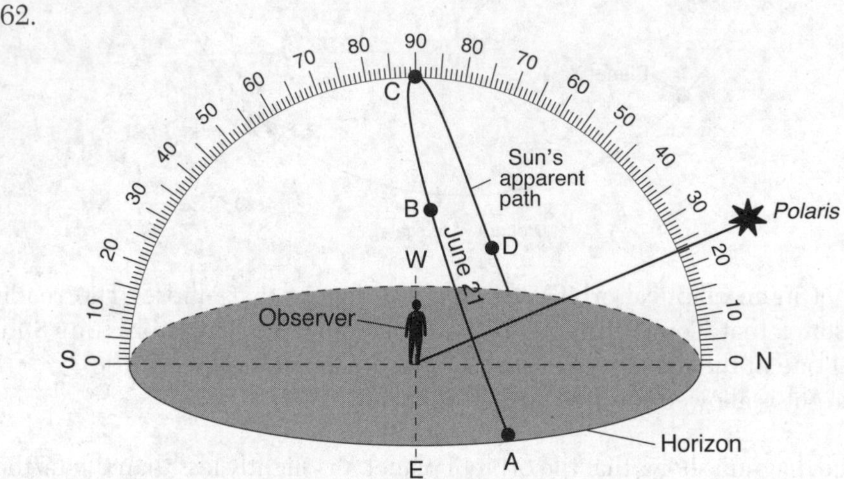

At sunrise at position *A*, the Sun would be low in the sky to an observer. The low angle of the Sun would cause a long shadow to be cast by the observer. At position *B*, the Sun would be higher in the sky and the shadow cast by the observer would be shorter. At position *C*, the Sun would be at an even higher in the sky and the shadow cast by the observer would be even shorter than at position *B*. At position *D*, the Sun would again be low in the sky and the observer would cast a long shadow. Thus, from position *A* to position *D*, the length of the observer's shadow will decrease from position *A* to position *C* and then increase from position *C* to position *D*.

One credit is allowed for an acceptable response. Acceptable responses include but are not limited to:

- **The length of the observer's shadow will decrease from *A* to *C*, then increase from *C* to *D*.**
- **The shadow will get shorter, then longer.**
- **The shadow was longest at *A*, became shortest at *C*, then became longer at *D*.**
- **decreases then increases**

63. The diagram shows that the altitude of *Polaris* at the observer's position is 23.5°. The altitude of *Polaris* is the same as an observer's latitude north of the equator. Therefore, the observer is located at 23.5° N latitude. Find the Surface Ocean Currents map in the *Reference Tables for Physical Setting/Earth Science.* Along the right edge of the map, note that 23.5° N is labeled "Tropic of Cancer." Thus, one piece of evidence that supports the inference that the observer is located at the Tropic of Cancer is the altitude of *Polaris*.

From the altitude of *Polaris*, it is clear that the observer is located in the Northern Hemisphere. It is given that the diagram represents the Sun's path on June 21, which is the summer solstice in the Northern Hemisphere. The diagram shows that at its highest point (noon), the altitude of the Sun is 90°. In other words, the Sun is directly overhead the observer. On June 21, the Sun is directly overhead (altitude 90°) at the Tropic of Cancer, which is located at latitude 23.5° N.

One credit is allowed for an acceptable response. Acceptable responses include but are not limited to:

- **The altitude of *Polaris* is at 23.5° *or* $23\frac{1}{2}$°.**
- **The noon Sun for June 21 is directly overhead.**
- **The highest Sun position is at 90° (zenith) on June 21.**
- **the angle/altitude of *Polaris***

Note: Credit is *not* allowed for $23\frac{1}{2}$° N or for 23.5° North, because a compass direction should not be included with an altitude value.

64. September 23 is the fall equinox. When Earth is at the equinox position, day and night are equal in length. Thus, an observer would experience 12 hours of daylight and 12 hours of darkness. Therefore, on September 23, the number of daylight hours at this location will be 12 hours.

One credit is allowed for **12 h.**

65. In 1851, French physicist Jean Foucault suspended a heavy iron ball on a long steel wire from the top of the dome of the Pantheon in Paris. As this pendulum swung back and forth, it appeared to change direction slowly, passing over different lines on the floor until eventually coming full circle to its original position after 24 hours. Since Foucault knew that the path of a freely swinging pendulum would not change on its own, he concluded that the apparent shift in the direction of swing of the pendulum was due to the floor (Earth's surface) rotating beneath the pendulum. Thus, the apparent change in direction of swing of a Foucault pendulum provides evidence that Earth is rotating. Therefore, the device that was used to first demonstrate that Earth rotates was the Foucault pendulum.

One credit is allowed for **Foucault pendulum *or* pendulum.**

PART C

66.

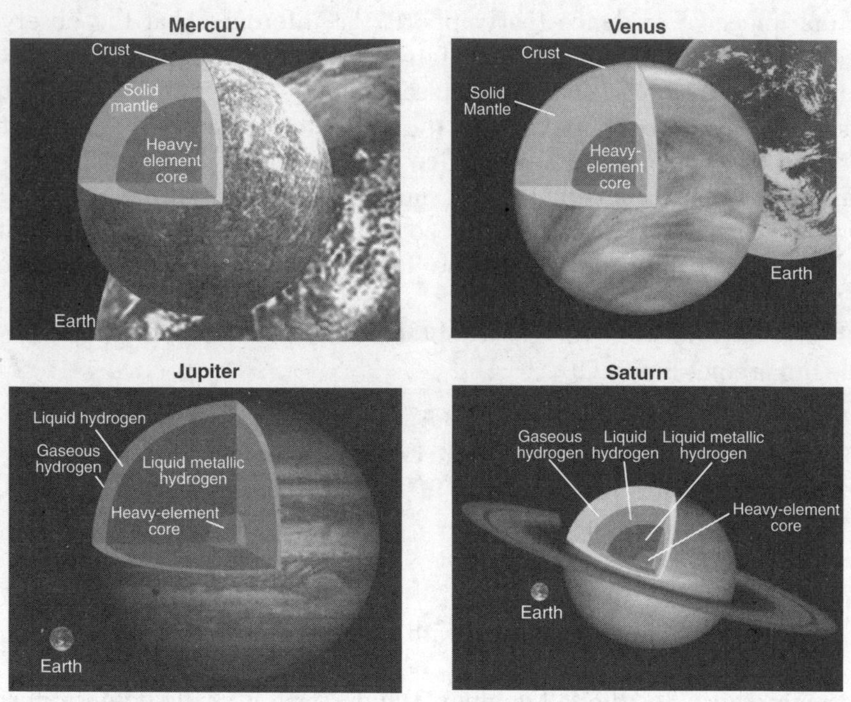

Seeds, Michael and Backman, Dana. 2011. *The Solar System.*

Find the Solar System Data chart in the *Reference Tables for Physical Setting/Earth Science.* In the column labeled "Celestial Object," locate Earth and Jupiter. Trace these rows right to the column labeled "Equatorial Diameter (km)." Note that Earth has an equatorial diameter of 12,756 km and

Jupiter has an equatorial diameter of 142,984 km. Therefore, compared to Earth's equatorial diameter, Jupiter's equatorial diameter is 11.2 times larger ($142{,}984 \div 12{,}756 = 11.2$).

One credit is allowed for any value from **11 to 12 times larger.**

67. Planets are visible from Earth by the light they reflect from the Sun. Planets vary greatly in apparent brightness, or how bright they appear in the sky. Planets may appear bright because they reflect more light from the Sun (greater albedo) or because they are closer to Earth. Larger planets reflect more light than do smaller planets. Find the Solar System Data chart in the *Reference Tables for Physical Setting/Earth Science.*

In the column labeled "Celestial Object," locate Mercury and Jupiter. Trace these rows right to the column labeled "Equatorial Diameter (km)." Note that Mercury has an equatorial diameter of 4,879 km and Jupiter has an equatorial diameter of 142,984 km. Therefore, the reason why Jupiter appears brighter in the night sky than Mercury, despite Jupiter's greater distance from Earth, is that Jupiter is so much larger than Mercury and reflects more sunlight than Mercury.

One credit is allowed for an acceptable response. Acceptable responses include but are not limited to:

- **Jupiter is much larger than Mercury.**
- **Larger objects appear brighter than smaller objects.**
- **Jupiter reflects more light/has a greater albedo.**

68. The planets can be divided by mass and density into the terrestrial (Earthlike) and the Jovian (Jupiter-like) planets. The terrestrial planets are small, dense, and rocky. They are composed mostly of metal silicates and iron. The terrestrial planets include the four innermost planets: Mercury, Venus, Earth, and Mars. The Jovian planets are large, low in density, and gaseous. They are composed mainly of hydrogen and helium. The Jovian planets include the outer planets: Jupiter, Saturn, Uranus, and Neptune. The models show three of the terrestrial planets: Mercury, Earth, and Venus.

One credit is allowed for *two* correct terrestrial planets and a correct explanation. Acceptable responses include but are not limited to:

Terrestrial planets:

- **Mercury**
- **Venus**
- **Earth**

Explanations:

- **Terrestrial planets are small and rocky.**
- **Their interior layers are not gaseous.**
- **Terrestrial planet densities are greater than Jovian planet densities.**
- **They are closer to the Sun.**
- **They have solid crusts.**

Note: Do *not* accept Mars as the terrestrial planet because it is *not* shown in the models.

69.

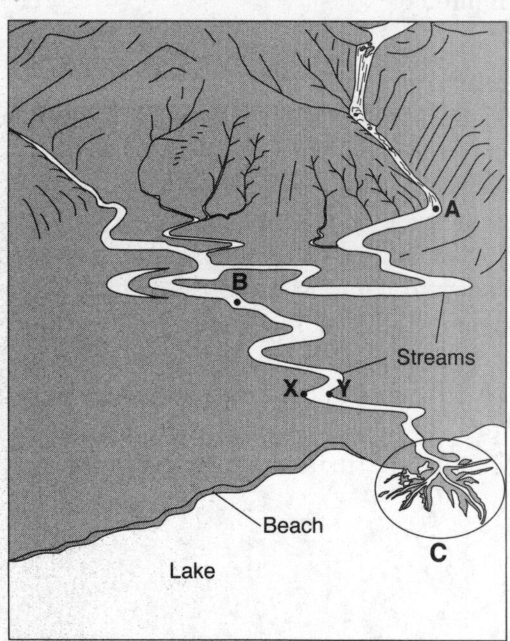

The diagram shows a large, fan-shaped deposit where a stream flows into the lake. When a stream flows into a standing body of water, the stream slows down and most of the stream's sediment is deposited. The resulting landform is a large, flat, fan-shaped pile of sediment at the mouth of the stream called a delta. Thus, depositional feature labeled *C* is a delta.

One credit is allowed for **delta.**

70. Note that in the diagram, points *X* and *Y* cut across a section of the stream where the stream channel curves in a meander. When a stream curves, or meanders, water velocity is higher along the outside of the curve and lower along the inside of the curve because the water is forced to cover a greater distance along the outside of the curve. Since stream erosion is greatest where the velocity of the water is highest, the outside of the curve of a meandering channel experiences more erosion than the inside of the curve. In the diagram, point *X* is located at the outside of a curve in the stream, which is where the velocity of the water and erosion are the greatest. Point *Y* is located at the inside of a curve in the stream, which is where the velocity of the water and erosion are the least. Therefore, the streambed will be more deeply eroded (the stream channel will be deeper) near point *X* than it is near point *Y*. Draw a line from *X* to *Y* on the cross section in your answer booklet showing that the stream channel is deepest near point *X*.

One credit is allowed if **the student's line is drawn from point X to point *Y* and shows that the stream channel is deepest near side X.**

Example of a 1-credit response:

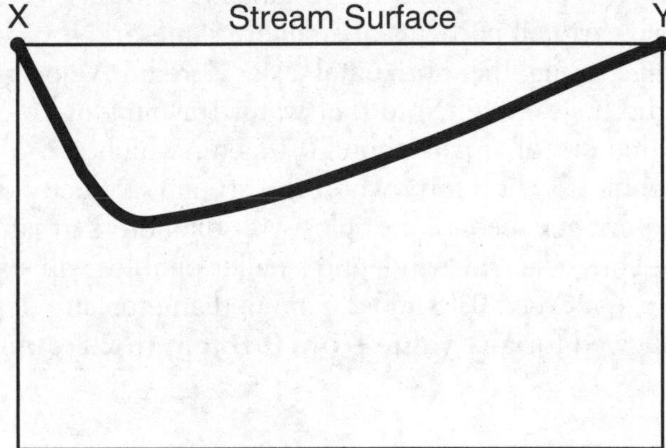

71. As sediments are carried along by the moving water in a stream, they bounce off of and rub against one another, a process called abrasion. When these sediments collide, smaller pieces break off sharp corners and edges on their surface, causing the sediments to become smaller and more rounded. The sediments are also abraded as they bounce and roll along the bottom of the streambed. Therefore, as sediments are transported downstream, their size will become smaller and their shape more rounded.

One credit is allowed for *two* correct responses. Acceptable responses include but are not limited to:

Size:

- **becomes smaller**
- **decrease**

Shape:

- **becomes more rounded**
- **less angular**
- **Sharp edges are worn down.**

Note: The term "smooth" is *not* accepted for the shape of the sediments, because smooth does not describe a shape but a texture.

72. Find the Relationship of Transported Particle Size to Water Velocity graph in the *Reference Tables for Physical Setting/Earth Science*. Along the horizontal axis "Stream Velocity (cm/s)," locate a stream velocity of 100 cm/s. Trace upward to the bold curve and then horizontally to the axis labeled "Particle Diameter (cm)." Note that water traveling at 100 cm/s can move particles with diameters of up to 2.5 cm, which are in the size range labeled "Pebbles." At that velocity, the stream can also carry all particles of smaller diameter. Next, locate a stream velocity of 10 cm/s along the horizontal axis "Stream Velocity (cm/s)," and trace upward to the bold curve. Note that water traveling at 10 cm/s can move particles with diameters of up to about 0.18 cm, which are also in the size range labeled "Pebbles." Therefore, when the stream's velocity decreases from 100 to 10 centimeters per second, particles with diameters greater than 0.2 cm will be deposited. Thus, clay, silt, sand, and smaller pebbles will stay in transport; some pebbles (those between 0.18 and 2.5 cm in diameter) are deposited.

One credit is allowed for **any value from 0.18 cm to 2.5 cm.**

73.

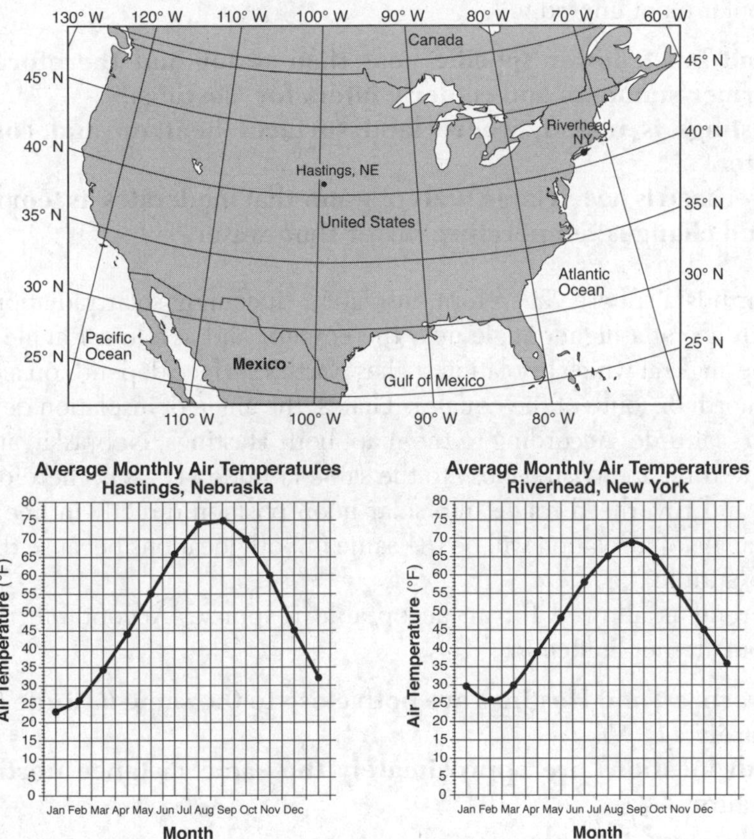

According to the map, Hastings, Nebraska, and Riverhead, New York, are located at nearly the same latitude. Therefore, both cities receive roughly the same intensity of insolation. However, land surfaces increase in temperature more than water surfaces when insolation strikes them. Land surfaces also cool off more rapidly than water surfaces. As a result, air temperatures are warmer in the summer and cooler in the winter over landmasses than they are over oceans at the same latitude. Therefore, large bodies of water tend to moderate the temperatures of nearby landmasses by warming them in winter and cooling them in summer. Thus, Hastings, Nebraska, most likely has a greater difference between maximum and minimum temperatures than Riverhead, New York, because Riverhead is located near a large body of water that moderates its temperatures. Therefore, Hastings has a greater annual temperature range than Riverhead because Hastings is located inland where land surfaces heat up and cool down faster. In contrast, Riverhead is near a large body of water that moderates its temperatures.

One credit is allowed for an acceptable response. Acceptable responses include but are not limited to:

- **Land has a lower specific heat than water and therefore causes warmer summers and cooler winters for Hastings.**
- **Hastings is inland, where land surfaces heat up and cool down faster.**
- **Riverhead is near a large body of water that moderates its temperature.**
- **Land changes temperature faster than water.**

74. Earth is a sphere. Therefore, insolation (incoming solar radiation) strikes Earth's surface at a higher angle near the equator and at a lower angle near the poles. The angle at which insolation strikes Earth's surface depends on a location's distance north or south of the equator. That is, the angle of insolation depends on a location's latitude. According to the map, both Hastings, Nebraska, and Riverhead, New York, are located close to the same latitude (41° N). Therefore, when Hastings and Riverhead rotate into solar noon position directly in line with the Sun, the angle of insolation will be the same in both locations because they are at the same latitude.

One credit is allowed for an acceptable response. Acceptable responses include but are not limited to:

- **Riverhead and Hastings are both close to the same latitude (approximately 41° N).**
- **Both locations are approximately the same distance north of the equator.**
- **at the same latitude**

75. Find the Planetary Wind and Moisture Belts in the Troposphere chart in the *Reference Tables for Physical Setting/Earth Science.* Note that the prevailing wind belt in the region between 30° N and 60° N is labeled "S.W. Winds." These winds are commonly called the prevailing southwesterlies or westerlies. Hastings, Nebraska, and Riverhead, New York, are both located close to the same latitude—41° N. Therefore, Hastings and Riverhead are located in this planetary wind belt. Thus, the planetary wind belt that primarily influences the climates of both Hastings and Riverhead are the S.W. winds.

One credit is allowed for an acceptable response. Acceptable responses include but are not limited to:

- **prevailing southwesterlies *or* westerlies**
- **S.W. Winds**

76. The map shows that Riverhead, New York, is located along the east coast of North America at about 41° N latitude. Find the Surface Ocean Currents map in the *Reference Tables for Physical Setting/Earth Science.* Locate the position corresponding to 41° N latitude along the east coast of North America. Note that the surface ocean current nearest this location is labeled the "Gulf Stream C." Therefore, the ocean current that most likely has the greatest effect on the climate of Riverhead is the Gulf Stream Current.

One credit is allowed for **Gulf Stream Current.**

77.

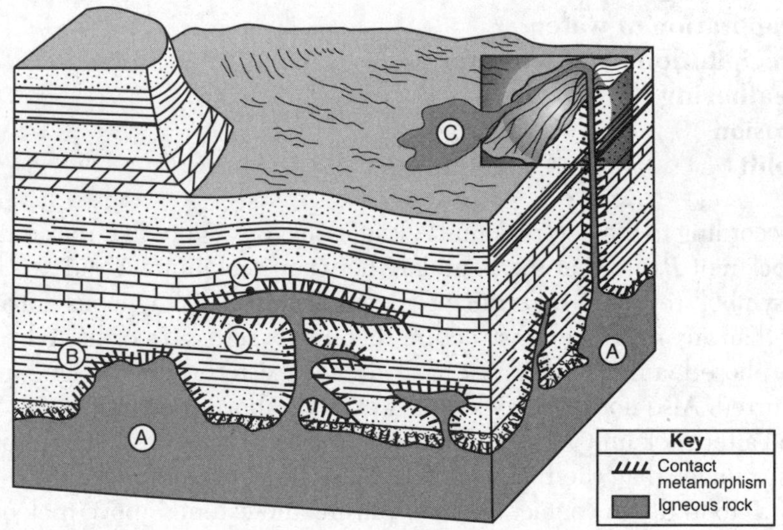

All sedimentary rock forms from sediments: either rock fragments (clastic sediments) or mineral crystals (precipitates or evaporites). Clastic sediments are formed when rock is uplifted, weathered, eroded, and then deposited. Sediments that are precipitates form when chemical reactions occur in water, forming insoluble substances that crystallize and then sink to the bottom. Sediments that are evaporites form when seawater evaporates. Removing water from a solution (dewatering) but not removing the dissolved minerals causes the solution to become more concentrated. When the solution of dissolved minerals in the remaining water becomes saturated, dissolved minerals start to crystallize out of the solution, and those crystals sink to the bottom.

Find the Rock Cycle in Earth's Crust diagram in the *Reference Tables for Physical Setting/Earth Science.* Locate the oval labeled "Sediments." Follow the arrow leading to the box labeled "Sedimentary Rock." Note the processes listed along the arrow: deposition, burial, compaction, and cementation. These processes are collectively called lithification, because they convert sediment to rock.

One credit is allowed for *two* correct responses. Acceptable responses include but are not limited to:

- **compaction**
- **cementation**
- **lithification**
- **deposition/sedimentation**
- **burial**
- **dewatering**
- **evaporation of water**
- **precipitation from seawater**
- **weathering**
- **erosion**
- **uplift**

78. According to the key, rock unit *A* is igneous rock. Note that rock unit *A* cuts across rock unit *B* and that the boundary between rock units *A* and *B* is marked with a symbol representing contact metamorphism. An igneous intrusion is younger than any rock layer or structure that it cuts across. In order to be contact metamorphosed, a rock layer must have already existed when the igneous intrusion occurred. Also note that rock unit *A* extends above rock unit *B*. If rock unit *B* formed after rock unit *A*, rock unit *B* would cover rock unit *A*. Therefore, rock unit *A* is younger than the rock unit *B* because it cuts across rock unit *B*, caused rock unit *B* to undergo contact metamorphism, and extends above rock unit *B*.

One credit is allowed for an acceptable response. Acceptable responses include but are not limited to:

- **Rock unit *A* cuts across (cross-cutting) rock unit *B*.**
- **There is contact metamorphism in rock unit *B*.**
- **Rock unit *A* intrudes into rock unit *B*.**
- **Parts of *A* extend above *B*.**

79. According to the key, both rock unit *A* and rock unit *C* are igneous rock. Note that rock unit *C* is an extension of rock unit *A* and has flowed out onto Earth's surface. Also note the broad, cone-shaped mound where rock unit *A* emerges at the surface—a volcano. Thus, rock unit *C* is an extrusive igneous rock that flowed out of a volcano as lava and then cooled and solidified at Earth's surface. When molten rock emerges onto Earth's surface, it is suddenly exposed to an environment that is hundreds or even thousands of degrees cooler than its temperature, so the molten rock cools rapidly. During rapid cooling, mineral

crystals may not have any time to form, in which case the solid formed is a noncrystalline glassy material. Alternatively, the mineral crystals may not able to grow very large before the rock solidifies.

Find the Scheme for Igneous Rock Identification in the *Reference Tables for Physical Setting/Earth Science*. In the chart at the top of the scheme, locate the section labeled "Extrusive (Volcanic)." Trace right to the column labeled "Crystal Size." Note that extrusive rocks are either noncrystalline or have crystals that are less than 1 mm in size. Thus, the igneous rock that formed at location *C* is composed of crystals less than 1 mm in size because the lava cooled quickly at Earth's surface.

One credit is allowed for an acceptable response. Acceptable responses include but are not limited to:

- **The rock at *C* has a fine texture because the lava cooled quickly.**
- **Rock *C* crystallized/solidified on Earth's surface.**
- **Extrusive igneous rock doesn't have time to grow large crystals.**
- **The environment of formation is extrusive/volcanic.**

80. Find the Scheme for Sedimentary Rock Identification in the *Reference Tables for Physical Setting/Earth Science.* In the column labeled "Map Symbol," locate the symbols corresponding to those of the sedimentary rocks in the block diagram at locations *X* and *Y*. Trace left to the column labeled "Rock Name." Note that the sedimentary rock layer at location *X* corresponds to limestone and the sedimentary rock layer at *Y* corresponds to sandstone.

Find the Scheme for Metamorphic Rock Identification in the *Reference Tables for Physical Setting/Earth Science.* In the column labeled "Type of Metamorphism," locate the row labeled "Contact (heat)." Trace right to the column labeled "Comments," and note the statement "Various rocks changed by heat from nearby magma/lava." Now trace right to the column labeled "Rock Name," and note that hornfels is a metamorphic rock that forms from various rocks by contact metamorphism. Thus, hornfels may form at both locations *X* and *Y*. In the column labeled "Type of Metamorphism," locate the row labeled "Regional or Contact." Trace right to the column labeled "Comments," and note the statements "Metamorphism of quartz sandstone" and "Metamorphism of limestone or dolostone." From these rows, trace right to the column labeled "Rock Name." Note that metamorphism of sandstone (location *Y*) forms quartzite and metamorphism of limestone or dolostone (location *X*) forms marble. Thus, metamorphic rocks that are most likely found in the zone of contact metamorphism at location *X* include marble and hornfels and at location *Y* include quartzite and hornfels.

One credit is allowed if *both* responses are correct. Acceptable responses include:

Location *X*:

- **marble**
- **hornfels**

Location *Y*:

- **quartzite**
- **hornfels**

Note: Credit is *not* allowed if hornfels is used for *both* locations *X* and *Y*.

81.

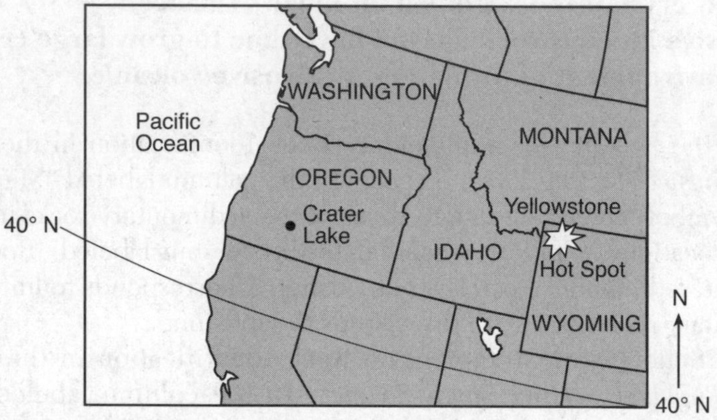

An isoline connects points of equal field value, and the field values shown on this map are depth. The 500-meter depth isoline connects all points that are located 500 meters below the lake surface. To construct the 500-meter depth isoline, connect all points that are at a depth of 500 meters with a smooth line. Several points on the map show a value of "500," however, these are not the only points on the map with that value. Begin with the point marked "500" to the lower left of point *X*. To the right of this point there is a point labeled "550" and a depth isoline labeled "400." It is logical to infer that between the depths "550" and "400," there is a point with the depth 500. Therefore, draw the isoline to curve south to pass between the 400-m isoline and the point labeled "550." Then curve north to reach the point labeled "500" to the lower right of point *X*. Now trace the isoline to the north between the points labeled "490" and "550." Then curve the isoline sharply to the south around the outside of the "550" and roughly parallel the 400-meter isoline. At the next point labeled "550" and extend the isoline between the “550” and the nearby “450.” Then extend the isoline south between the “450” and the “570.” Continue extending the line to the south between the points labeled “460” and “560” to the next point labeled “500.” From there, extend the isoline west between the points labeled “450” and “560.” Then turn north to pass between the points labeled “450” and “565.” Then curve to the northwest between the point labeled “540” and the 400-meter isoline to the next point labeled “500.” From here, extend the isoline west between the points labeled “460” and “530” to the next point labeled “500.” From there, extend the line east between the points labeled “450” and “540” to connect with the point labeled “500” at which you began. A correctly drawn 500-meter depth isoline is shown below.

One credit is allowed if **the 500-meter depth isoline is correctly drawn. If additional isolines are drawn, all isolines must be correct to receive credit.**

Example of a 1-credit response:

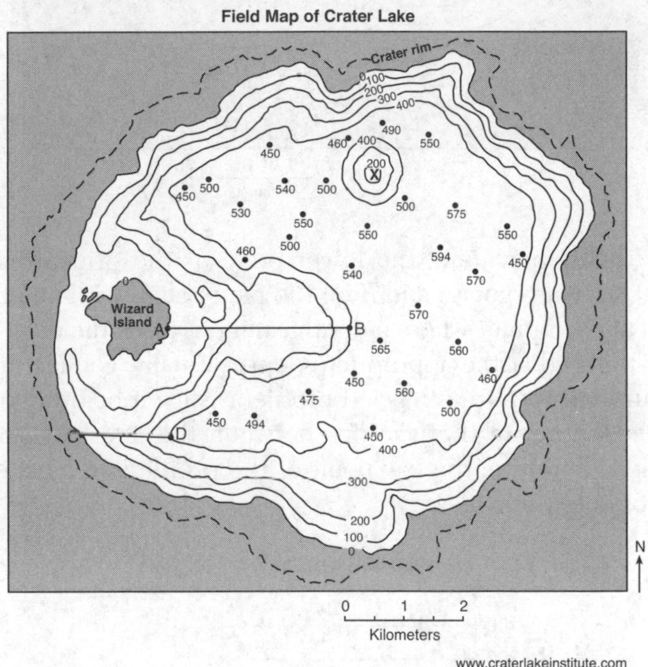

Note: The isoline must touch *all five* points for the 500-meter depth.

82. To construct a profile of the depth of Crater Lake along line *AB*, proceed as follows. Place the straight edge of a piece of scrap paper along the solid line connecting point *A* to point *B*. Mark the edge of the paper at points *A* and *B* and

wherever the paper intersects a depth isoline. Wherever the paper intersects a depth isoline, label the mark with the depth of the isoline as shown below.

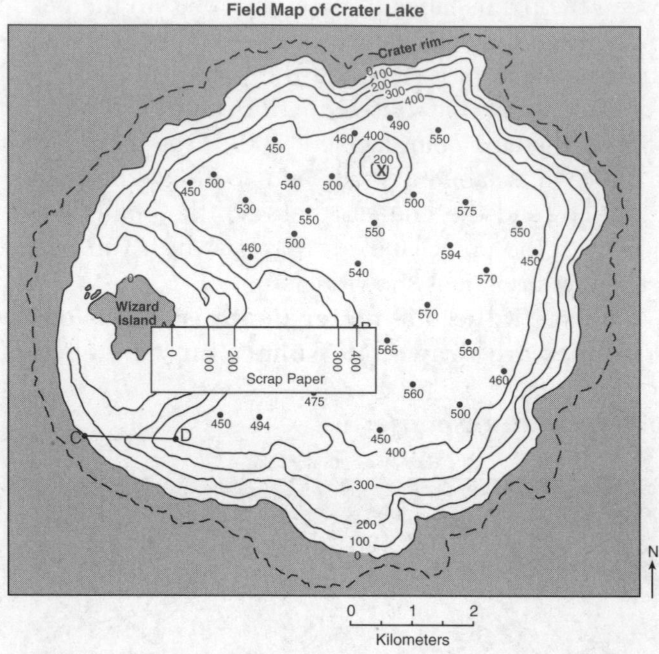

Now place this paper along the lower edge of the grid provided in your answer booklet so that points *A* and *B* on the paper align with points *A* and *B* on the grid. Then at each point where a depth isoline crosses the edge of the paper, draw a dot on the grid at the appropriate depth. Finally, connect all of the dots in a smooth curve to form the finished profile as shown below. Note that along the profile, the bottom of the lake lies between the 400-foot isoline and the 500-foot isoline. Therefore, the low point of the profile line must extend below 400 feet but not below 500 feet.

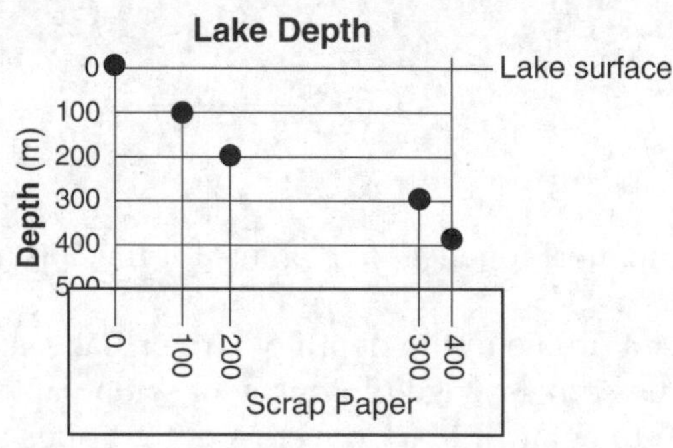

One credit is allowed if ***all five*** **plots are located within or touch the rectangles shown and are correctly connected with a line that passes within or touches each rectangle.**

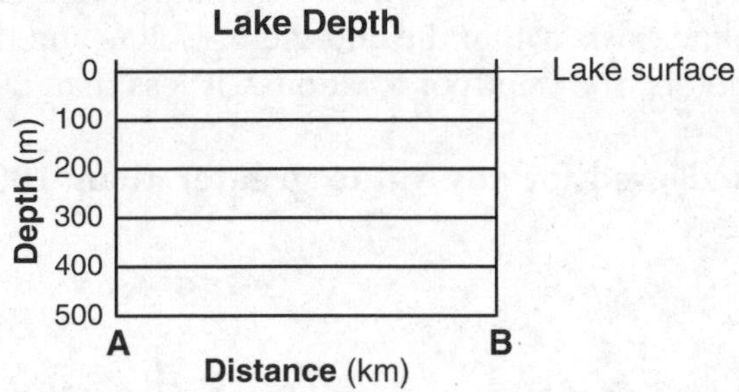

83. Find the Equations chart of the *Reference Tables for Physical Setting/Earth Science*, and note the equation for gradient:

$$\text{Gradient} = \frac{\text{change in field value}}{\text{distance}}$$

The map shown is a lake depth map. The field value is depth. Note that point *C* is located directly on the 0-meter depth isoline and that point *D* is located directly on the 400-meter depth isoline. Thus, the change in field value (depth) from *C* to *D* is 400 meters.

On a piece of scrap paper, mark off the distance along the straight line between points *C* and *D*. Compare the distance marked off on the scrap paper to the scale printed beneath the map to determine the distance between points *C* and *D*. The distance is about 1.5 kilometers. Substitute these values in the equation and solve:

$$\text{Gradient} = \frac{400 \text{ m}}{1.5 \text{ km}} = 266.6 \text{ m/km}$$

Thus, the gradient between location *A* and location *B* is approximately 267 m/km.

One credit is allowed for **any value from 250 to 286 m/km *or* –250 to –286 m/km.**

84. Note that the interval between isolines on the map is 100 meters. The map shows that *X* is located within the isoline labeled "200." The concentric isolines surrounding the 200-meter isoline decrease in value as you approach location *X*. Therefore, the depth within the 200-meter isoline is less than 200 meters. Since no additional isoline exists within the 200-meter isoline and the interval between isolines is 100 meters, the depth of location *X* is less than 200 meters but more than 100 meters.

One credit is allowed for **any value greater than 100 meters but less than 200 meters**.

85.

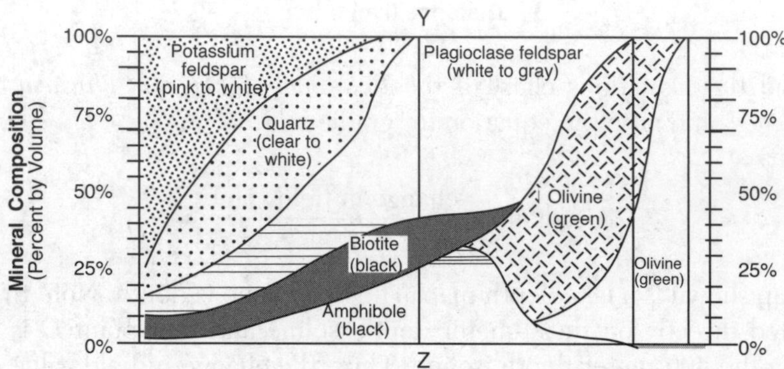

According to the diagram, line *YZ* extends through the regions of three minerals on the chart: amphibole, biotite, and plagioclase feldspar. The total distance from *Z* to *Y* on the line corresponds to 100% on the scale along the left edge of the chart labeled "Mineral Composition (Percentage by Volume)." The distance along line *YZ* that extends through each of the three mineral regions corresponds to the percentage by volume of each of those minerals in the rock represented by line *YZ*. To determine the percentage of each of these three minerals in the rock, first mark off the distance along line *YZ* that extends through each region. Then compare each distance to the scale labeled "Mineral Composition (Percent by Volume)" along the left side of the diagram. For example using a piece of scrap paper, mark off the distance that line *YZ* passes through the region labeled "Amphibole (black)." Now place the scrap paper so that one of these marks aligns with 0% on the scale labeled "Mineral Composition (Percentage by Volume)." Read the value corresponding to the other mark. Note that percentage of line *YZ* that passes through the amphibole region corresponds to about 27% on the percentage by volume scale. Thus, the rock represented by line *YZ* is 27% amphibole by volume. Repeat this procedure for each of the other two mineral regions through which line *YZ* passes on the diagram. Note that the rock represented by line *YZ* also contains about 16% biotite and 57% plagioclase feldspar by volume.

One credit is allowed if ***all three* percentages are within the ranges shown below.**

Plagioclase feldspar: **any value from 54% to 60%**

Biotite: **any value from 13% to 19%**

Amphibole: **any value from 24% to 30%**

Topic	Question Numbers (Total)	Wrong Answers (x)	Grade
Standards 1, 2, 6, and 7: Skills and Application			
Skills			
Standard 1 Analysis, Inquiry, and Design	4, 5, 7–11, 13, 14, 19, 21, 25–27, 30, 32–37, 40, 43, 45–47, 49–53, 55–68, 71–77, 79–81, 83, 85		$\frac{100(57-x)}{57} = \%$
Standard 2 Information Systems			
Standard 6 Interconnectedness, Common Themes	1–4, 7, 12, 14–16, 19, 21–23, 27–29, 32, 36–46, 50–52, 54–56, 59, 62–64, 66–71, 73–85		$\frac{100(57-x)}{57} = \%$
Standard 7 Interdisciplinary Problem Solving			
Standard 4: The Physical Setting/Earth Science			
Astronomy			
The Solar System (MU 1.1a, b; 1.2d)	3, 38–40, 58–61, 64		$\frac{100(9-x)}{9} = \%$
Earth Motions and Their Effects (MU 1.1c, d, e, f, g, h, i)	1, 6, 9, 36, 37, 41, 62, 63, 65		$\frac{100(9-x)}{9} = \%$
Stellar Astronomy (MU 1.2b)			
Origin of Earth's Atmosphere, Hydrosphere, and Lithosphere (MU 1.2e, f, h)			
Theories of the Origin of the Universe and Solar System (MU 1.2a,c)	2, 5, 66–68		$\frac{100(5-x)}{5} = \%$

Topic	Question Numbers (Total)	Wrong Answers (x)	Grade
Meteorology			
Energy Sources for Earth Systems (MU 2.1a, b)	8		$\frac{100(1-x)}{1} = \%$
Weather (MU 2.1c, d, e, f, g, h)	10, 14, 42–44, 54–57		$\frac{100(9-x)}{9} = \%$
Insolation and Seasonal Changes (MU 2.1i; 2.2a, b)	17, 74		$\frac{100(2-x)}{2} = \%$
The Water Cycle and Climates (MU 1.2g; 2.2c, d)	11, 12, 16, 18, 20, 73, 75, 76		$\frac{100(8-x)}{8} = \%$
Geology			
Minerals and Rocks (MU 3.1a, b, c)	30, 33–35, 50, 77, 79, 80, 85		$\frac{100(9-x)}{9} = \%$
Weathering, Erosion, and Deposition (MU 2.1s, t, u, v, w)	28, 29, 31, 49, 69, 70–72		$\frac{100(8-x)}{8} = \%$
Plate Tectonics and Earth's Interior (MU 2.1j, k, l, m, n, o)	7, 15, 19, 25–27, 45, 46, 48		$\frac{100(9-x)}{9} = \%$
Geologic History (MU 1.2i, j)	4, 13, 21, 22, 32, 47, 51–53, 78		$\frac{100(10-x)}{10} = \%$
Topographic Maps and Landscapes (MU 2.1p, q, r)	23, 24, 81–84		$\frac{100(6-x)}{6} = \%$
ESRT			
2011 Edition Reference Tables for Physical Setting/Earth Science	4, 5, 7, 8, 10, 11, 13, 14, 19, 21, 25–27, 30, 33–35, 40, 45–47, 50, 52, 53, 55, 56, 59, 63, 66–68, 72, 75–77, 79, 80, 83, 85		$\frac{100(39-x)}{39} = \%$

To further pinpoint your weak areas, use the Topic Outline in the front of the book.
MU = Major Understanding (see Topic Outline)

Examination August 2019

Physical Setting/Earth Science

PART A

Answer all questions in this part.

Directions (1–35): For *each* statement or question, choose the word or expression that, of those given, best completes the statement or answers the question. Some questions may require the use of the *2011 Edition Reference Tables for Physical Setting/Earth Science*. Record your answers in the space provided.

1 Most asteroids in our solar system are located in the asteroid belt between the orbits of Mars and Jupiter. What is the approximate distance from the Sun to the asteroid belt?

(1) 129 million km
(2) 210 million km
(3) 403 million km
(4) 1103 million km

1 ____

2 The table below shows tide data for a location on the north shore of Long Island, in New York State.

Day	Tide	Time
Tuesday	High Tide	12:11 a.m.
	Low Tide	6:23 a.m.
	High Tide	12:36 p.m.
	Low Tide	6:49 p.m.
Wednesday	High Tide	1:02 a.m.
	Low Tide	7:15 a.m.
	High Tide	1:27 p.m.

Based on these data, what is the most likely time of the next high tide?

(1) 1:53 a.m.
(2) 1:53 p.m.
(3) 7:40 a.m.
(4) 7:40 p.m.

2 ____

3 Which planet takes longer to complete one rotation on its axis than it does to complete one orbit of the Sun?

(1) Mercury
(2) Venus
(3) Earth
(4) Mars

3 ____

4 Jupiter is a Jovian planet. Compared to the terrestrial planets, Jupiter has a

(1) shorter period of revolution
(2) shorter distance to the Sun
(3) greater density
(4) greater mass

4 ____

5 The entire constellation of Orion is visible in the night sky in January to an observer in New York State. Which statement explains why this constellation is *not* visible in the night sky to this observer in June?

(1) Earth rotates on its axis.
(2) Earth revolves around the Sun.
(3) The constellation Orion orbits the Sun.
(4) The tilt of Earth's axis changes throughout the year.

5 ____

6 The approximate age of the universe is estimated to be

(1) 4.6 million years
(2) 4.6 billion years
(3) 13.8 million years
(4) 13.8 billion years

6 ____

7 The photograph below shows the apparent path of the Sun photographed at 20-minute intervals and combined into one photograph.

Baker, Robert H. *An Introduction to Astronomy*

Which motion is responsible for the apparent path of the Sun shown in the photograph?

(1) Earth's rotation
(2) Earth's revolution
(3) Sun's rotation
(4) Sun's revolution

7 _____

8 The mass extinction of the dinosaurs, approximately 65.5 million years ago, is inferred by most scientists to have been caused by

(1) a large energy surge from the surface of the Sun
(2) the occurrence of a major ice age
(3) an impact event occurring on Earth's surface
(4) earthquakes occurring along crustal plate boundaries

8 _____

9 The diagram below represents Earth as viewed from space. Letter *A* represents the angle of tilt of Earth's rotational axis. Line *XY* is perpendicular to the plane of Earth's orbit.

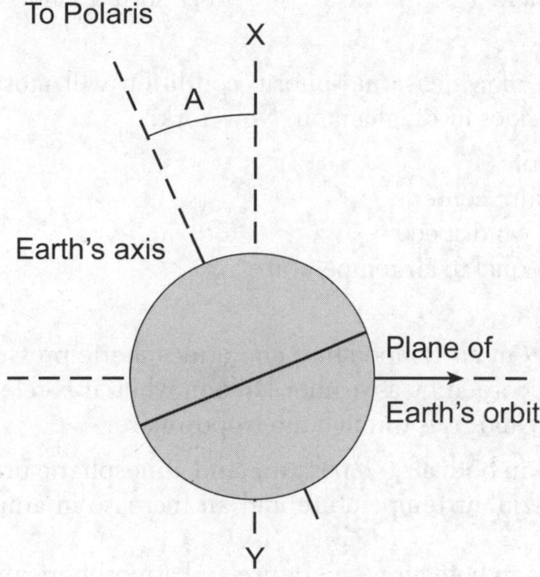

What is the value of the angle represented by letter *A*?

(1) 15.0°
(2) 23.5°
(3) 66.5°
(4) 90.0°

9 _____

10 In the spring, when snow cover on the land melts, the water will most likely infiltrate Earth's surface where that land surface is

(1) still frozen with a steep slope
(2) still frozen with a gentle slope
(3) no longer frozen with a steep slope
(4) no longer frozen with a gentle slope

10 _____

11 Water is returned from Earth's surface to the atmosphere by

(1) condensation and transpiration
(2) condensation and precipitation
(3) evaporation and transpiration
(4) evaporation and precipitation

11 _____

12 Which type of sediment sample normally has the greatest permeability rate?

(1) unsorted pebbles
(2) unsorted sand
(3) sorted pebbles
(4) sorted sand

12 ____

13 The presence of which atmospheric condition will most likely result in clear skies in Binghamton, New York?

(1) high humidity
(2) high-pressure center
(3) increasing wind speed
(4) dewpoint equal to air temperature

13 ____

14 Which changes in air temperature and atmospheric pressure will normally be recorded by a weather balloon when it is released at Earth's surface and rises through the troposphere?

(1) a decrease in both air temperature and atmospheric pressure
(2) a decrease in air temperature and an increase in atmospheric pressure
(3) an increase in both air temperature and atmospheric pressure
(4) an increase in air temperature and a decrease in atmospheric pressure

14 ____

15 Which gas is most effective in absorbing incoming harmful ultraviolet radiation in Earth's stratosphere before that radiation reaches Earth's surface?

(1) nitrogen
(2) hydrogen
(3) oxygen
(4) ozone

15 ____

16 The major cause of monsoon rains in India and southeast Asia is seasonal

(1) shifts in the prevailing wind belts
(2) shifts in ocean currents
(3) changes in the energy radiated from the Sun
(4) changes in worldwide atmospheric temperatures

16 ____

17 The cross section below represents the windward and leeward sides of a mountain. The arrows show the direction of air movement over the mountain. Points *X* and *Y* represent locations on Earth's surface.

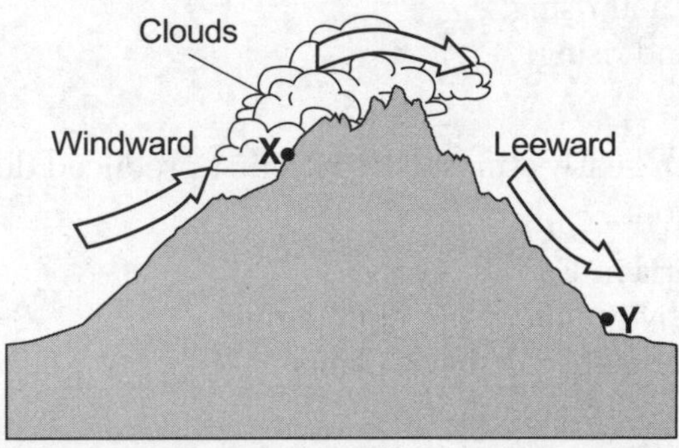

Compared to the temperature and water vapor content of the air at location *X*, the temperature and water vapor content at location *Y* are most likely

(1) warmer and wetter
(2) warmer and drier
(3) cooler and wetter
(4) cooler and drier

17 ____

18 Which surface ocean current transports cool water to lower latitudes?

(1) Gulf Stream
(2) Peru Current
(3) West Greenland Current
(4) East Australia Current

18 ____

19 Which land surface characteristics produce the greatest amount of absorption of insolation at Earth's surface?

(1) dark color and rough texture
(2) dark color and smooth texture
(3) light color and rough texture
(4) light color and smooth texture

19 ____

20 The planetary surface winds and air currents near Earth's equator are usually

(1) converging and sinking
(2) diverging and sinking
(3) converging and rising
(4) diverging and rising 20 _____

21 Oxygen in Earth's early atmosphere was first produced during the Precambrian from

(1) cyanobacteria in Earth's oceans
(2) volcanic activity along plate boundaries
(3) the absorption of sunlight by plants
(4) evaporation of ocean water 21 _____

22 Which life-form existed on Earth for the *shortest* period of time?

(1) dinosaurs
(2) trilobites
(3) ammonoids
(4) placoderm fish 22 _____

23 During the late Silurian epoch, very salty seas extended from New York State to Michigan. These environmental conditions resulted in the formation of halite layers. At the same time, *Eurypterus remipes* lived in nearby environments. Both the halite layers and *Eurypterus remipes* fossils can now be used to identify a specific geologic time interval because both formed over a

(1) large geographic area and in a short geologic time
(2) large geographic area and in a long geologic time
(3) small geographic area and in a short geologic time
(4) small geographic area and in a long geologic time 23 _____

24 If it takes a *P*-wave five minutes to travel from the epicenter of an earthquake to a seismic station, approximately how long will it take an *S*-wave to travel that same distance?

(1) 15 minutes
(2) 12 minutes
(3) 9 minutes
(4) 4 minutes 24 _____

25 Earthquake *S*-waves do *not* pass through which two interior Earth layers?

(1) rigid mantle and asthenosphere
(2) asthenosphere and stiffer mantle
(3) stiffer mantle and outer core
(4) outer core and inner core

25 _____

26 The cross section below represents the bedrock structure of a section of Earth's crust. Letters *A* through *H* represent rock units. Line *XY* represents a fault. The rock layers have *not* been overturned.

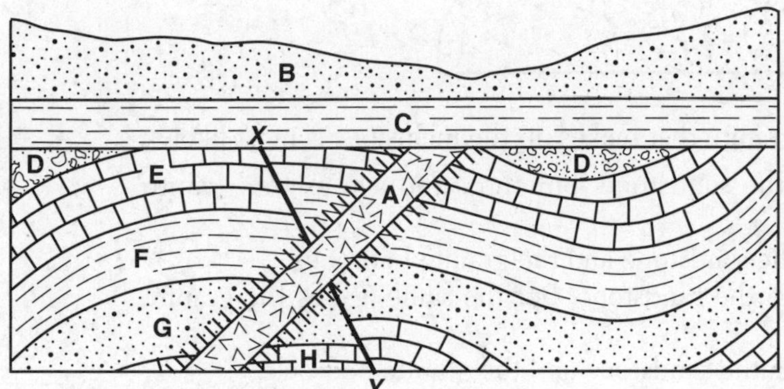

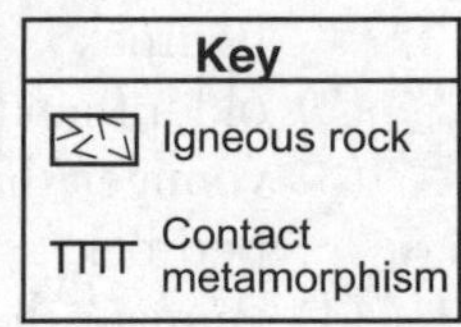

When did faulting along line *XY* occur?

(1) after the intrusion of rock unit *A*
(2) after the deposition of rock units *B*, *C*, and *D*
(3) after the formation of rock units *E* and *F*
(4) before the formation of rock units *G* and *H*

26 _____

27 The cross section below represents rock units within Earth's crust.

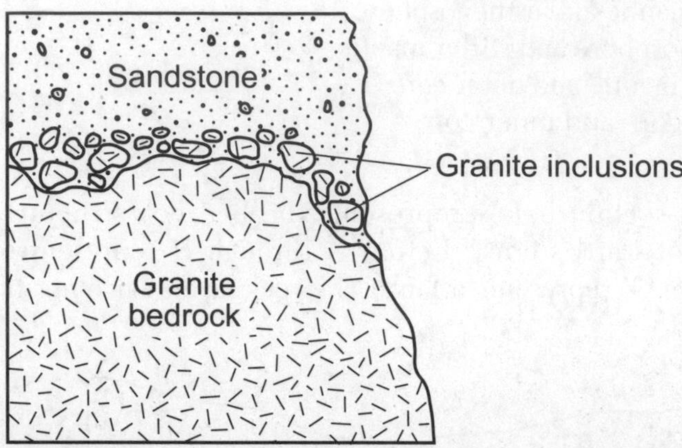

The age of the granite that makes up the inclusions is most likely

(1) older than the sandstone, but the same age as the granite bedrock
(2) older than the sandstone and the granite bedrock
(3) younger than the sandstone, but the same age as the granite bedrock
(4) younger than the sandstone and the granite bedrock 27 ____

28 A portion of the Generalized Landscape Regions of New York State map below shows the location of the Catskill Aqueduct that flows into the Kensico Reservoir, which supplies the residents of New York City with drinkable water.

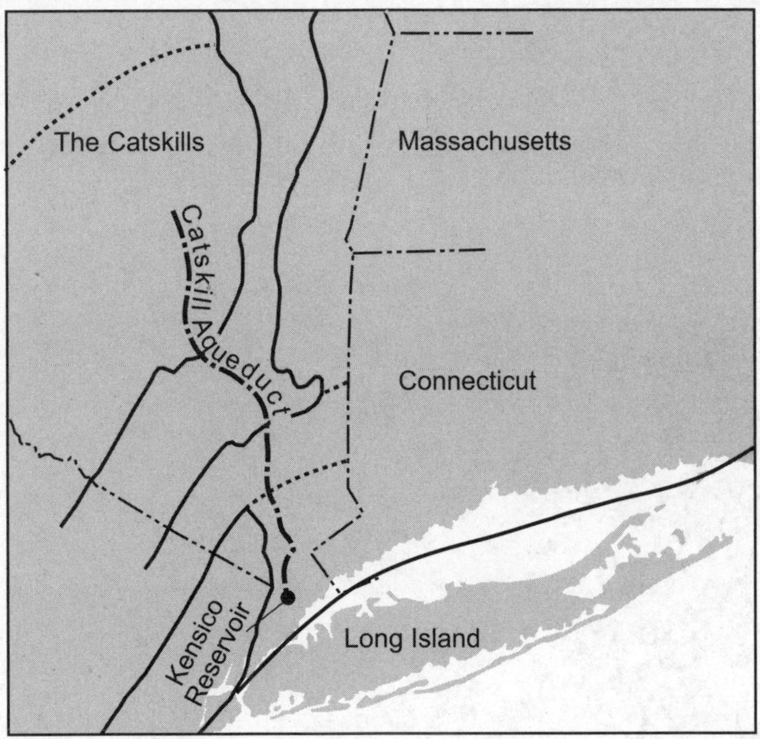

Which sequence shows the order of landscape regions that are crossed as water flows through the Catskill Aqueduct?

(1) Allegheny Plateau, Hudson-Mohawk Lowlands, Taconic Mountains, Newark Lowlands
(2) Allegheny Plateau, Hudson-Mohawk Lowlands, Hudson Highlands, Manhattan Prong
(3) Atlantic Coastal Plain, Newark Lowlands, Hudson Highlands, Hudson-Mohawk Lowlands
(4) Atlantic Coastal Plain, Manhattan Prong, Hudson Highlands, Allegheny Plateau

28 _____

29 The map below shows the land area in New York State drained by the Oswego River and its tributaries.

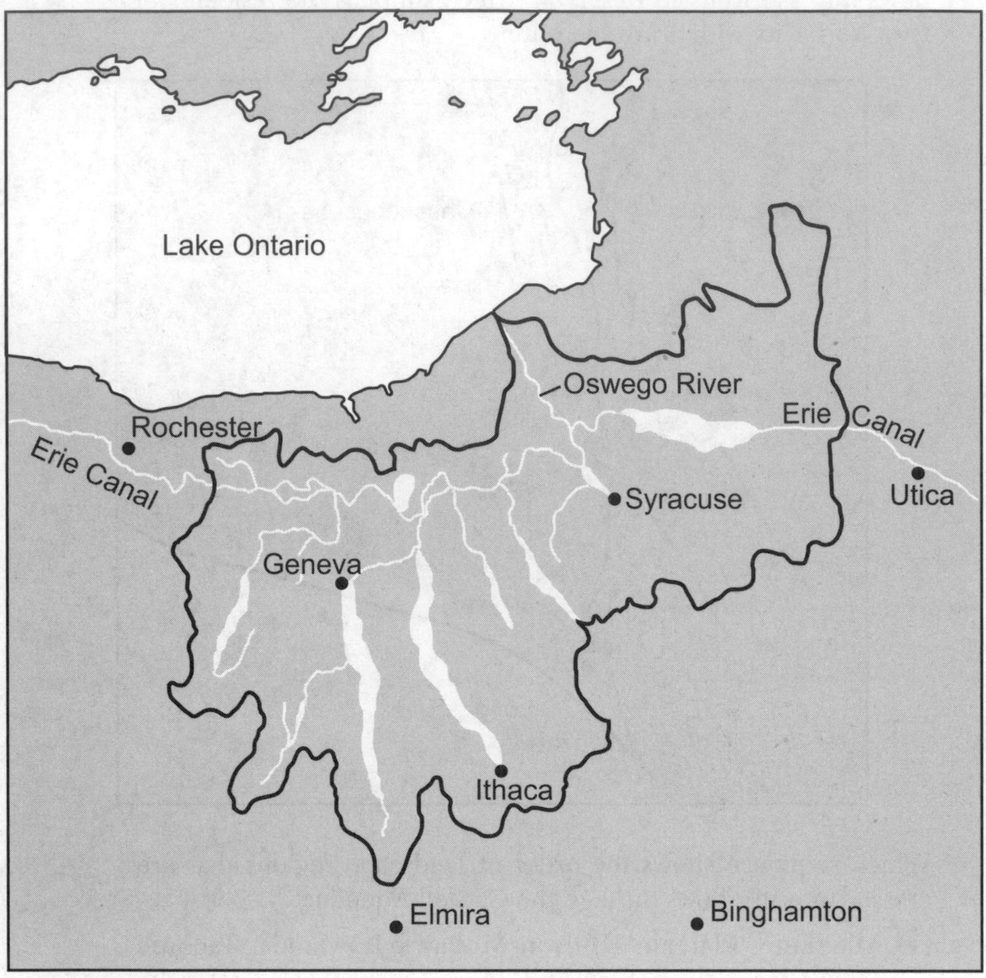

The land area drained by the Oswego River and its tributaries is called a

(1) delta
(2) watershed
(3) water table
(4) floodplain

29 _____

30 The photograph below shows a portion of the Genesee River in western New York. Letters *A*, *B*, *C*, and *D* are locations in the river.

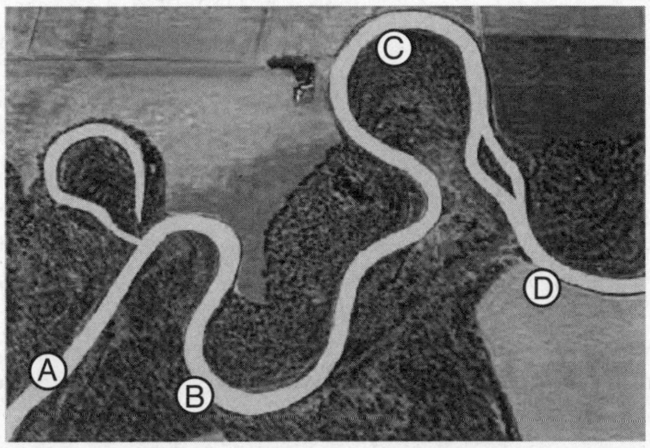

At which location would deposition of sediments most likely be greater than erosion?

(1) *A*
(2) *B*
(3) *C*
(4) *D*

30 ____

31 The rate of soil development in tropical areas is usually greater than the rate of soil development in arctic areas because tropical areas have

(1) less chemical weathering and a scarcity of living organisms
(2) less chemical weathering and an abundance of living organisms
(3) more chemical weathering and a scarcity of living organisms
(4) more chemical weathering and an abundance of living organisms

31 ____

32 New York State's Finger Lakes exist today because

(1) U-shaped valleys were dammed by glacial sediments
(2) V-shaped valleys are being eroded by streams
(3) a drop in sea level occurred, leaving the lakes
(4) a rise in sea level occurred, flooding the region

32 ____

33 Which igneous rock could physically weather to beach sand that contains the minerals pyroxene, plagioclase feldspar, and olivine?

(1) dunite
(2) granite
(3) peridotite
(4) gabbro

33 ____

34 Which two physical properties of graphite make it a good mineral for use in pencils?

(1) luster and fracture
(2) cleavage and color
(3) hardness and streak
(4) greasy feel and composition

34 ____

35 Which mineral is produced when two atoms of iron chemically combine with three atoms of oxygen?

(1) garnet
(2) pyrite
(3) magnetite
(4) hematite

35 ____

PART B–1

Answer all questions in this part.

Directions (36–50): For *each* statement or question, choose the word or expression that, of those given, best completes the statement or answers the question. Some questions may require the use of the *2011 Edition Reference Tables for Physical Setting/Earth Science*. Record your answers in the space provided.

Base your answers to questions 36 through 38 on the map below and on your knowledge of Earth science. The map shows the locations of Jamestown, Watertown, and Kingston in New York State.

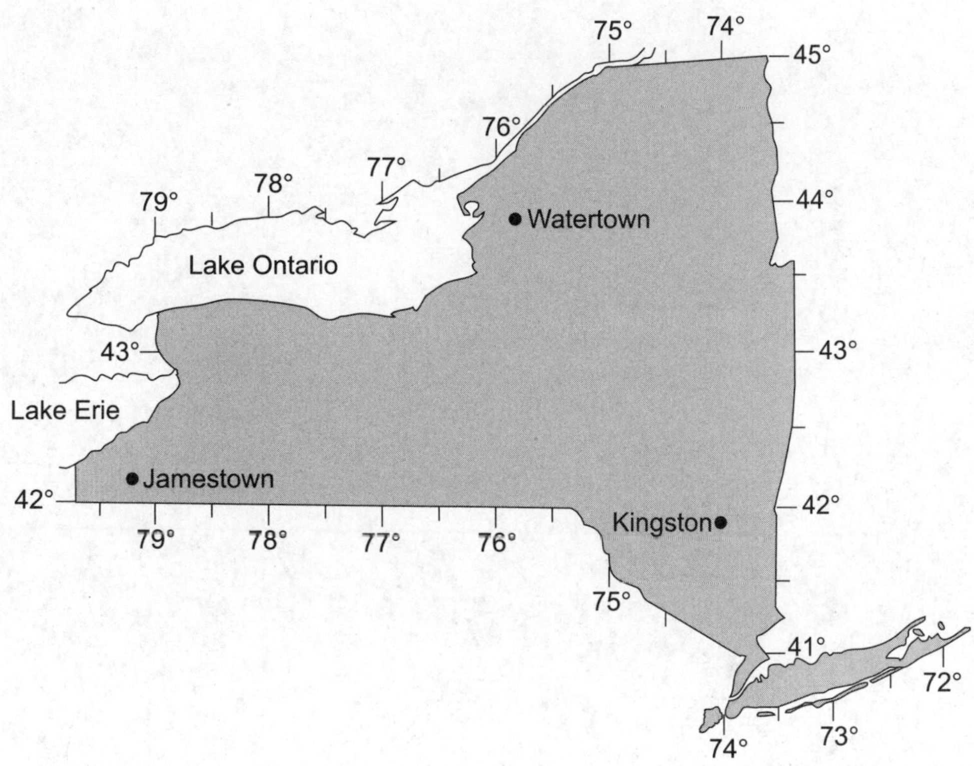

36 Compared to the time of sunrise in Kingston, the time of sunrise in Jamestown would be approximately

(1) 1 hour earlier
(2) 1 hour later
(3) 20 minutes earlier
(4) 20 minutes later

36 _____

37 An observer in Watertown, New York, would see the star *Polaris* at an altitude of approximately

(1) 44°
(2) 45°
(3) 75°
(4) 76°

37 ____

38 Earth's system of longitude is based on which motion?

(1) Earth's rotation
(2) the Sun's rotation
(3) Earth's revolution around the Sun
(4) the Sun's revolution around Earth

38 ____

Base your answers to questions 39 through 41 on the passage and photograph below and on your knowledge of Earth science. The photograph shows a portion of the Patagonia Marble Caves found in South America.

Patagonia Marble Caves

The Patagonia Marble Caves of South America are found on the shores of General Carrera Lake at a location of 46.5° S 72° W. Most of the water in the lake comes from the melting of nearby glaciers. Many small particles carried by the glacier remain suspended in the meltwater that fills this lake, causing a distinct blue color. Over the last 6200 years, the water of the lake has been weathering and eroding the marble bedrock found along the shores and within the lake itself. The marble dissolved faster at the water surface, where the moving water is interacting with the marble bedrock, producing countless caves, mazes, columns, and tunnels in the marble.

http://www.dont-complain.com/2015/01/17/marble-caves-chile/

39 The map below shows four locations in South America, labeled *A*, *B*, *C*, and *D*.

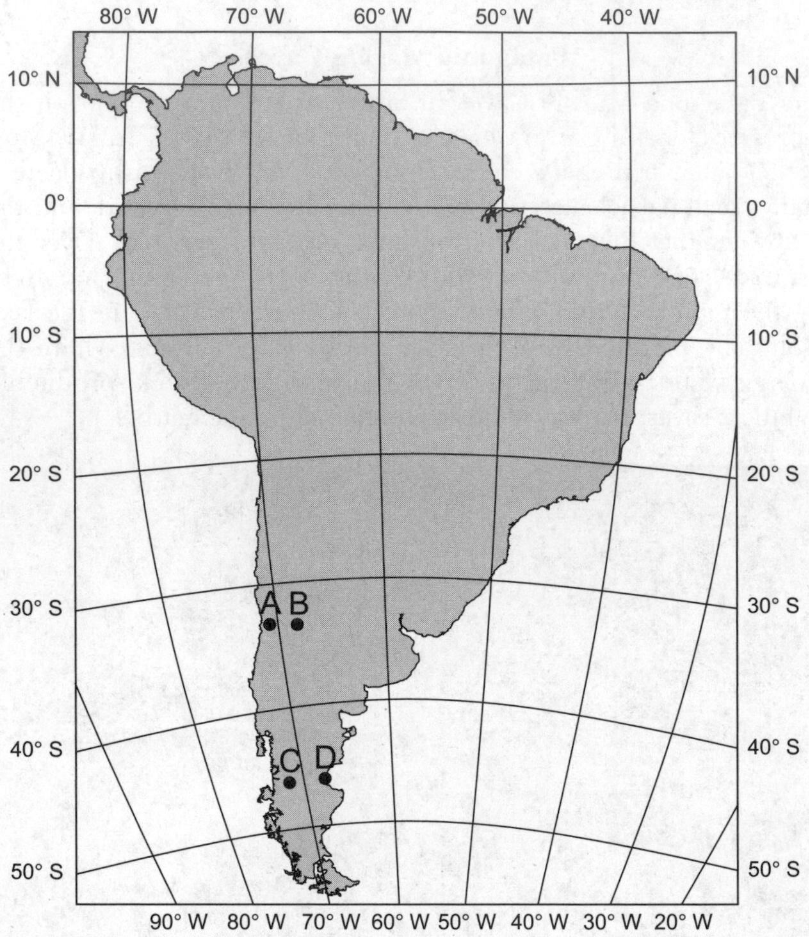

Which lettered point on the map best represents the location of the Patagonia Marble Caves?

(1) *A*
(2) *B*
(3) *C*
(4) *D*

39 _____

40 Which two processes formed the marble bedrock shown in the photograph?

(1) heat and pressure
(2) compaction and cementation
(3) melting and solidification
(4) precipitation and evaporation 40 _____

41 Most marble is composed primarily of

(1) quartz and/or potassium feldspar
(2) calcite and/or dolomite
(3) halite and/or olivine
(4) pyroxene and/or plagioclase feldspar 41 _____

Base your answers to questions 42 through 44 on the photographs below and on your knowledge of Earth science. One photograph shows a digital device that recorded several weather variables. The second photograph shows two weather instruments, labeled *A* and *B*.

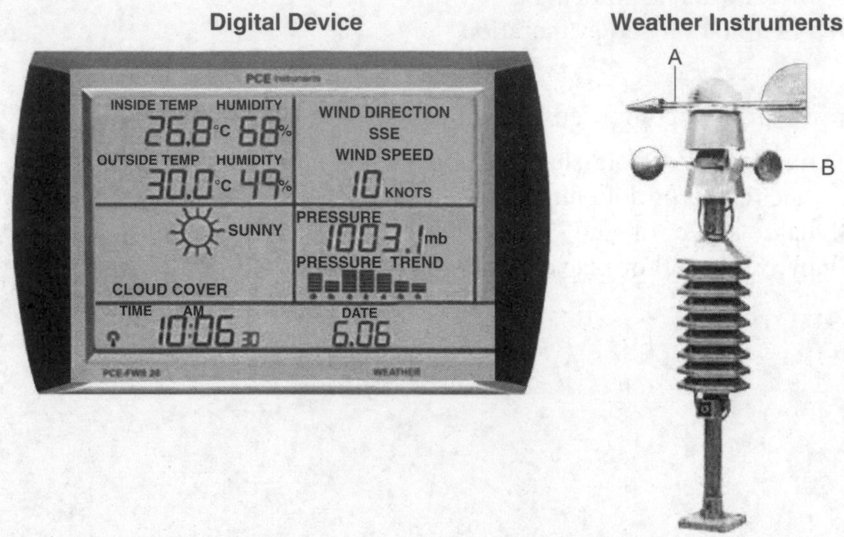

Adapted from: https://www.pce-instruments.com/us/measuring-instruments/

42 The barometric pressure was recorded on the digital device in millibars. What is the equivalent air pressure in inches of mercury (Hg)?

(1) 29.59 in of Hg
(2) 29.60 in of Hg
(3) 29.62 in of Hg
(4) 29.65 in of Hg

42 ____

43 Based on the outside air temperature and relative humidity shown on the digital device, what is the approximate dewpoint for the time shown on the device?

(1) 8°C
(2) 12°C
(3) 18°C
(4) 31°C

43 ____

44 Which table correctly identifies weather instruments *A* and *B* and the weather variable that each measures?

	Weather Instrument	Weather Variable Measured
A	wind vane	wind direction
B	anemometer	wind speed

(1)

	Weather Instrument	Weather Variable Measured
A	wind vane	wind speed
B	anemometer	wind direction

(3)

	Weather Instrument	Weather Variable Measured
A	anemometer	wind speed
B	wind vane	wind direction

(2)

	Weather Instrument	Weather Variable Measured
A	anemometer	wind direction
B	wind vane	wind speed

(4)

44 ____

Base your answers to questions 45 through 47 on the diagrams below and on your knowledge of Earth science. Diagram *A* represents a Foucault pendulum set up at a particular location on Earth. The line on the floor marked 0 is the path of the pendulum when it was first set in motion. Diagram *B* represents the curving of planetary winds located in two areas of Earth's surface due to the Coriolis effect.

Diagram A: Foucault's Pendulum

Swing
Wire 200 feet long
Steel ball
0 2 4 6 8 hours
Floor

Diagram B: Coriolis Effect on Planetary Winds

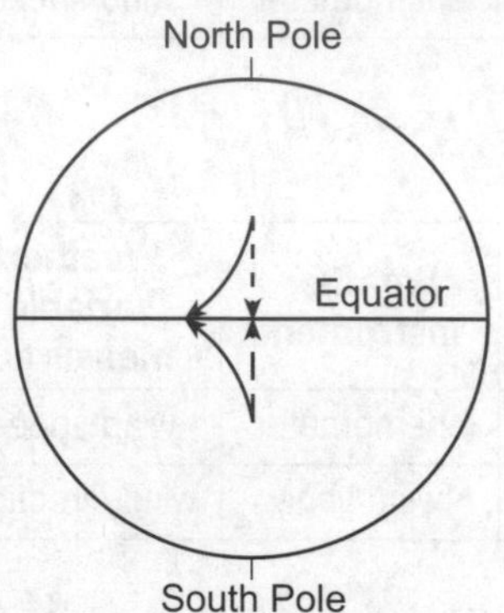

(Not drawn to scale)

Key
- - → Original direction of wind
——→ Deflected path of wind

45 The apparent shift in the direction of swing of a Foucault pendulum and the curving of winds by the Coriolis effect are both evidence of Earth's

(1) tilt
(2) shape
(3) rotation
(4) revolution

45 ____

46 In diagram *A*, the apparent shift of the path of the pendulum is shown at two-hour intervals for the first eight hours of pendulum motion. During the eight hours, the pendulum's path was displaced 60 degrees from where it started. How many degrees did the apparent path of the pendulum change each hour (h)?

(1) 7.5°/h
(2) 15°/h
(3) 60°/h
(4) 120°/h

46 ____

47 In diagram *B*, what are the names of the planetary wind belts formed by the deflected winds shown in the Northern and Southern hemispheres?

Hemisphere	Wind Belt
Northern	Southeast winds
Southern	Northwest winds

(1)

Hemisphere	Wind Belt
Northern	Southwest winds
Southern	Northwest winds

(3)

Hemisphere	Wind Belt
Northern	Northeast winds
Southern	Southeast winds

(2)

Hemisphere	Wind Belt
Northern	Northwest winds
Southern	Southeast winds

(4)

47 ____

Base your answers to questions 48 through 50 on the cross section below and on your knowledge of Earth science. The cross section shows the boundary between the Juan de Fuca Plate and the North American Plate.

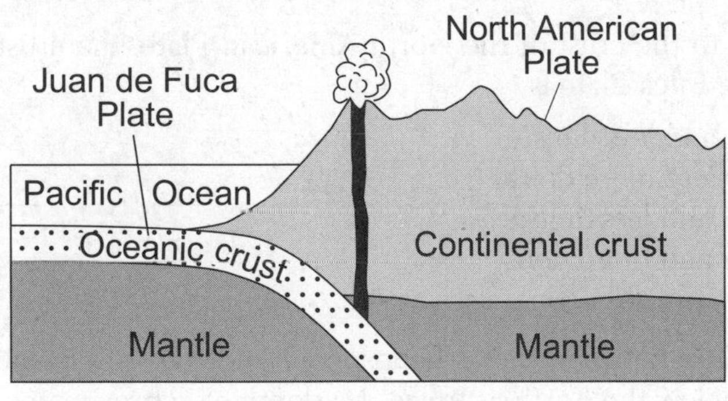

(Not drawn to scale)

48 In which diagram do the arrows best represent the relative motion of the upper mantle at this plate boundary?

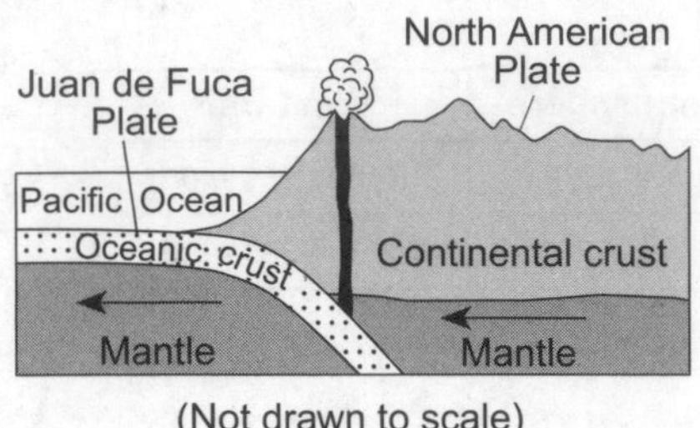

(Not drawn to scale)

(1)

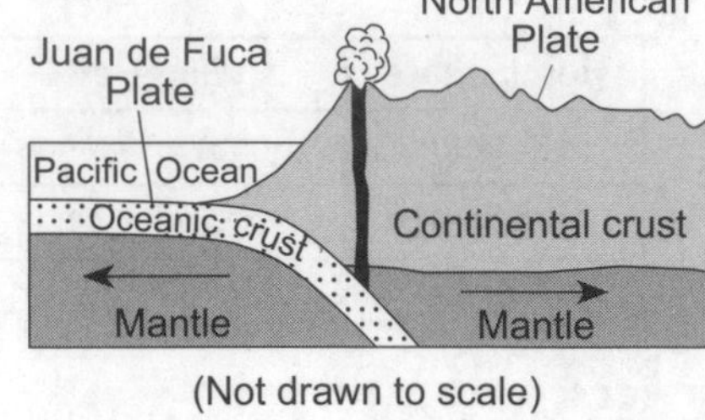

(Not drawn to scale)

(3)

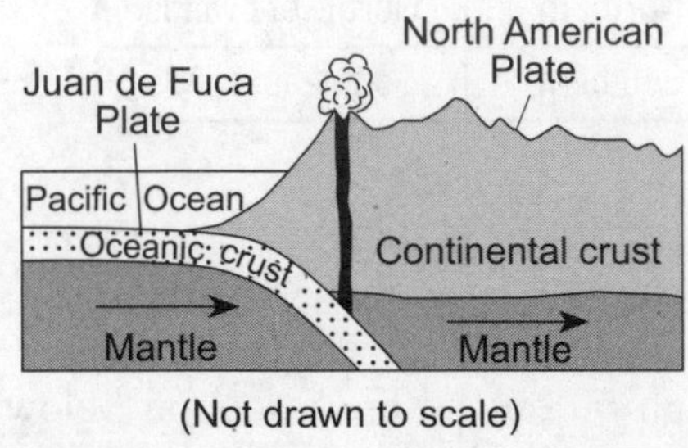

(Not drawn to scale)

(2)

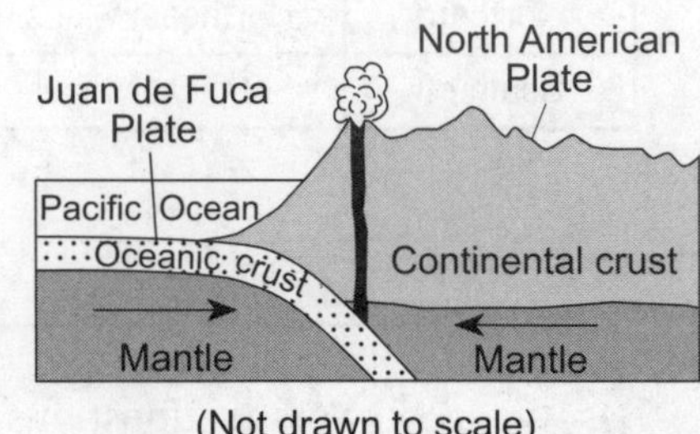

(Not drawn to scale)

(4)

48 _____

49 Compared to the crust of the North American Plate, the crust of the Juan de Fuca Plate is

(1) thicker and less dense
(2) thicker and more dense
(3) thinner and less dense
(4) thinner and more dense

49 _____

50 The boundary between the asthenosphere and the stiffer mantle is located below the Earth's surface at a depth of approximately

(1) 100 km
(2) 700 km
(3) 2500 km
(4) 3000 km

50 _____

PART B–2

Answer all questions in this part.

Directions (51–65): Record your answers in the spaces provided. Some questions may require the use of the *2011 Edition Reference Tables for Physical Setting/Earth Science*.

Base your answers to questions 51 through 53 on the passage below, the cross section below, and on your knowledge of Earth science. The passage describes the discovery of ocean floor magnetism. The cross section represents a pattern of normal and reverse polarity of the magnetic field preserved in the igneous bedrock of the oceanic crust west of the Mid-Atlantic Ridge. The magnetic polarity pattern of the bedrock on the east side of the ridge has been left blank.

Ocean Floor Magnetism

Scientists in the early 1960s were surprised to find there was a pattern in the ocean floor magnetism preserved in the bedrock of the Atlantic Ocean floor. They found that the magnetism in the bedrock was arranged in an orderly pattern parallel to the Mid-Atlantic Ridge. This mountain ridge, often marked by earthquakes and volcanic eruptions, runs roughly north-south. Earth is currently in a period of normal polarity. However, the magnetic record preserved in the rocks indicates that Earth's magnetic poles have reversed positions many times in the past. Since the initial discovery of this ocean floor magnetism, similar magnetic patterns have also been found parallel to the mid-ocean ridges in all of the other oceans.

51 Complete the diagram *below* by shading in the pattern of normal polarity on the east side of the Mid-Atlantic Ridge center. Assume that the rate of plate movement was constant on both sides of the ridge center. Your answer must show the correct width and placement of each normal polarity section. [1]

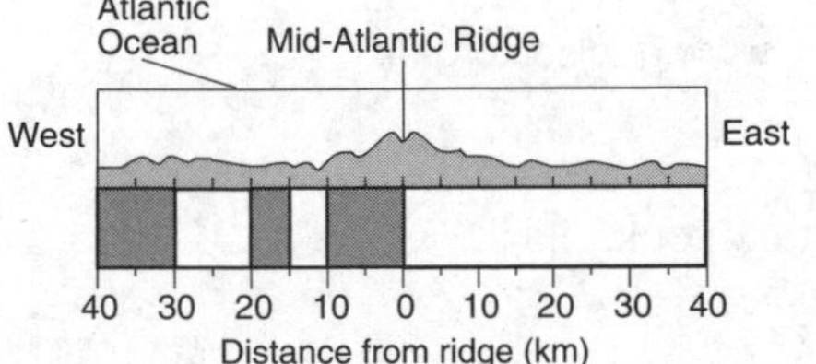

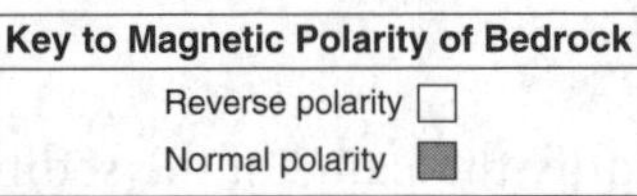

52 Identify the type of tectonic plate boundary at the mid-ocean ridges where these magnetic patterns were produced. [1]

________________________________ **plate boundary**

53 On the set of axes *below*, draw a line to represent the relationship between the distance from the Mid-Atlantic Ridge and the age of the ocean floor bedrock. [1]

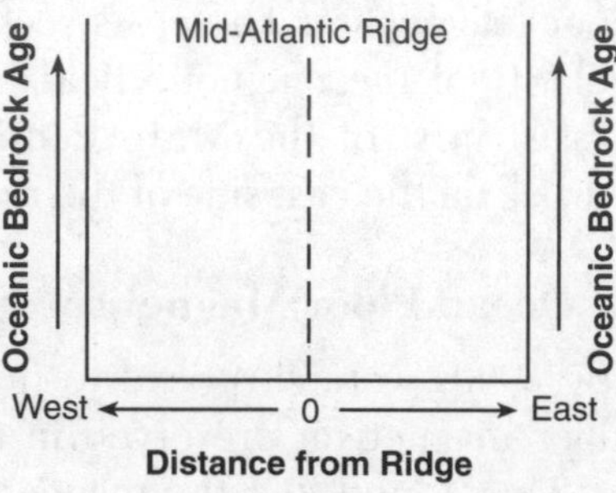

Base your answers to questions 54 through 56 on the photograph below and on your knowledge of Earth science. The photograph is of a rock sample composed of pebbles that have been cemented together.

www.geolsoc.org.uk

54 Identify the name of the sedimentary rock. [1]

____

55 Identify the total possible range of particle diameters, in centimeters, for a particle to be classified as a pebble. [1]

The range is from ____________ **cm** to ____________ **cm.**

56 Describe the evidence from the photograph that supports the inference that the particles in this rock were most likely transported by running water. [1]

____

____

____

Base your answers to questions 57 through 59 on the diagrams below and on your knowledge of Earth science. The diagrams represent a model of the disintegration of a sample of the radioactive isotope carbon-14 (^{14}C).

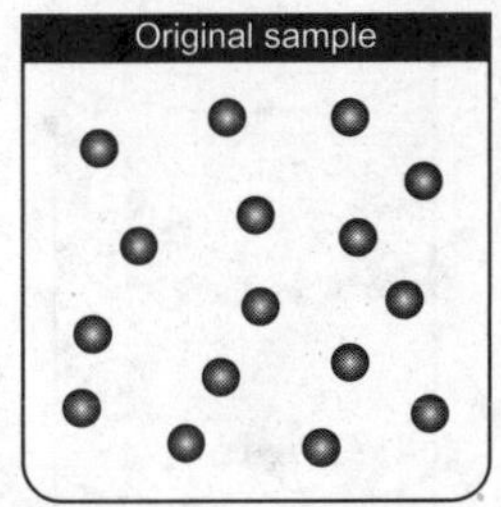

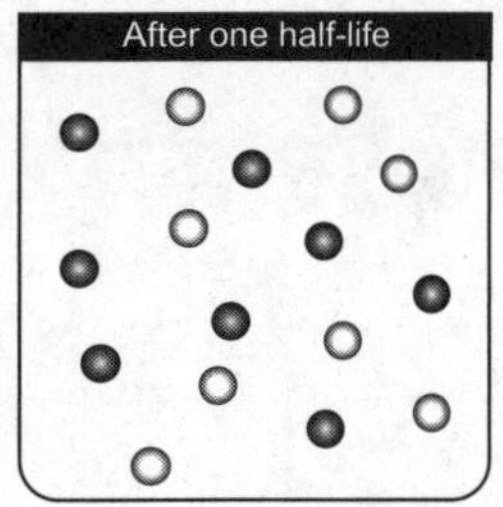

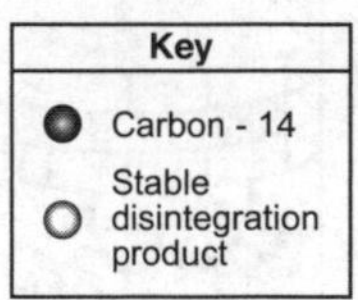

57 Determine the number of carbon-14 atoms that would most likely remain at the end of the second half-life. [1]

____

58 Identify the stable disintegration product that is produced when carbon-14 decays. [1]

____

59 Identify the radioactive isotope that has a half-life that is approximately the same as the estimated age of Earth, by using the Radioactive Decay Data table in the *Earth Science Reference Tables*. [1]

____

____

Base your answers to questions 60 through 62 on the diagrams below and on your knowledge of Earth science. The diagrams represent drill core samples from two different locations (I and II). A drill core is a cylinder of rock material removed from the bedrock. Letters *A* through *J* represent different rock layers. Some layers contain index fossils. The rock layers shown have *not* been overturned.

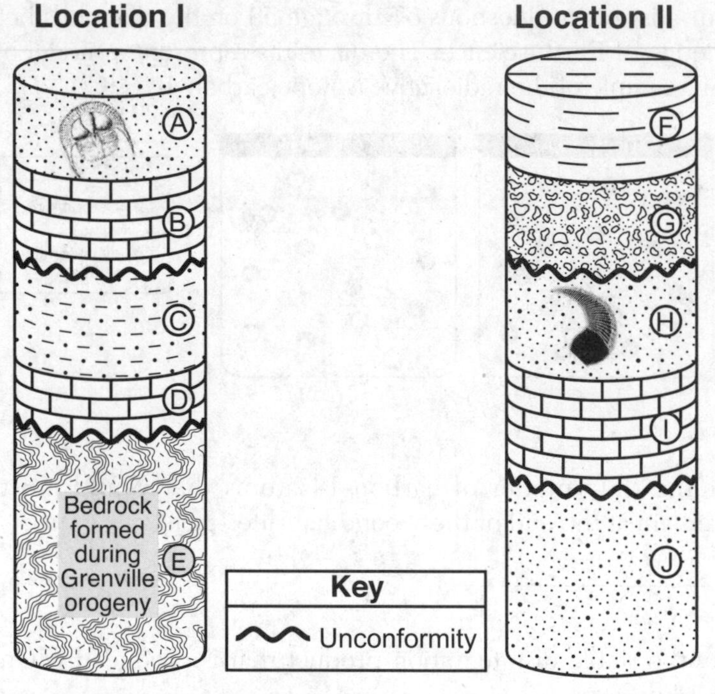

60 Identify the geologic period when rock layer *A* was deposited. [1]

______________________ **Period**

61 List the relative ages of rock layers *D*, *E*, *F*, and *G* by listing the letters from oldest to youngest. [1]

__________ → __________ → __________ → __________

Oldest ————————————————→ Youngest

62 Identify *two* processes that formed the unconformities in these drill cores. [1]

Process 1: ____________________

Process 2: ____________________

Base your answers to questions 63 through 65 on the graph below and on your knowledge of Earth science. The graph shows the discharge of the Delaware River at Barryville, New York, for a one-week period during March 2004.

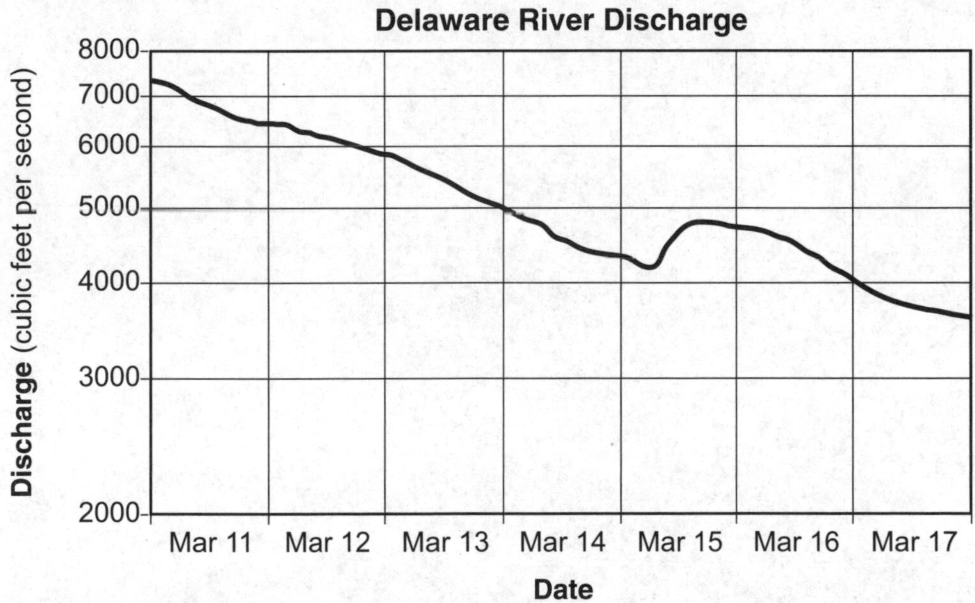

63 Identify the date shown on the graph when the river was most likely carrying the greatest amount of sediment. [1]

____

64 State *one* possible cause for the increase in stream discharge on March 15. [1]

____

65 The Delaware River flows out of the Catskills and into the Atlantic Ocean. Identify the general compass direction toward which the Delaware River flows along the Pennsylvania–New York State border. [1]

____

PART C

Answer all questions in this part.

Directions (66–85): Record your answers in the spaces provided. Some questions may require the use of the *2011 Edition Reference Tables for Physical Setting/Earth Science*.

Base your answers to questions 66 through 68 on the topographic map below and on your knowledge of Earth science. The map shows an eroded drumlin at Chimney Bluffs State Park along the shoreline of Lake Ontario east of Rochester, New York. Lines *AB* and *CD* are reference lines. Elevations are shown in feet.

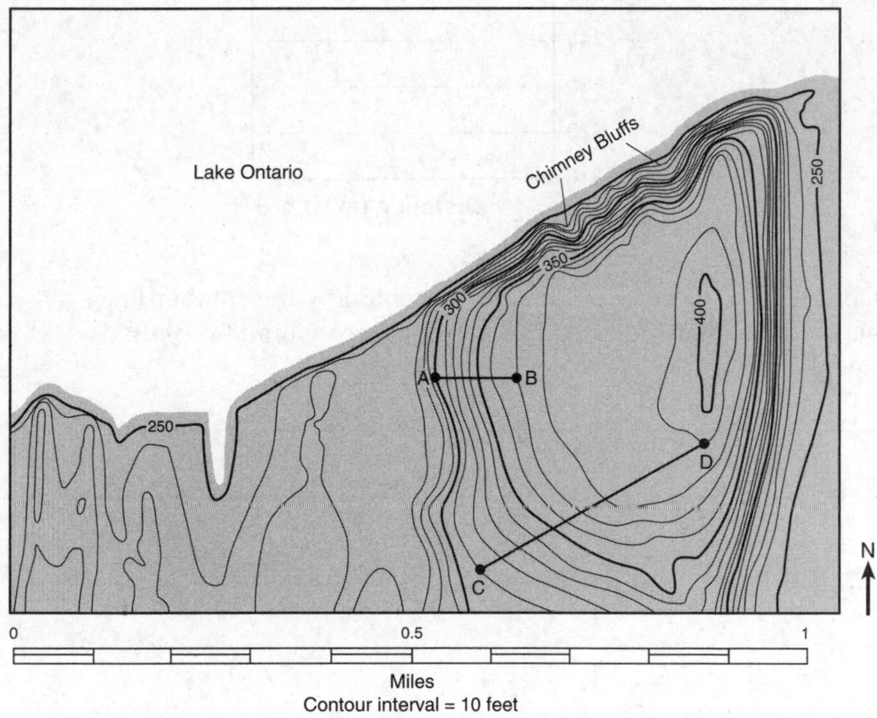

66 Calculate the gradient along the line between points *A* and *B* in feet per mile. [1]

______________ **ft/mile**

67 On the grid *below*, construct a topographic profile along line *CD* by plotting the elevation of each contour line that crosses line *CD*. The elevations for points *C* and *D* have been plotted on the grid. Connect *all nine* plots with a line from point *C* to point *D* to complete the profile. [1]

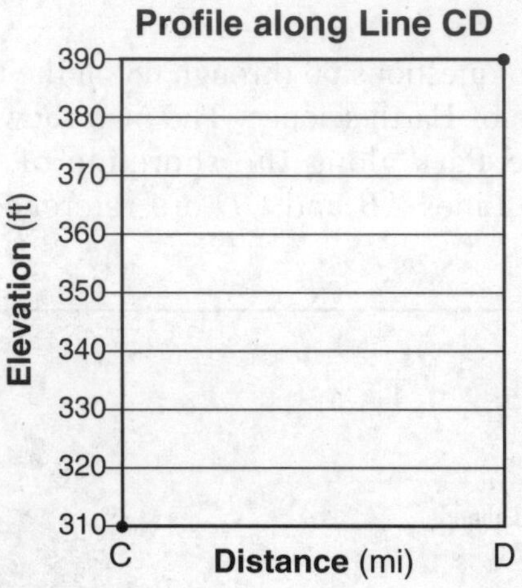

68 Describe *one* piece of evidence represented by the contour lines on the map that indicates the north side of Chimney Bluffs is steep. [1]

____

____

Base your answers to questions 69 through 71 on the weather map below and on your knowledge of Earth science. The map shows the center of a low-pressure system (**L**). The location for Albany, New York, is shown.

130° W 120° W 110° W 100° W 90° W 80° W 70° W 60° W
55° N
50° N
45° N
40° N
35° N
30° N
25° N
L
Canada
Albany
Atlantic Ocean
Pacific Ocean
Mexico
Gulf of Mexico
110° W 100° W 90° W 80° W

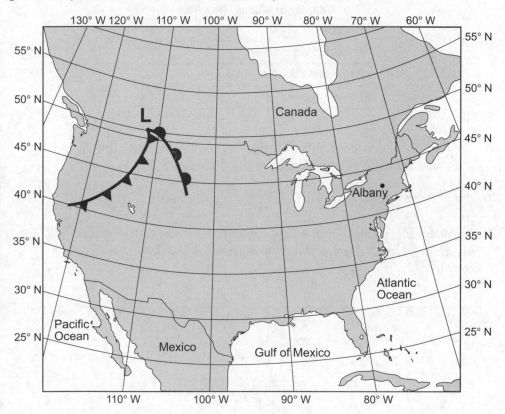

69 The station model below represents the weather conditions for Albany, New York.

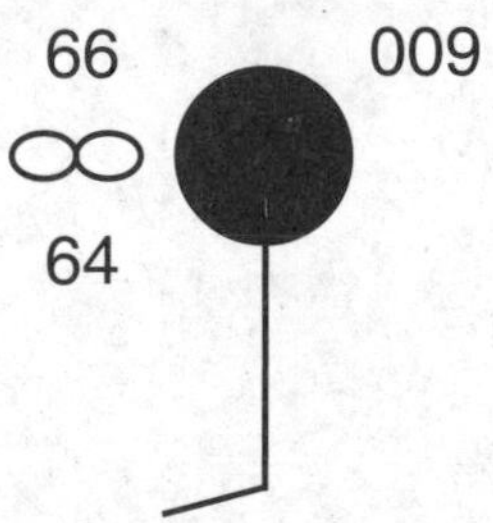

Complete the table *below* by indicating the weather conditions represented on the station model for Albany, New York. [1]

Weather in Albany, New York	
Dewpoint	°F
Amount of cloud cover	%
Barometric pressure	mb
Present weather	

70 Write the standard two-letter air-mass symbol to identify the air mass that normally forms over the Gulf of Mexico. [1]

____

71 Identify *one* process that causes clouds to form in the moist air rising along the frontal boundaries of the low-pressure system. [1]

____

Base your answers to questions 72 through 74 on the passage and data table below and on your knowledge of Earth science. The data table shows the monthly average high air temperatures, in degrees Fahrenheit (°F), and monthly average snowfall, in inches (in), at the summit (top) of Mount Washington in New Hampshire.

Mount Washington

Mount Washington, located in the state of New Hampshire, is one of the highest mountains east of the Mississippi River. This mountain, as well as many mountains across upstate New York and the northeast, has deep-cut glacial valleys formed by continental glaciers approximately 1.7 to 1.8 million years ago. Hurricane-force wind gusts are observed at the summit of Mount Washington on the average of 110 days per year, including a record wind speed of 231 miles per hour. It also receives very high levels of snow, averaging 282 inches (23.5 feet) of snow per year.

Mount Washington Monthly Average High Air Temperatures and Average Snowfall

Month	Jan	Feb	Mar	Apr	May	Jun	Jul	Aug	Sep	Oct	Nov	Dec
Average High Air Temperature (°F)	14	15	21	30	41	50	54	53	47	36	28	18
Average Snowfall (in)	44	40	45	36	12	1	Trace	0.1	2	18	38	46

72 On the grid *below*, construct a line graph by plotting the average high air temperatures for each month listed in the data table. Connect *all twelve* plots with a line. [1]

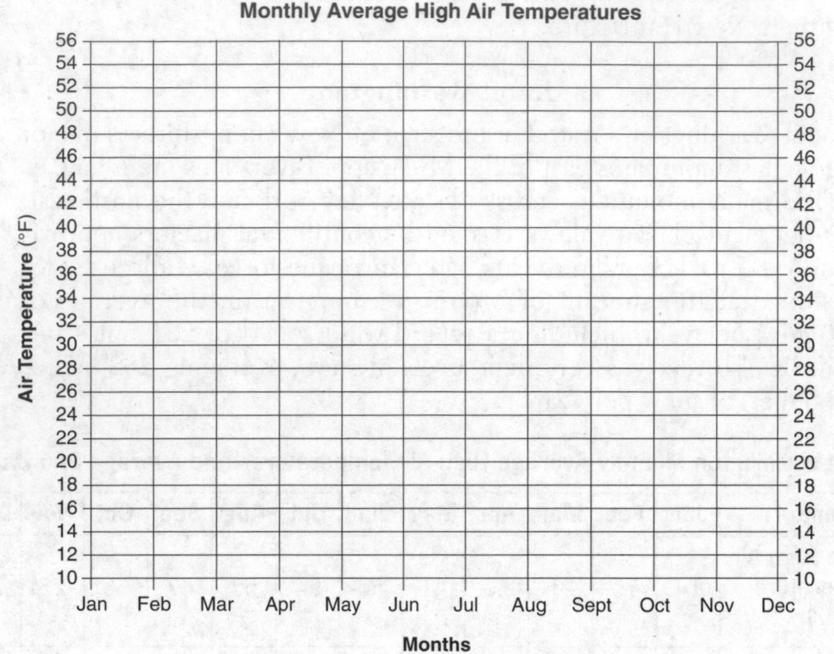

73 Identify the climate factor that best explains why Mount Washington has relatively low air temperatures throughout the year. [1]

____

74 State the name of the geologic epoch when continental glaciers last carved the valleys around Mount Washington and across upstate New York. [1]

____________________ **Epoch**

Base your answers to questions 75 through 78 on the table below and on your knowledge of Earth science. The table shows the velocities, in kilometers per second (km/s), for several galaxies, represented by letters *A*, *B*, *C*, *D*, and *E*, that are moving away from Earth. The vast majority of stars and galaxies in the universe are moving away from our solar system. Scientific evidence indicates that the farther away a galaxy is, the faster it is moving away.

Velocities of Galaxies Moving Away From Earth

Galaxy	Velocity (km/s)
A	61,000
B	15,000
C	1200
D	39,000
E	22,000

75 List the galaxies in order from closest to Earth to farthest from Earth. [1]

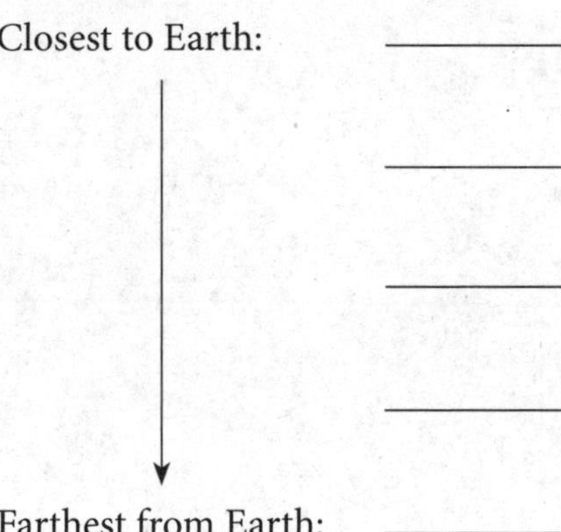

76 Identify the name of the theory for the formation of the universe that scientists developed after observing that most galaxies are moving away from each other. [1]

____

77 Identify the evidence scientists use to determine that a galaxy is moving away from Earth. [1]

____

____

78 A star in one of these galaxies has a surface temperature of 8000 K and a luminosity of 10. Identify the stage and color of this star. [1]

Stage: ____________________

Color: ____________________

Base your answers to questions 79 through 81 on the graph below and on your knowledge of Earth science. The graph shows the altitude of the Sun for a 24-hour period on a certain day of the year at four different latitudes.

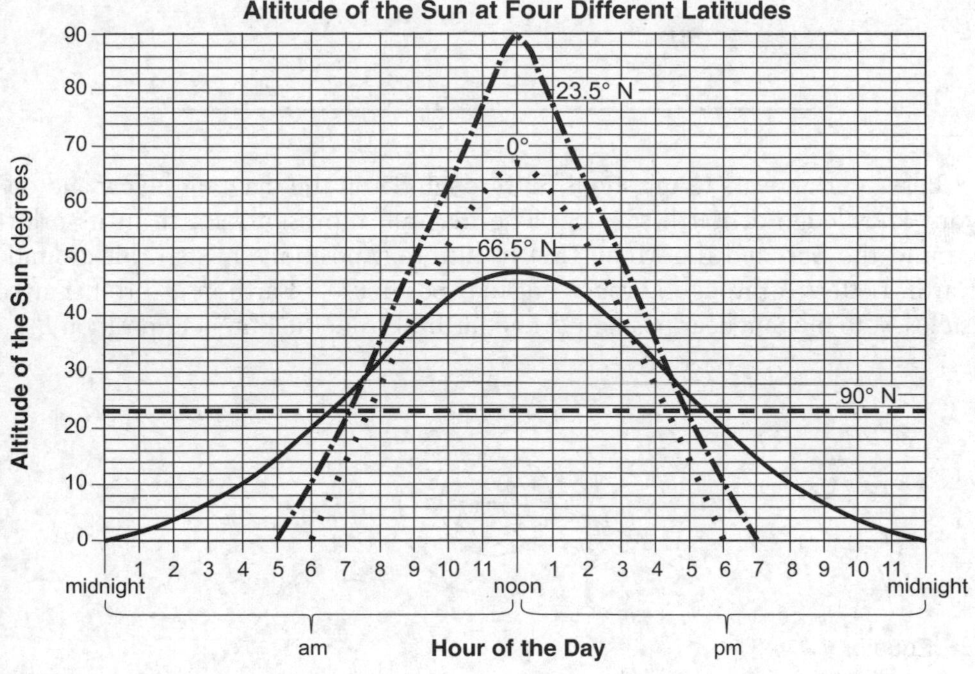

79 Identify the date represented by the data shown on the graph. [1]

____

80 Identify the latitude that has the greatest intensity of insolation at approximately noon on this date, and describe the evidence shown on the graph to support your answer. [1]

Latitude: ______________________

Evidence: ____

____

81 Identify the compass direction toward which an observer's shadow would extend if the observer is located exactly at 90° N at noon on this date. [1]

____

Base your answers to questions 82 through 85 on the diagram below and on your knowledge of Earth science. The diagram represents Earth in its orbit around the Sun and the Moon (M) in different positions in its orbit around Earth. Letters *A* through *D* represent four positions of Earth in its orbit. Earth is closest to the Sun near position *D* and farthest from the Sun near position *B*.

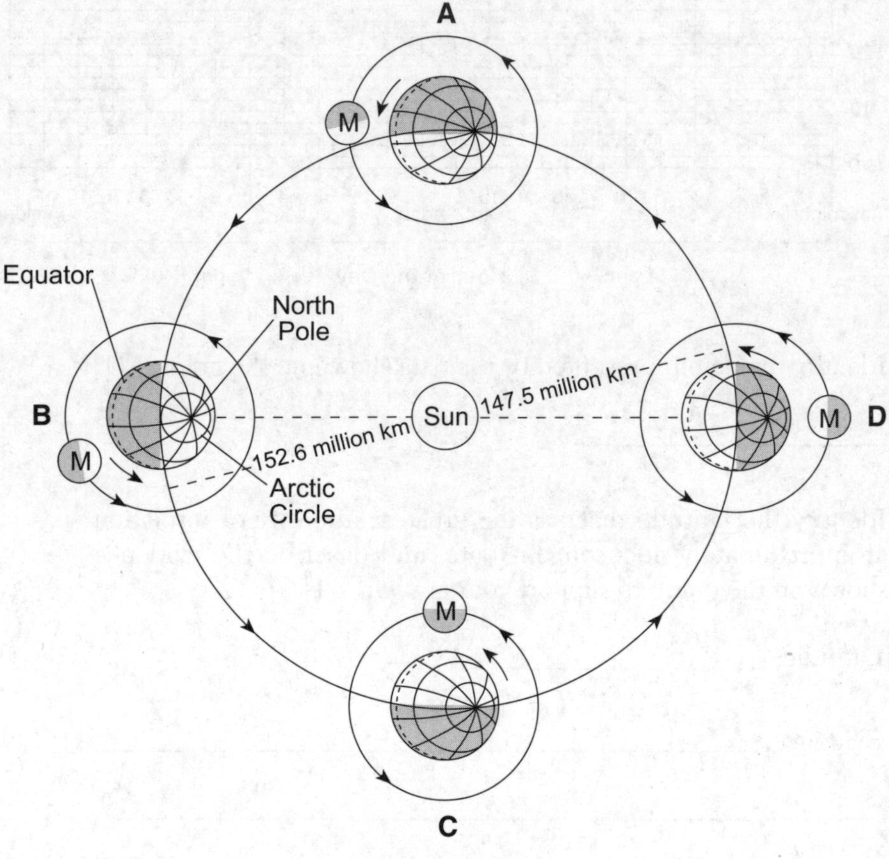

(Not drawn to scale)

82 The photograph below shows the surface of the Moon as seen from Earth during the full moon phase.

Identify the lettered position of Earth and the Moon when this full moon phase can be observed, and state the name of the eclipse that could occur at this position. [1]

Lettered position: ______________________________

Name of eclipse: ______________________________

83 Identify the Northern Hemisphere season that is beginning at position *A*. [1]

____

84 Explain why the gravitational attraction between the Sun and Earth increases as Earth moves from position *C* to position *D*. [1]

____

____

85 Describe *one* piece of evidence shown by the diagram that indicates this is a heliocentric model. [1]

____

____

Answers
August 2019
Physical Setting/Earth Science

Answer Key

PART A

1. 3	**8.** 3	**15.** 4	**22.** 4	**29.** 2
2. 1	**9.** 2	**16.** 1	**23.** 1	**30.** 3
3. 2	**10.** 4	**17.** 2	**24.** 3	**31.** 4
4. 4	**11.** 3	**18.** 2	**25.** 4	**32.** 1
5. 2	**12.** 3	**19.** 1	**26.** 3	**33.** 4
6. 4	**13.** 2	**20.** 3	**27.** 1	**34.** 3
7. 1	**14.** 1	**21.** 1	**28.** 2	**35.** 4

PART B–1

36. 4	**39.** 3	**42.** 3	**45.** 3	**48.** 4
37. 1	**40.** 1	**43.** 3	**46.** 1	**49.** 4
38. 1	**41.** 2	**44.** 1	**47.** 2	**50.** 2

PART B–2 and **PART C**. *See* **Answers Explained**.

Answers Explained

PART A

1. **3** It is given that the asteroid belt is between the orbits of Mars and Jupiter. Find the Solar System Data table in the *Reference Tables for Physical Setting/Earth Science.* Locate both "Mars" and "Jupiter" in the column labeled "Celestial Object." Follow both rows to the right, and note the column labeled "Mean Distance from Sun (million km)." Note that Mars and Jupiter orbit the Sun at distances of 227.9 and 778.4 million kilometers, respectively. Therefore, most asteroids are located between 227.9 and 778.4 million kilometers from the Sun. Of the choices given, only choice (3) 403 million km is within this range.

2. **1** Ocean tides are cyclic and predictable. By using subtraction, it can be determined that the time from the last high tide on Tuesday until the next high tide Wednesday morning is 12 hours 26 minutes (e.g., 12:36 p.m. – 1:02 a.m. = 12:26). Therefore, the next high tide after the last high tide on Wednesday will occur about 12 hours and 26 minutes after 1:27 p.m., or at 1:53 a.m.

3. **2** Find the Solar System Data chart in the *Reference Tables for Physical Setting/Earth Science.* Locate "Venus" in the column labeled "Celestial Object." Follow the row for Venus to the right, and note in the column labeled "Period of Revolution (d=days) (y=years)" the value 224.7 days. Continue following the row for Venus to the right. In the column labeled "Period of Rotation at the Equator," note the value 243 days. Thus, Venus takes more time to complete one rotation on its axis than to complete one revolution around the Sun.

WRONG CHOICES EXPLAINED:

(1) Find the Solar System Data chart in the *Reference Tables for Physical Setting/Earth Science.* Locate "Mercury" in the column labeled "Celestial Object." Note that Mercury's period of revolution is 88 days and its period of rotation is 59 days. Thus, it takes *less* time for Mercury to complete one rotation on its axis than to complete one revolution around the Sun.

(3) Find the Solar System Data chart in the *Reference Tables for Physical Setting/Earth Science.* Locate "Earth" in the column labeled "Celestial Object." Note that Earth's period of revolution is 365.26 days and its period of rotation is 23 hours 56 minutes 4 seconds. Thus, it takes *less* time for Earth to complete one rotation on its axis than to complete one revolution around the Sun.

(4) Find the Solar System Data chart in the *Reference Tables for Physical Setting/Earth Science.* Locate "Mars" in the column labeled "Celestial Object." Note that the period of revolution for Mars is 687 days and its period of rotation is 24 hours 37 minutes 23 seconds. Thus, it takes *less* time for Mars to complete one rotation on its axis than to complete one revolution around the Sun.

4. **4** The planets can be divided by mass and density into the terrestrial (Earth-like) and Jovian (Jupiter-like) planets. The terrestrial planets include the four innermost planets: Mercury, Venus, Earth, and Mars. The Jovian planets include the four outermost planets: Jupiter, Saturn, Neptune, and Uranus. Find the Solar System Data chart in the *Reference Tables for Physical Setting/Earth Science.* Locate the column labeled "Mass (Earth = 1)." Note that the four terrestrial planets range in mass from 0.06 to 1.00, while Jupiter has a mass of 317.83. Thus, compared to the terrestrial planets, Jupiter has a greater mass.

WRONG CHOICES EXPLAINED:

(1) Find the Solar System Data chart in the *Reference Tables for Physical Setting/Earth Science.* Locate the column labeled "Period of Revolution (d=days) (y=years)." Note that the periods of revolution for the terrestrial planets range from 88 days to 687 days. Note that the period of revolution of Jupiter is 11.9 years. Thus, compared to the terrestrial planets, Jupiter has a longer period of revolution.

(2) Find the Solar System Data chart in the *Reference Tables for Physical Setting/Earth Science.* Locate the column labeled "Mean Distance from Sun (million km)." Note that the distance from the Sun for the terrestrial planets ranges from 57.9 to 227.9 million km. Note that Jupiter's distance from the Sun is 778.4 million km. Thus compared to the terrestrial planets, Jupiter has a longer distance to the Sun.

(3) Find the Solar System Data table in the *Reference Tables for Physical Setting/Earth Science.* Locate the column labeled "Density (g/cm^3)." Note that the densities of the terrestrial planets range from 3.9 g/cm^3 to 5.5 g/cm^3 while the density of Jupiter is 1.3 g/cm^3. Thus, compared to the terrestrial planets, Jupiter has a lower density.

5. **2** At different times of the year, Earth is at different points along its orbit, and the night side of Earth faces different portions of the universe. Since stars are visible only at night, only constellations in the direction of Earth's night side are visible from Earth. Since the constellation Orion is visible in New York State at night in January, the night side of Earth faces Orion in January. Five months later, in June, Earth has revolved to a position in its orbit nearly opposite the Sun from its January position. Thus, the night side of Earth faces away from Orion in June.

Orion is not visible in New York State in June because the Sun lies between Earth and Orion.

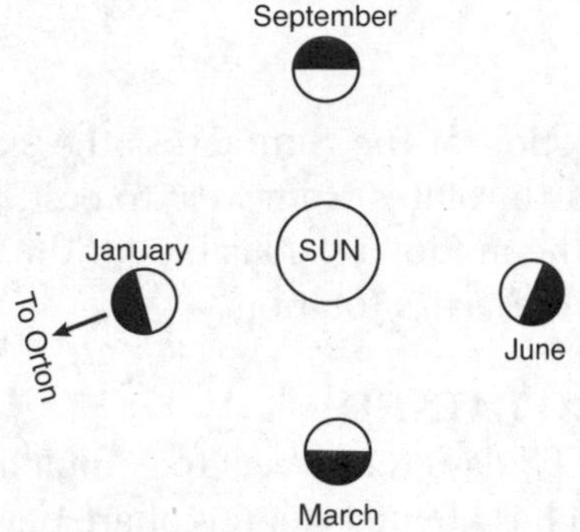

Therefore, the best explanation for why this constellation is not visible in the night sky to this observer in June is that Earth revolves around the Sun.

WRONG CHOICES EXPLAINED:

(1) Earth rotates on its axis once every day. Earth's daily rotation causes a daily apparent motion of constellations, not a seasonal change in constellations visible from Earth.

(3) The Sun is located too far from the stars in constellations to exert enough gravitational attraction to hold the constellations in orbit and cause them to revolve around the Sun.

(4) The side of Earth that is in darkness depends on only Earth's position in relation to the Sun. In whatever direction Earth is tilted, night always occurs on the side of the planet located directly opposite the Sun. Thus, the tilt of Earth's axis does not affect the direction Earth faces at night (nor does it affect the constellations that are visible in that direction).

6. **4** According to the Big Bang theory, the universe started out with all of its matter and energy in a small volume and then expanded outward in all directions, a motion very much like an explosion. The explosion associated with the Big Bang theory and the formation of the universe occurred *before* the formation of Earth and the solar system. In other words, the universe is older than either Earth or the solar system. Find the Geologic History of New York State chart in the *Reference Tables for Physical Setting/Earth Science.* Locate the column labeled "Eon." Note that the entry "Estimated time of origin of Earth and solar system" corresponds to 4600 million years ago (4.6 billion years ago) on the time scale. Thus, the explosion associated with the Big Bang theory took place

more than 4.6 billion years ago. Based on the current expansion rate of the universe, scientists extrapolate back to when the universe was all concentrated in a small volume and infer that the Big Bang occurred approximately 13.8 billion years ago.

7. **1** This apparent motion of the Sun across the sky is the result of Earth's rotation on its axis. As Earth rotates from west to east, the Sun appears to move from east to west. Thus, the motion responsible for the apparent path of the Sun shown in the photograph is Earth's rotation.

WRONG CHOICES EXPLAINED:

(2) Earth takes about 365 days (one year) to complete one 360-degree revolution around the Sun, which is about 1 degree of arc per day. The Sun has clearly moved through more than 1 degree of arc during the time shown in the photograph. Thus, it is unlikely that Earth's revolution around the Sun is responsible for the apparent path of the Sun shown in the photograph.

(3) In order for the Sun to change position in the sky, it must change position relative to an observer on Earth. The Sun's rotation on its axis does not change its position relative to an observer on Earth. The Sun could remain stationary to an observer on Earth and still rotate. Therefore, the apparent path of the Sun through the sky cannot be caused by the Sun's rotation on its axis.

(4) The Sun does not revolve around Earth. Therefore, this motion could not cause the apparent path of the Sun shown in the photograph.

8. **3** Evidence suggests that an impact event was responsible for the extinction of many life-forms at the end of the Cretaceous period approximately 65.5 million years ago. The dividing line between the Cretaceous and the Tertiary is a thin layer of clay that has been identified in sediments worldwide. This boundary clay contains numerous indications of a massive impact event: levels of iridium higher than those found in Earth's crust but similar to those found in meteorites, shocked quartz grains, melt spherules, soot from the widespread forest fires ignited by the impact, as well as evidence of large waves such as would be caused by an ocean impact. The impact of asteroids or large meteorites would generate a massive shock wave and spew large amounts of dust high into the atmosphere. The resulting decrease in sunlight would cause global temperatures to decrease markedly and have devastating effects on photosynthetic life and all of the life-forms that depend on them for food. Thus, some scientists have inferred that the mass extinction of the dinosaurs approximately 65.5 million years ago was caused by an impact event occurring on Earth's surface.

WRONG CHOICES EXPLAINED:

(1) Measurements of the Sun's energy output indicate that the Sun is relatively stable. Although there are fluctuations associated with the sunspot cycle, these do not occur only at times of mass extinctions. Thus, there does not appear to be a cause and effect relationship between large energy surges from the surface of the Sun and mass extinctions.

(2) About five major ice ages have occurred during Earth's history. The most recent, the Quaternary, began about 2.8 million years ago. The ice age before that, the Karoo, occurred between 360 and 260 million years ago. Thus, the mass extinction of dinosaurs about 65.5 million years ago did not occur during a major ice age. Therefore, there is no basis to infer that the mass extinction of dinosaurs was caused by the occurrence of a major ice age.

(4) Earthquakes are unlikely to have caused mass extinctions, because they occur almost continuously, not solely at times of mass extinctions.

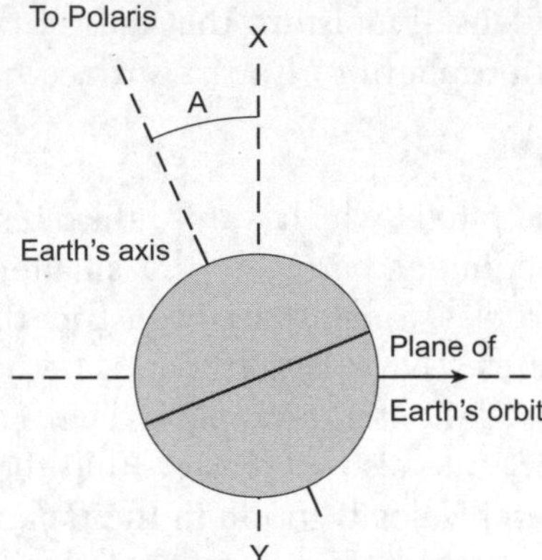

9. **2** It is given that line *XY* is perpendicular to the plane of Earth's orbit. Earth's axis of rotation is tilted 23.5° from a line perpendicular to the plane of its orbit. Thus, the value of the angle represented by letter *A* is 23.5°.

10. **4** The rate at which meltwater can infiltrate soil depends on a number of factors. One factor is the surface slope. The gentler the slope, the more slowly the water will flow downhill and the more time the water has to infiltrate before it runs off. Another factor is the permeability of the surface. If the land surface is still frozen, ice fills the pore spaces between particles, making the surface impermeable. However, if the surface is no longer frozen, the spaces between particles are not blocked by ice, allowing water to infiltrate. Thus, meltwater will most likely infiltrate where the land surface is no longer frozen with a gentle slope.

11. **3** Water is returned to Earth's atmosphere in the form of water vapor. To form water vapor, water must change from either the solid or the liquid phase to the gaseous phase. Since most of Earth's surface is liquid water, most water vapor enters the atmosphere by changing from a liquid to a gas, a process called evaporation. Water vapor also enters the atmosphere directly from the leaves of plants that cover much of Earth's land surfaces, a process called transpiration. Thus, water is returned from Earth's surface to the atmosphere by evaporation and transpiration.

WRONG CHOICES EXPLAINED:

(1) Condensation refers to the change from the gaseous to the liquid phase of matter. Thus, condensation does not return water vapor to the atmosphere.

(2) Condensation refers to the change from the gaseous to the liquid phase of matter. Precipitation is condensed moisture that falls to the ground. Neither process returns water from Earth's surface to the atmosphere.

(4) Precipitation is condensed moisture that falls to the ground. This process returns water from the atmosphere to Earth's surface, not from Earth's surface to the atmosphere.

12. **3** Permeability is the rate at which water infiltrates (trickles through) sediment. The smaller the sediment particles, the smaller the empty spaces are between the sediment particles. Small spaces constrict the flow of water through soil. Large sediment particles have larger spaces between them, which allow water to flow through the sediment more rapidly. Therefore, as sediment particle size increases, permeability rate also increases. Find the Relationship of Transported Particle Size to Water Velocity graph in the *Reference Tables for Physical Setting/Earth Science*. Note the dashed lines labeled with the size ranges corresponding to clay, silt, sand, pebbles, cobbles, and boulders. Note that pebbles are larger in diameter than sand. Thus, pebbles have a greater permeability rate than sand. Sorting also affects permeability rate. Unsorted pebbles have smaller spaces through which water can infiltrate than do sorted pebbles. In unsorted pebbles, smaller pebbles fill the spaces between larger pebbles. Therefore, sorted pebbles have the greatest permeability rate.

13. **2** High-pressure centers are usually associated with air sinking toward the surface. When air sinks, it goes through a warming and drying process. Therefore, the relative humidity of the air decreases, and condensation and cloud formation become less likely. Thus, high-pressure centers are usually associated with clear skies.

WRONG CHOICES EXPLAINED:

(1) Clouds form when air is cooled to its dewpoint and water vapor condenses into water droplets. The closer the air temperature is to the dewpoint, the higher the relative humidity is. Therefore, cloudy skies are associated with higher relative humidity than are clear skies.

(3) Most clouds and precipitation occur at or around frontal boundaries. Frontal boundaries mark the transition from higher pressure cool air to lower pressure warm air. The greatest pressure differences, and thus the highest wind speeds, occur at or near these boundaries. Thus, increasing wind speed is associated with frontal boundaries and cloudy skies, not with clear skies.

(4) When the air temperature reaches the dewpoint, moisture begins to condense, clouds form, and precipitation results. Thus, when the dewpoint equals the air temperature, cloudy skies are the most likely result, not clear skies.

14. **1** Find the Selected Properties of Earth's Atmosphere chart in the *Reference Tables for Physical Setting/Earth Science*. Locate the vertical axis labeled "Altitude." Note that an altitude of "0" corresponds to ground level. Now find "Troposphere" in the portion of the chart labeled "Temperature Zones." Note the dashed line labeled "Tropopause" marking the top of the troposphere. On the graph labeled "Temperature (°C)," note that from ground level to the tropopause, temperature decreases. On the graph labeled "Pressure (atm)," note that from ground level to the tropopause, atmospheric pressure decreases. Thus, as a weather balloon released at Earth's surface rises through the troposphere, there will be a decrease in both air temperature and atmospheric pressure.

15. **4** The ozone layer is a portion of the upper stratosphere that contains high concentrations of the gas ozone. One property of ozone is that it absorbs certain wavelengths of ultraviolet radiation. The short wavelength and high energy of ultraviolet radiation enables it to damage cells. Therefore, the gas most effective in absorbing incoming harmful ultraviolet radiation in Earth's stratosphere before that radiation reaches Earth's surface is ozone.

16. **1** A monsoon is a seasonally reversing system of surface winds caused by temperature differences between continents and neighboring oceans. A monsoon is a large-scale version of the type of *daily* changes in circulation we see in land and sea breezes. In a monsoon, we see *seasonal* changes in wind direction between a large landmass and surrounding oceans.

For example during the summer, the large landmass of Asia warms up and a low-pressure center develops in the southern part of the continent. This low-pressure center results in winds that blow inland from the surrounding oceans.

These moist summer monsoon winds bring clouds and heavy rains. During the winter, intense cooling produces a mass of very cold, very dry air over the northern part of the continent. The result is a high-pressure center in the northern part of the continent that causes the winds to reverse direction and blow from the continental interior outward to the coast. Although these winter monsoon winds may warm up as they descend toward the coast, they pick up little moisture as they blow over the land and bring clear, dry weather. Thus, the major cause of monsoon rains in India and Southeast Asia is seasonal shifts in the prevailing wind belts.

WRONG CHOICES EXPLAINED:

(2) A monsoon is a seasonally reversing system of surface winds caused by temperature differences between land and ocean, not shifts in ocean currents.

(3) Measurements of the Sun's energy output indicate that it is relatively stable. Although fluctuations are associated with the sunspot cycle, these do not occur on a seasonal basis and are therefore unlikely to be the major cause of monsoons.

(4) Monsoons are regional, not worldwide, occurrences. Changes in worldwide temperatures would affect all parts of Earth, not just certain regions.

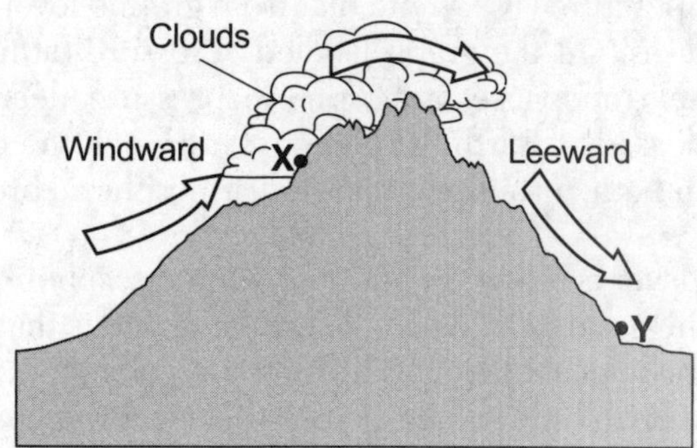

17. **2** The cross-section shows prevailing winds forcing air to rise along the windward side of the mountain range. As the air rises to higher elevations, air pressure and temperature decrease, causing the air to expand and cool. When the air cools to its dewpoint temperature, water vapor begins to condense, forming clouds. If enough water vapor condenses and if large water droplets form, precipitation will occur on the windward side. Thus, location *X* will be cool and wet.

As the air sinks on the leeward side of the mountains, air pressure increases, causing the air to be compressed and, therefore, warm. The air is warmed above its dewpoint, and any available moisture evaporates. Therefore, clouds and precipitation are much less likely to form on the leeward side of the mountain, and location *Y* will be warm and dry. Therefore, compared to location *X*, the temperature and water vapor content at location *Y* are most likely warmer and drier.

18. **2** Find the Surface Ocean Currents map in the *Reference Tables for Physical Setting/Earth Science*. In the key, note the symbols for warm and cool ocean currents. Locate the Gulf Stream, Peru Current, West Greenland Current, and East Australia Current. Note that of these, the Peru Current and West Greenland Current are cool currents. Note that the Peru Current transports cool water toward the equator and the West Greenland Current transports cool water toward the pole. Therefore, the surface ocean current that transports cool water to lower latitudes is the Peru Current.

19. **1** Color is a fairly good indicator of whether a substance absorbs more light than it reflects. The lighter the color of a substance, the more light is being reflected by it; the darker the color, the more light is absorbed by it.

If a surface is smooth, light is more likely to be immediately reflected away from the surface than if it is rough. The irregularities on a rough surface can cause some of the light to hit the surface several times before leaving. Each time the light strikes the surface, a little more of its energy is absorbed. Thus, the greatest amount of absorption of insolation at Earth's surface is produced by a dark color and rough texture.

20. **3** Note that the question refers to *surface* winds and air currents. Find the Planetary Wind and Moisture Belts in the Troposphere chart in the *Reference Tables for Physical Setting/Earth Science.* Note the arrows showing air movements at the surface near the equator. As you can see, the surface winds north and south of the equator are both blowing toward the equator. In other words, the surface winds are converging at the equator. The arrows also show air currents at the equator moving upward from the surface toward the tropopause, or rising. Thus, the planetary surface winds and air currents near Earth's equator are usually converging and rising.

21. **1** Earth's early atmosphere most likely formed by outgassing, which was the release of water and trace gases trapped in rocks in the interior of the hot, young Earth. Comets that collided with Earth were another likely source of gases. Comets are rich in ices of water, carbon dioxide, carbon monoxide, ammonia, and organic compounds. Thus, Earth's early atmosphere contained virtually no free oxygen. However, over time, life-forms evolved that were able to carry out photosynthesis. Photosynthetic life-forms use energy from sunlight to convert carbon dioxide and water into carbohydrates (food) and oxygen. Therefore, once photosynthetic life evolved, these organisms released oxygen into the environment.

Find the Geologic History of New York State chart in the *Reference Tables for Physical Setting/Earth Science*. In the column labeled "Era," locate the statements "Oceanic oxygen produced by cyanobacteria combines with iron, forming iron oxide layers on ocean floor" and "Oceanic oxygen begins to enter the atmosphere." Cyanobacteria, or blue-green algae, are single-celled organisms that produce gaseous oxygen as a by-product of photosynthesis. Oxygen is highly reactive and combines with substances dissolved in seawater. Some of these substances, such as carbonates, were soluble and remained in solution in the oceans. Others, such as iron oxides, were insoluble and settled out of the oceans to form layers of iron oxide on the ocean floor. Eventually, oxygen levels grew high enough in the oceans that excess oxygen began to escape into the atmosphere. As photosynthetic organisms multiplied over time, photosynthesis steadily increased. Over millions of years, the oxygen released by these photosynthetic organisms steadily increased the oxygen in Earth's atmosphere to its present level. Thus, scientists infer that oxygen in Earth's early history was first produced during the Precambrian from cyanobacteria in Earth's oceans.

WRONG CHOICES EXPLAINED:

(2) Volcanic activity along plate boundaries occurred throughout Earth's history and was occurring long before oxygen was first produced in Earth's atmosphere. Therefore, it is unlikely that volcanic activity first produced the oxygen in Earth's atmosphere.

(3) Find the Geologic History of New York State chart in the *Reference Tables for Physical Setting/Earth Science*, and locate the "Life on Earth" column. Note in this column that the earliest land plants and animals appeared during the Silurian, well after oxygen in Earth's early atmosphere was first produced during the Precambrian. Therefore, it is unlikely that the absorption of sunlight by plants first produced the oxygen in Earth's atmosphere.

(4) The evaporation of ocean water involves the change from liquid water to water vapor. Water vapor is not oxygen. Furthermore, oceans existed and ocean water evaporated long before oxygen in Earth's early atmosphere was produced. Therefore, it is unlikely that evaporation of ocean water first produced the oxygen in Earth's atmosphere.

22. **4** Find the Geologic History of New York State chart in the *Reference Tables for Physical Setting/Earth Science*, and locate the column labeled "Time Distribution of Fossils." Note the vertical gray bars labeled with the names of different groups of organisms. The vertical gray bars indicate the range of time during which each group of organisms existed on Earth. The bottom of the gray bar indicates when the group first appeared on Earth, and the top of the bar indicates when the group became extinct or, if it extends to the top of the column,

that the group still exists. Locate the bars labeled "dinosaurs," "trilobites," "ammonoids," and "placoderm fish." Note that of these four bars, the bar labeled "placoderm fish" is the shortest. Therefore, of these four choices, the life-form that existed on Earth for the shortest period of time was placoderm fish.

23. **1** Find the Geologic History of New York State chart in the *Reference Tables for Physical Setting/Earth Science*. Note that *Eurypterus* is fossil "**M**" in the illustrations of specific important index fossils printed along the bottom of the chart. In the column labeled "Time Distribution of Fossils," locate the circled letter "**M**" along the black line labeled "eurypterids." Trace horizontally to the left to the column labeled "Period," and note that the approximate time of the existence of *Eurypterus* was during the Silurian. Index fossils are remains of organisms that had distinctive body features, were common and abundant, and had a broad, even worldwide range yet existed for only a short period of time. Their distinctive body features and broad distribution make index fossils easy to find in widely separated rock layers. Their short existence pinpoints the time period during which the rock layer was formed. Thus, organisms that later became good index fossils lived over a wide geographic area and existed for a short time. Evaporites, such as halite, that formed during a short period of time when very salty inland seas existed in an area allows one to roughly date the age of the rocks. Abundance over a wide geographic area also allows rocks in widely separated places to be correlated. Thus, both the halite layers and the *Eurypterus remipes* fossils can now be used to identify a specific geologic time interval because both formed over a large geographic area and in a short geologic time.

24. **3** Find the Earthquake *P*-Wave and *S*-Wave Travel Time graph in the *Reference Tables for Physical Setting/Earth Science.* On the vertical axis labeled "Travel Time (min)," locate "5." From this point, trace right until you intersect the bold curve labeled "P." From this intersection, trace downward vertically to the horizontal axis labeled "Epicenter Distance ($\times$ 10^3 km)." Note that you intersect the axis at "2600 km." Thus, a *P*-wave takes 5 minutes to travel 2600 km. Now, trace vertically upward from 2600 km until you intersect the bold curve labeled "S." From this intersection, trace horizontally to the left to the vertical axis labeled "Travel Time (min)." Note that an *S*-wave takes 9 minutes to travel 2600 km. Thus, an *S*-wave takes 9 minutes to travel the same distance a *P*-wave travels in 5 minutes.

25. **4** *P*-waves are compression waves that compress and expand rock in the direction of wave travel. Solids, liquids, and gases are all compressible; therefore, *P*-waves can be transmitted through solids, liquids, and gases. On the other hand, *S*-waves are transverse waves that twist rock back and forth, deforming their shape

in a direction perpendicular to that of wave travel (laterally). *S*-waves can be transmitted only through solids. *S*-waves cannot travel through liquids and gases, because when liquids and gases are deformed laterally, they do not return to their original shape. Find the Inferred Properties of Earth's Interior in the *Reference Tables for Physical Setting/Earth Science*. Locate the outer core in the cross section at the top of the diagram. From the outer core, trace downward between the dashed lines to the "Temperature (°C)" graph, and note that the actual temperature of the outer core is higher than its melting temperature. Thus, the outer core is a liquid. Note, too, that the outer core is the only place on the "Temperature (°C)" graph where this is the case, so the outer core is the only layer of Earth's interior that is a liquid. Therefore, *S*-waves from an earthquake cannot pass through the liquid outer core and as a result will also not reach the inner core. Thus, earthquake *S*-waves do not pass through the outer core and the inner core.

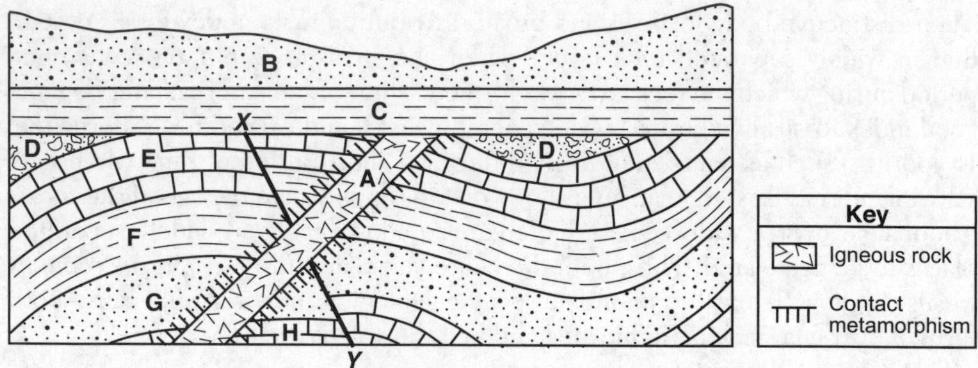

26. **3** Note that the fault along line *XY* has displaced rock units *E*, *F*, *G*, and *H*. Any rock unit displaced by a fault is older than the fault because it already had to exist in order to be displaced. Thus, the faulting along *XY* occurred after the formation of rock units *E*, *F*, *G*, and *H*. Therefore, the faulting along line *XY* occurred after the formation of rock units *E* and *F*.

WRONG CHOICES EXPLAINED:

(1) Note that the igneous rock intrusion *A* cuts across the fault and is not displaced by the fault. If the intrusion had existed before faulting occurred, the fault would cut across the intrusion and the intrusion would have been displaced by the fault. Therefore, the faulting along line *XY* occurred before, not after, the intrusion of rock unit *A*.

(2) Note that rock units *B* and *C* have not been displaced by faulting along line *XY*. Thus, the faulting along line *XY* occurred before rock units *B* and *C* were deposited.

(4) Note that the fault along line *XY* has displaced rock units *G* and *H*. Any rock unit displaced by a fault is older than the fault because it already had to exist in order to be displaced. Thus, the faulting along line *XY* occurred after the formation of rock units *G* and *H*.

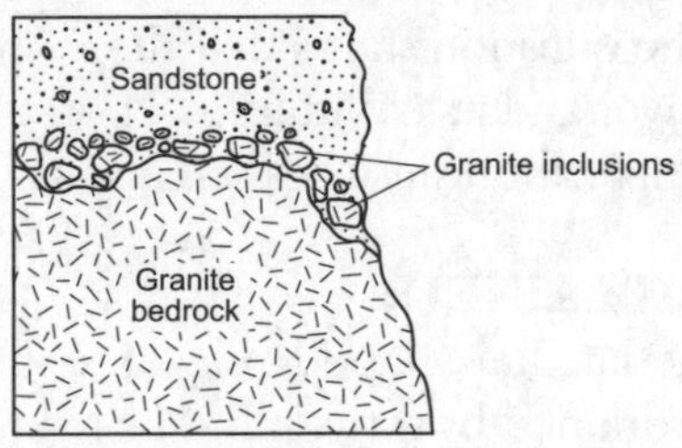

27. **1** An inclusion is a solid fragment, liquid globule, or pocket of gas enclosed in a mineral or rock. The geologic principle of inclusions states that the inclusions in a rock are older than the rock itself. Thus, the granite inclusions are older than the sandstone. Note that the granite inclusions lie near the granite bedrock beneath the sandstone. Therefore, it is reasonable to infer that the inclusions shown near the bottom of the sandstone are pieces of the granite bedrock broken loose by weathering before the sandstone was deposited. Thus, the age of the granite that makes up the inclusions is most likely older than the sandstone, but the same age as the granite bedrock.

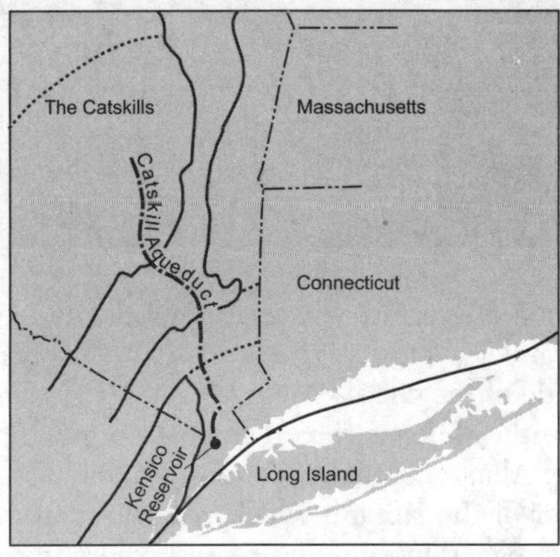

28. **2** It is given that the Catskill Aqueduct flows into the Kensico Reservoir. Find the Generalized Landscape Regions of New York State map in the *Reference Tables for Physical Setting/Earth Science*. Locate the region corresponding to the location of the Catskill Aqueduct. Note that the Catskills are part of the

Allegheny Plateau. Therefore, the landscape regions that are crossed as water flows through the Catskill Aqueduct, in order, are the Allegheny Plateau, Hudson-Mohawk Lowlands, Hudson Highlands, and Manhattan Prong.

29. **2** By definition, a watershed is the area drained by a stream and its tributaries. A river is a stream. The question states that the map represents the land area drained by the Oswego River and its tributaries. Therefore, the land area drained by the Oswego River and its tributaries is called a watershed.

WRONG CHOICES EXPLAINED:

(1) A delta is a large, flat, fan-shaped pile of sediment at the mouth of a stream, not the land area drained by a river and its tributaries.

(3) A water table is the upper surface of groundwater filling underground pore spaces, not the land area drained by a river and its tributaries.

(4) A floodplain is a low plain adjacent to a stream that is likely to be flooded if a stream overflows, not the area drained by a river and its tributaries.

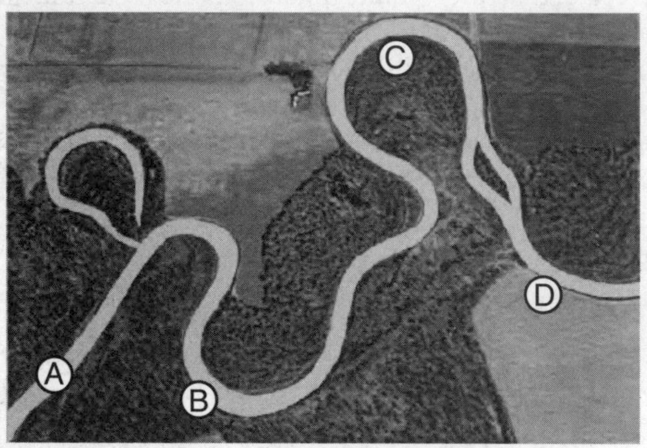

30. **3** Stream erosion is greatest where water velocity is greatest. When the stream curves, or meanders, velocity is greatest along the outside of the curve and lowest along the inside of the curve. Water velocity is greatest along the outside of the curve because the water is forced to cover a greater distance along the outside of the curve. Along a straight section of a stream's channel, velocity is greatest in the center of the stream, which is farthest from the friction of the banks or bed. Since stream deposition is greatest where the velocity of the water is slowest, the inside of the curve of a meandering channel experiences the greatest amount of deposition. Of the four locations on the Genesee River, only location *C* is on the inside curve of a meander. Therefore, deposition of sediments would most likely be greater than erosion at location *C*.

31. **4** Soil is the accumulation of loose, weathered material that covers much of Earth's land surface. The main component of soil is weathered rock. However, a true soil also contains water, air, bacteria, and decayed plant and animal material (humus).

Since weathered rock is the main component of soil, one of the most important factors controlling the development of soil is the multitude of weathering processes that convert the bedrock to soil. The single most important factor affecting weathering processes is climate. Climate is typically expressed in terms of two factors: temperature and precipitation. Both of these factors influence the type and rate of weathering. Warm climates favor chemical weathering; cold climates favor physical weathering. In both cases, the more moisture that is present, the more pronounced is the weathering. Tropical climates tend to have high temperatures and abundant rainfall. The combination of heat and water accelerates chemical weathering, leading to the rapid breakdown of rock. Living organisms can cause both physical and chemical weathering of rock. For example, tree roots can grow into crevices in rock and pry the rock apart, causing mechanical breaking apart of the rock. Moss and fungi can grow on the surface of rock. Their roots produce a weak acid that can chemically break down the rock. Ants and earthworms bring rock fragments to the surface, where the fragments are exposed to air and water that further weather them. Arctic areas tend to have cold temperatures and low rainfall. These two factors result in living organisms being less abundant in arctic areas than in tropical areas. Therefore, the rate of soil development is usually greater in tropical areas than in arctic areas because tropical areas have more chemical weathering and an abundance of living organisms.

32. **1** During the Pleistocene, huge continental glaciers formed in arctic and subarctic regions and flowed outward (south), merging with smaller valley glaciers that had already formed in mountainous regions. These processes covered most of New York State with a layer of ice thousands of feet thick. Nearly all of New York State displays evidence of glaciation. Glacial lakes, such as the Finger Lakes, formed when the glaciers retreated and meltwater filled depressions whose openings were blocked by moraines (piles of sediment) deposited by the glaciers. Since glaciers often flow through existing valleys, the Finger Lakes' basins are probably old stream valleys that were widened and deepened by the glacier into U-shaped valleys and later filled with meltwater when the ends of the valleys were dammed by moraines. Thus, New York State's Finger Lakes exist today because U-shaped valleys were dammed by glacial sediments.

WRONG CHOICES EXPLAINED:

(2) If V-shaped valleys are being eroded by streams, the water is flowing in the stream, not forming a lake.

(3) If the Finger Lakes were bodies of water stranded by falling sea levels, the water in the lakes would be saltwater, not freshwater.

(4) If the Finger Lakes were bodies of water created by rising sea levels flooding the region, the water in the lakes would be saltwater, not freshwater.

33. **4** Physical weathering breaks down rocks without changing their chemical composition. Therefore, the chemical composition of the rock fragments (beach sand) is the same as the composition of the rock from which they formed. Find the Scheme for Igneous Rock Identification in the *Reference Tables for Physical Setting/Earth Science*. On the graph labeled "Mineral Composition (relative by volume)," locate the regions labeled pyroxene, plagioclase feldspar, and olivine. Find the section of the graph in which a vertical line would pass through the regions for all three of these minerals. From this section, trace upward to the chart in the scheme labeled "Igneous Rocks." Note that it intersects the rocks gabbro, diabase, basalt, vesicular basalt, scoria, and basaltic glass. Thus of the choices given, the igneous rock that could physically weather into beach sand that contains the minerals pyroxene, plagioclase feldspar, and olivine is gabbro.

WRONG CHOICES EXPLAINED:

(1) Find the Scheme for Igneous Rock Identification in the *Reference Tables for Physical Setting/Earth Science*. In the part of the Scheme labeled "Igneous Rocks," locate dunite. From there, trace vertically downward to the graph labeled "Mineral Composition (relative by volume)." Note that dunite contains only the mineral olivine. Therefore, dunite could not physically weather to form beach sand that contained pyroxene or plagioclase feldspar.

(2) Find the Scheme for Igneous Rock Identification in the *Reference Tables for Physical Setting/Earth Science*. In the part of the Scheme labeled "Igneous Rocks," locate granite. From there, trace vertically downward to the graph labeled "Mineral Composition (relative by volume)." Note that granite contains no pyroxene or olivine. Therefore, granite could not physically weather to form beach sand that contained pyroxene or olivine.

(3) Find the Scheme for Igneous Rock Identification in the *Reference Tables for Physical Setting/Earth Science*. In the part of the Scheme labeled "Igneous Rocks," locate peridotite. From there, trace vertically downward to the graph labeled "Mineral Composition (relative by volume)." Note that peridotite contains no plagioclase feldspar. Therefore, peridotite could not physically weather to form beach sand that contained plagioclase feldspar.

34. **3** Graphite is used in pencils because of its ability to leave dark marks on paper. The dark marks are the powdered graphite left behind on the paper. To leave a streak on paper, the graphite must be very soft. In order for the streak to be easily visible, it must be dark in color. Find the Properties of Common Minerals table in the *Reference Tables for Physical Setting/Earth Science.* In the column labeled "Mineral Name," locate graphite. From graphite, trace left to the column labeled "Hardness," and note that graphite has a hardness of 1–2. Thus, graphite is among the softest of minerals. Now trace right to the column labeled "Distinguishing Characteristics," and note that graphite has a black streak. Thus, the two physical properties of graphite that make it a good mineral to use in pencils are hardness and streak.

35. **4** Find the Properties of Common Minerals table in the *Reference Tables for Physical Setting/Earth Science.* At the bottom of the table, note that the chemical symbol for iron is "Fe" and the chemical symbol for oxygen is "O." Thus, the chemical formula for a mineral that is a chemical combination of two atoms of iron and three atoms of oxygen is written as Fe_2O_3. Now find the column labeled "Composition," and locate Fe_2O_3. From there, trace right to the column labeled "Mineral Name." Note that the mineral with the chemical formula Fe_2O_3 is hematite. Thus, the mineral produced when two atoms of iron chemically combine with three atoms of oxygen is hematite.

WRONG CHOICES EXPLAINED:

(1) Find the Properties of Common Minerals table in the *Reference Tables for Physical Setting/Earth Science.* In the column labeled "Mineral Name," locate garnet. From garnet, trace left to the column labeled "Composition." Note that the composition of garnet is $Fe_3Al_2Si_3O_{12}$. In other words, garnet is formed by chemically combining three atoms of iron, two atoms of aluminum, three atoms of silicon, and twelve atoms of oxygen, not two atoms of iron and three atoms of oxygen.

(2) Find the Properties of Common Minerals table in the *Reference Tables for Physical Setting/Earth Science.* In the column labeled "Mineral Name," locate pyrite. From pyrite, trace left to the column labeled "Composition." Note that the composition of pyrite is FeS_2. In other words, pyrite is formed by chemically combining one atom of iron and two atoms of sulfur, not two atoms of iron and three atoms of oxygen.

(3) Find the Properties of Common Minerals table in the *Reference Tables for Physical Setting/Earth Science.* In the column labeled "Mineral Name," locate magnetite. From magnetite, trace left to the column labeled "Composition." Note that the composition of magnetite is Fe_3O_4. In other words, magnetite is formed by chemically combining three atoms of iron and four atoms of oxygen, not two atoms of iron and three atoms of oxygen.

PART B–1

36. **4** Earth rotates through 360° of longitude in 24 hours, or 15° of longitude per hour (360/24 = 15). Thus, every 15° of longitude that separates two locations corresponds to a difference of 1 hour (60 minutes) in solar time. Find Jamestown on the map and note that it is located at 79.2° W longitude. Find Kingston on the map and note that it is located at 74° W longitude. Thus, Kingston and Jamestown are separated by 5.2° of longitude. A 5.2° difference in longitude is equivalent to a difference in solar time of 0.35 hours (5.2°/15°/h = 0.35 h), or about 21 minutes (60 min/h × 0.35 h = 21 min). Thus, Jamestown and Kingston will experience sunrise 21 minutes apart. As Earth rotates from west to east, Kingston in the eastern part of New York State will be carried from darkness into daylight (experience sunrise) before Jamestown in the western part of New York State. Therefore, compared to the time of sunrise in Kingston, the time of sunrise in Jamestown will be approximately 20 minutes later.

37. **1** Find Watertown on the map, and note that it is located at 44° N latitude. The altitude of *Polaris* is the same as an observer's latitude north of the equator. Therefore, an observer in Watertown, New York, would see the star *Polaris* at an altitude of approximately 44°.

38. **1** Earth's axis and poles provide the reference points for lines of longitude. Lines of longitude are great circles that pass through the north and south poles and that cross the equator at 90°. These great circles are also called meridians. The longitude of a point is the angular distance between two meridians: a fixed reference meridian and the meridian passing through the point. Since the angular distance in any full circle is 360° and the angular distance covered in one Earth rotation is 360°, it was logical to link Earth's rotation to the longitude coordinate system. In one rotation, Earth moves through 360° of angular distance and 360° of longitude. Thus, Earth's system of longitude is based on Earth's rotation.

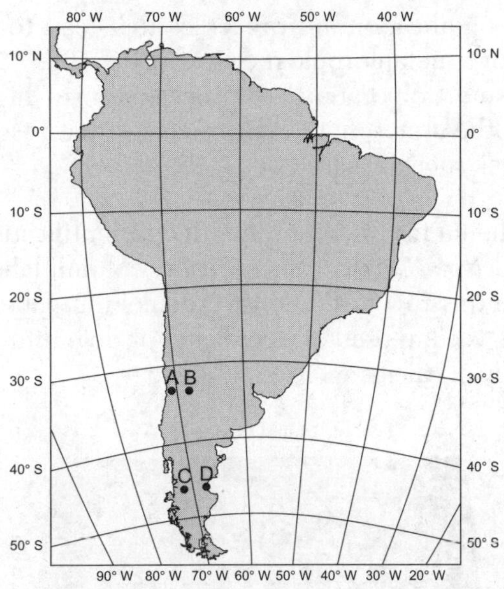

39. **3** The reading passage states "The Patagonia Marble Caves . . . at a location of 46.5° S 72° W." Locate the vertical longitude lines labeled "70° W" and "80° W." Longitude 72° W is located between longitudes 70° W and 80° W, that is, to the left of the 70° W longitude line. Locate the latitude lines labeled "40° S" and "50° S." Latitude 46.5° S is located between latitudes 40° S and 50° S, that is, below the 40° S latitude line. Therefore, the lettered point on the map that best represents the location of the Patagonia Marble Caves is left of the 70° W line and below the 40° S line, or *C*.

40. **1** Find the Scheme for Metamorphic Rock Identification in the *Reference Tables for Physical Setting/Earth Science.* In the column labeled "Rock Name," locate marble. Trace this row to the left to the "Comments" column, and note that the metamorphism of limestone or dolostone forms marble. Metamorphism is the change in the mineral composition or texture of a rock due to heat, pressure, or chemical activity without the rock undergoing melting into liquid magma. Therefore, two processes that formed the marble bedrock shown in the photograph are heat and pressure.

WRONG CHOICES EXPLAINED:

(2) Compaction and cementation are processes that transform loose particles of sediment into solid sedimentary rocks, not processes that change an existing rock into a metamorphic rock such as marble.

(3) Melting and solidification are processes that lead to the formation of a new igneous rock, not a metamorphic rock such as marble.

(4) Precipitation and evaporation are processes involving the movement of water out of or into the atmosphere, not processes that result in the formation of a metamorphic rock such as marble.

41. **2** Find the Scheme for Metamorphic Rock Identification in the *Reference Tables for Physical Setting/Earth Science.* In the column labeled "Rock Name," locate marble. Trace this row to the left to the column labeled "Composition," and note that marble is composed of calcite and/or dolomite. Thus, most marble is composed primarily of calcite and/or dolomite.

Digital Device

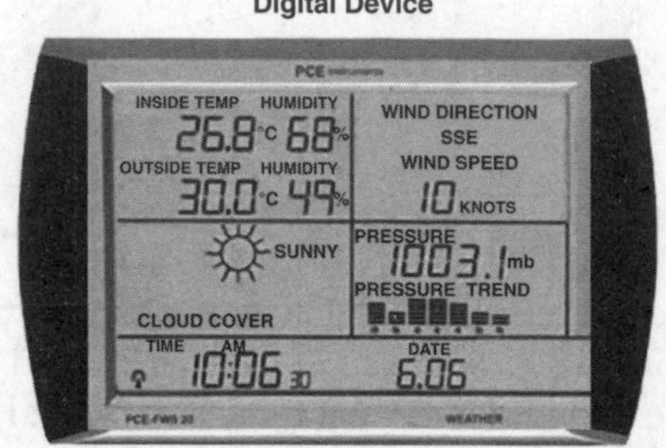

42. **3** The barometric pressure shown on the digital device is 1003.1 mb. Find the Pressure scale in the *Reference Tables for Physical Setting/Earth Science*. Locate "1003.1" on the left side of the scale labeled "millibars (mb)." Trace right to the corresponding scale labeled "inches (in of Hg*)." Note that 1003.1 millibars corresponds to 29.62 inches of Hg. Note at the bottom of the scale that "*Hg = mercury." Thus, 1003.1 millibars corresponds to 29.62 inches of mercury.

43. **3** According to the digital device, the outside air temperature is 30.0°C and the outside relative humidity is 49%. Find the Relative Humidity (%) chart in the *Reference Tables for Physical Setting/Earth Science*. Find the "Dry-Bulb Temperature (°C)" scale along the left side of the chart, and locate the row labeled "30." Trace this row to the right to the cell containing the value "49." From this cell, trace vertically upward to the "Difference Between Wet-Bulb and Dry-Bulb Temperatures (°C)" scale along the top of the chart. Note the value listed there is "8." Thus, when the dry-bulb temperature is 30°C and the relative humidity is 49%, the difference between the wet-bulb and dry-bulb temperatures is 8°C.

Now find the Dewpoint (°C) chart in the *Reference Tables for Physical Setting/ Earth Science*. Locate the column headed "8" in the "Difference Between Wet-Bulb and Dry-Bulb Temperatures (°C)" scale along the top of the chart.

Locate the row corresponding to 30°C on the "Dry-Bulb Temperature (°C)" scale along the left side of the chart. Trace to the right from the row labeled "30" until you intersect the column headed "8." At this intersection, note the value of "18." Thus, when the dry-bulb temperature is 30°C and the relative humidity is 49%, the dewpoint is 18°C.

Weather Instruments

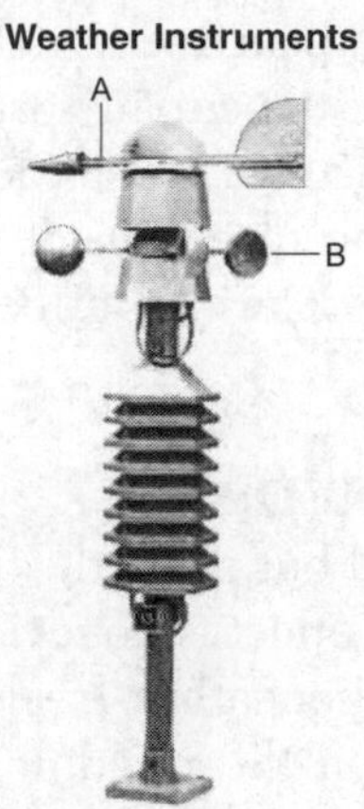

44. **1** In the photograph, weather instrument *A* consists of an arrow mounted on a rod in such a way that it can rotate freely. Wind exerts more force on the wide tail of the arrow than on the narrow head of the arrow. This causes the arrow to swing so that its head points in the direction from which wind is blowing, which is then registered on the digital device. This instrument is called a wind vane. Thus, the weather variable measured by a wind vane is wind direction.

Weather instrument *B* consists of a series of cups mounted on a shaft. The cups cause the shaft to spin when wind blows against them. The faster the wind blows, the faster the cups spin the shaft and the higher the wind speed that registers on the digital device. This instrument is called an anemometer. Thus, the weather variable measured by an anemometer is wind speed.

Therefore, *A* is a wind vane that measures wind direction, and *B* is an anemometer that measures wind speed.

45. **3** In 1851, the French physicist Jean Foucault devised a conclusive proof of Earth's rotation by means of a pendulum. He suspended a heavy iron ball on a long steel wire from the top of the dome of the Pantheon in Paris. As the pendulum swung back and forth, it appeared to change direction slowly, passing over different lines on the floor until eventually coming full circle to its original position after 24 hours. Since he knew that the path of a freely swinging pendulum would not change on its own, Foucault concluded that the apparent shift in the direction

of swing of the pendulum was due to the floor (Earth's surface) rotating beneath the pendulum. Thus, a Foucault pendulum is evidence of Earth's rotation.

Gaspard Coriolis successfully predicted the behavior of objects moving near a rotating Earth. In a stationary system, the behavior of fluids, such as the atmosphere, would be to move directly in a straight line from regions of high pressure to regions of low pressure. However, that is not what occurs on Earth. Large-scale movements of the atmosphere appear to be deflected from a straight line and instead follow a curved path. Since the curving path of winds would not occur on a stationary Earth, this apparent deflection is considered the result of Earth's rotation.

Thus, the apparent shift in the direction of swing of a Foucault pendulum and the curving of winds by the Coriolis effect are both evidence of Earth's rotation.

WRONG CHOICES EXPLAINED:

(1) If Earth's axis was not tilted but if Earth still rotated, there would still be a Coriolis effect. Thus, the Coriolis effect is not evidence of the tilt of Earth's axis.

(2) The shape of the surface beneath a freely swinging pendulum (whether flat or spherical) has no effect on the pendulum's direction of swing because there is no contact between the pendulum and the surface. Thus, a Foucault pendulum provides no evidence of Earth's shape.

(4) Since Earth's axis remains pointed in the same direction (parallel to itself) as Earth orbits the Sun, the direction of swing of a Foucault pendulum would remain unchanged in relation to Earth as the planet revolves around the Sun. Thus, the apparent shift in the direction of swing of a Foucault pendulum does not provide evidence of Earth's revolution. Furthermore, if Earth only rotated but did not revolve around the Sun, the curving path of winds would occur. Thus, the Coriolis effect is not considered evidence of Earth's revolution.

46. **1** Find the Equations chart in the *Reference Tables for Physical Setting/Earth Science*. Locate the equation for rate of change:

$$\text{rate of change} = \frac{\text{change in value}}{\text{time}}$$

It is given that the pendulum's path was displaced 60 degrees during eight hours. Substitute these values into the rate of change equation and solve:

$$\text{rate of change} = \frac{60°}{8\text{ h}} = 7.5°/\text{h}$$

Thus, the apparent path of the pendulum changed 7.5°/h.

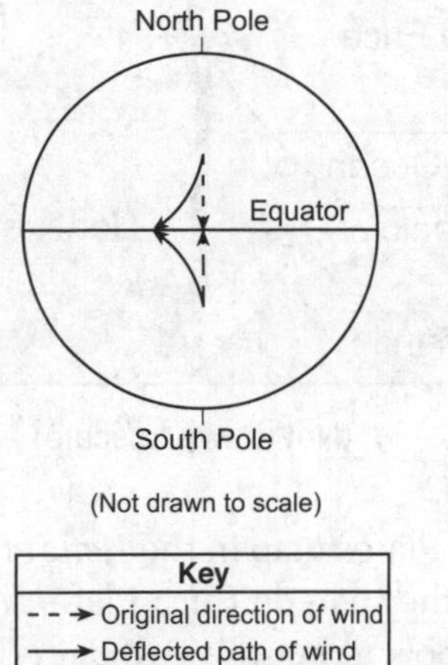

47. **2** In diagram *B*, the Northern Hemisphere is the half of Earth extending from the equator to the North Pole; the Southern Hemisphere is the half of Earth extending from the equator to the South Pole.

By convention, when north is toward the top of a diagram and south is toward the bottom, east is to the right and west is to the left. Thus, in diagram *B*, the deflected path of wind in the Northern Hemisphere is from the north to the west, and the deflected path of wind in the Southern Hemisphere is from the south to the west. Winds are named by the direction *from* which they blow, not the direction *toward* which they blow. Therefore, to an observer in the path of these winds, the deflected winds in the Northern Hemisphere *come from* the northeast and the deflected winds in the Southern Hemisphere *come from* the southeast. Now find the Planetary Wind and Moisture Belts in the Troposphere chart in the *Reference Tables for Physical Setting/Earth Science.* Note that the planetary wind belt just north of the equator is labeled "N.E. Winds," and the planetary wind belt just south of the equator is labeled "S.E. Winds." Thus, the names of the planetary wind belts formed by deflected winds shown in the Northern and Southern Hemispheres is best shown by the table in choice (2).

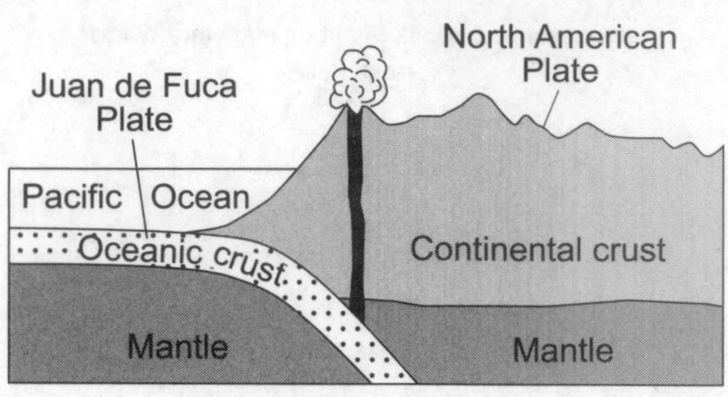

(Not drawn to scale)

48. **4** Find the Tectonic Plates map in the *Reference Tables for Physical Setting/Earth Science,* and locate the Juan de Fuca Plate. According to the key at the bottom of the map, the symbols along the eastern edge of the Juan de Fuca Plate correspond to a convergent plate boundary (subduction zone). Note that the arrows in the key indicate that the plates on either side of a convergent boundary are moving toward one another. Convection currents in the mantle cause lithospheric plates to move. Therefore, the relative motion of the upper mantle on either side of the boundary is toward the boundary. This relative motion is best shown by the diagram in choice (4).

49. **4** According to the diagram, the Juan de Fuca Plate consists of oceanic crust, and the North American Plate consists of continental crust. By visual inspection of the diagram, it can be seen that the Juan de Fuca Plate is thinner than the North American Plate. Find the Inferred Properties of Earth's Interior chart in the *Reference Tables for Physical Setting/Earth Science.* Locate the section labeled "Density (g/cm^3)" along the right-hand side of the cross section. Note that continental crust has a density of 2.7 g/cm^3 and that oceanic crust has a density of 3.0 g/cm^3. Thus, oceanic crust is denser than continental crust. Therefore, compared to the crust of the North American Plate, the Juan de Fuca Plate is thinner and more dense.

50. **2** Find the Inferred Properties of Earth's Interior chart in the *Reference Tables for Physical Setting/Earth Science*. Locate the asthenosphere and stiffer mantle in the cross section of Earth's interior at the top of the chart. Locate the dashed line corresponding to the boundary between the asthenosphere and the stiffer mantle. Follow this dashed line downward until it intersects the horizontal scale labeled "Depth (km)." Note that the dashed line intersects the scale at approximately 700 km. Thus, the boundary between the asthenosphere and the stiffer mantle is located below Earth's surface at a depth of approximately 700 km.

PART B–2

51. When lava is extruded in the rifts in mid-ocean ridges, crystals of minerals affected by magnetism align with Earth's magnetic field while the lava cools. When solidified, the rock contains a record of Earth's magnetic field locked in its crystals. Studies of ancient rocks show that Earth's magnetic field has reversed many times in the past. Research ships carrying sensitive instruments that can measure small changes in magnetism discovered that magnetism in rocks on the ocean floors reveals a striking pattern of strips of normal and reversed magnetic fields. The pattern on one side of the ridge is the mirror image of the pattern on the other side of the ridge. This suggested that rocks were forming at the ridges and then moving sideways away from the ridges. Thus, to complete the diagram, shade the pattern of normal polarity on the east side of the Mid-Atlantic Ridge center so that the width and placement of each normal polarity section is the mirror image of the pattern on the west side of the ridge, as shown below.

One credit is allowed if **the width and placement of the shaded areas are correctly indicated, as shown below.**

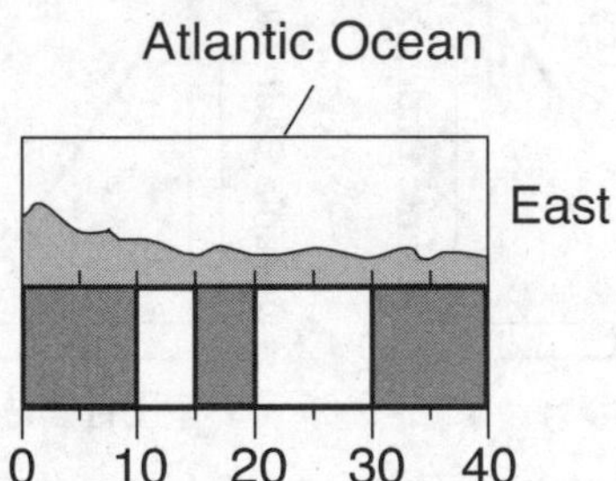

52. Find the Tectonic Plates map in the *Reference Tables for Physical Setting/Earth Science*. Note the symbol representing most mid-ocean ridges. In the key, note that the symbol corresponding to the one representing most mid-ocean ridges is a divergent plate boundary. Also note the statement "usually broken by transform faults along mid-ocean ridges." Thus, the type of tectonic plate boundary at the mid-ocean ridges where these magnetic patterns were produced is a divergent plate boundary.

One credit is allowed for **divergent/diverging**.

53. Find the Tectonic Plates map in the *Reference Tables for Physical Setting/Earth Science*, and locate the Mid-Atlantic Ridge. Note that according to the key, the Mid-Atlantic Ridge is a divergent boundary. Along divergent boundaries, adjacent plates move apart and open rifts through which magma can rise to the surface and solidify to form new ocean floor. As the plates continue to move apart, this process is repeated, constantly creating new ocean floor and pushing the older floor on either side of the mid-ocean ridge away from the ridge. Thus, the age of ocean floor bedrock increases with distance away from the mid-ocean ridge. Therefore, the line representing the relationship between the distance from the Mid-Atlantic Ridge and the age of the ocean floor bedrock should form a U-shape or a V-shape, with the youngest age at the Mid-Atlantic Ridge and the oldest ages farthest from the ridge as shown below.

One credit is allowed for **a U-shape or a V-shape, with the youngest age at the Mid-Atlantic Ridge and the oldest ages farthest from the ridge**.

Examples of 1-credit responses:

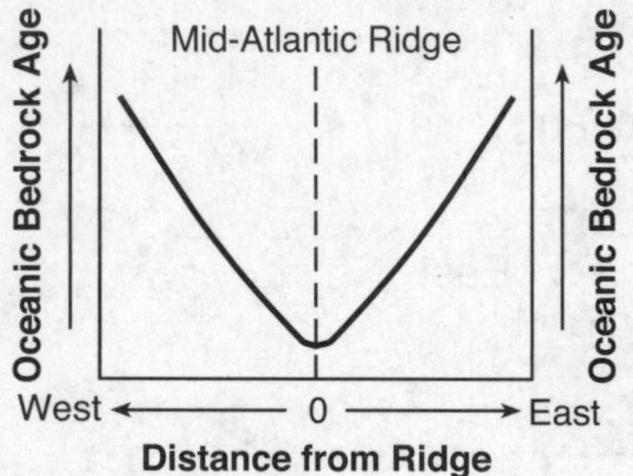

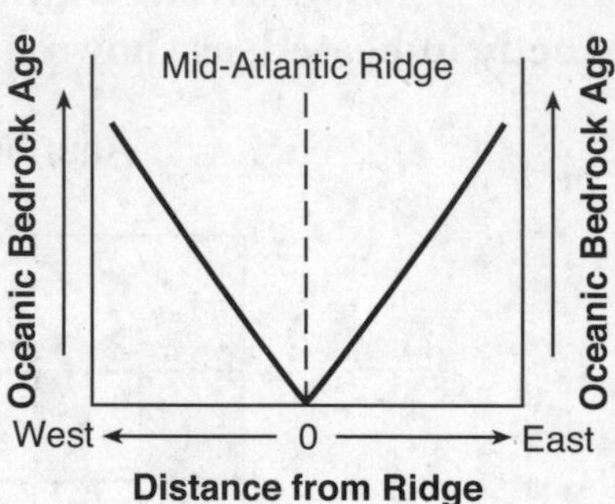

Note: One credit is allowed even if the oldest ages are *not* exactly at the same "oceanic bedrock age" level.

www.geolsoc.org.uk

54. It is given that the rock in the photograph is composed of pebbles that have been cemented together. Note that the pebbles in the photograph of the rock are rounded in shape.

Find the Scheme for Sedimentary Rock Identification in the *Reference Tables for Physical Setting/Earth Science*. In the "Grain Size" column, note the entry "pebbles, cobbles, and/or boulders embedded in sand, silt, and/or clay." Trace this double row right to the "Comments" column, and note the single row corresponding to "rounded fragments." Trace this row right to the "Rock Name" column, and note that a sedimentary rock composed of rounded pebbles is "conglomerate." Trace left to the "Texture" column, and note that conglomerate has a "clastic (fragmental)" texture. Clastic sedimentary rocks are composed of fragments of rock or sediments. Find the Rock Cycle in Earth's Crust diagram in the *Reference Tables for Physical Setting/Earth Science*. Note along the arrow leading from "Sediments" to "Sedimentary Rock" the label "compaction and/or cementation." Thus, a rock sample composed of rounded pebbles that have been cemented together is conglomerate.

One credit is allowed for **conglomerate**.

55. Find the Relationship of Transported Particle Size to Stream Velocity graph in the *Reference Tables for Physical Setting/Earth Science*. Along the right side of the graph, locate the region labeled "pebbles" among the listed particle size names. Note that the dashed line marking the upper limit of the pebble range is labeled "6.4" and the dashed line marking the lower limit of the pebble range is labeled "0.2." Also note that these dashed lines intersect the vertical axis labeled "Particle Diameter (cm)." Therefore, these values represent centimeters. Particles in the range of 0.2 cm to 6.4 cm in diameter are classified as pebbles. Thus, the total possible range of particle diameters, in centimeters, is 0.2 cm to 6.4 cm.

One credit is allowed for **0.2 cm** to **6.4 cm**.

56. Note that the particles in the rock mostly have a smooth, rounded shape. As sediments are carried along by running water, they bounce off of and rub against one another. These collisions cause smaller pieces to break off of sharp corners and edges on the surface of the sediments, particularly at corners that protrude—a process called abrasion. The sediments are also abraded as they bounce, roll, and scrape against the bottom of the streambed. As a result, sediments eroded by running water change in size and shape, becoming smaller, smoother, and more rounded. Therefore, the smooth, rounded shape of the pebbles in the conglomerate in the photograph supports the inference that the particles in this rock were likely transported by running water.

One credit is allowed. Acceptable responses include but are not limited to:

- **Rounded pebbles usually indicate transportation by running water.**
- **Sediments tumbling or bouncing in a stream produce the round shapes.**
- **The particles are smooth and have round shapes.**

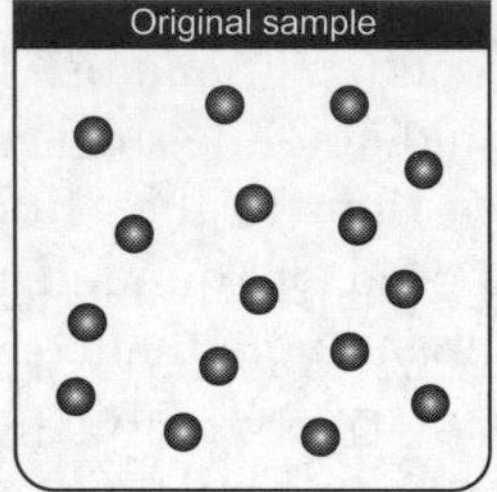

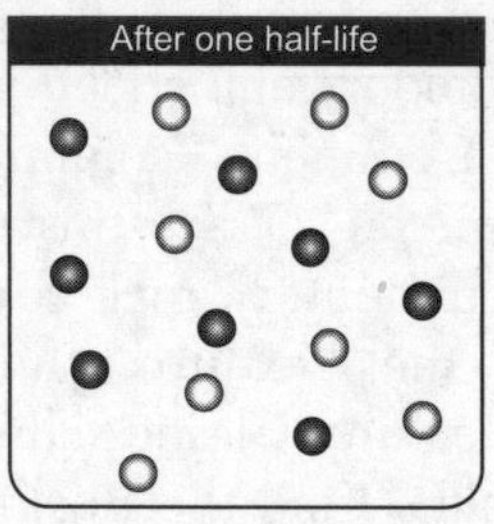

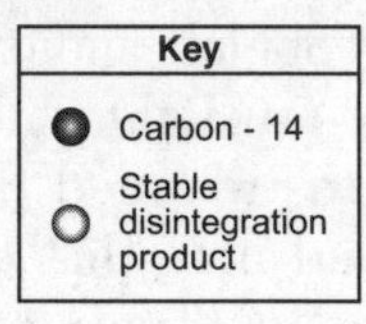

57. According to the key, carbon-14 atoms are shaded on the diagram. Count the number of shaded particles in the original sample and note that it contains sixteen carbon-14 atoms. Count the number of shaded particles in the sample after one half-life and note that it contains eight carbon-14 atoms. The decay of a radioactive isotope occurs at a statistically predictable rate known as its half-life. Half-life is the time required for one-half of the unstable radioactive isotope (such as carbon-14) to change into a stable disintegration product. Every half-life, one-half of the radioactive isotope atoms in the sample decay, forming atoms of stable disintegration product. Thus, at the end of the second half-life, one-half of the eight carbon-14 atoms that remained in the sample after the first half-life will decay, leaving four carbon-14 atoms.

One credit is allowed for **4 *or* four**.

58. Find the Radioactive Decay Data chart in the *Reference Tables for Physical Setting/Earth Science*. Locate "Carbon-14" in the "Radioactive Isotope" column. Trace right to the column labeled "Disintegration" and note that $^{14}C \rightarrow {}^{14}N$. Find the Properties of Common Minerals table in the *Reference Tables for Physical Setting/Earth Science.* In the key labeled "*Chemical symbols" at the bottom of the chart, note that "C = carbon" and "N = nitrogen." Thus, carbon-14 disintegrates into the stable disintegration product nitrogen-14.

One credit is allowed. Acceptable responses include but are not limited to:

- **nitrogen-14**
- $\mathbf{^{14}N}$
- $\mathbf{^{14}C \rightarrow {}^{14}N}$

59. Find the Geologic History of New York State chart in the *Reference Tables for Physical Setting/Earth Science*. In the column labeled "Era," find the statement "Estimated time of origin of Earth and solar system." Trace left to the scale labeled "Million years ago" and note that the origin of Earth occurred about 4600 million years ago (4.6 billion years ago).

Locate the Radioactive Decay Data chart in the *Reference Tables for Physical Setting/Earth Science.* Note that uranium-238 (^{238}U) has a half-life of 4.5×10^9 years; that is, 4.5 billion years. Thus, the radioactive isotope that has a half-life that is approximately the same as the estimated age of Earth is uranium-238 (^{238}U).

One credit is allowed. Acceptable responses include, but are not limited to:

- **uranium-238**
- $\mathbf{^{238}U}$

Note: The response "uranium" alone is *not* accepted because there are different isotopes of uranium.

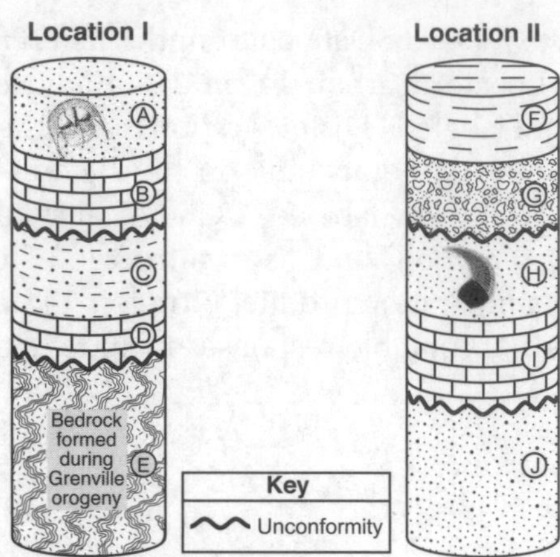

60. Index fossils are remains of organisms that had distinctive body features, were common and abundant, and had a broad, even worldwide range yet existed for only a short period of time. Their distinctive body features and broad distribution make index fossils easy to find in widely separated rock layers. Their short existence pinpoints the time period during which the rock layer was formed.

Find the Geologic History of New York State chart in the *Reference Tables for Physical Setting/Earth Science*. Locate the diagrams of index fossils along the bottom of the chart. Note that the index fossil that most closely resembles the one shown in rock layer *A* is "B–*Cryptolithus*." In the column labeled "Time Distribution of Fossils," locate the circled letter "**B**" and note that it lies on the gray bar labeled "trilobites." Note the explanation at the top of the column, "The center of each lettered circle indicates the approximate time of existence of a specific index fossil." From the center of the circled letter "**B**" representing *Cryptolithus*, trace left to the column labeled "Period." Note that *Cryptolithus* lived during the Ordovician Period. Therefore, the geologic period when rock layer *A* was deposited was the Ordovician Period.

One credit is allowed for **Ordovician Period**.

Note: "Middle Ordovician" is *not* accepted because that is an epoch.

61. It is given that the rock layers at these locations have not been overturned. The principle of superposition states that the bottom layer of a sedimentary series is the oldest, unless it was overturned or had older rock thrust over it, because the bottom layer was deposited first. Similarly, a rock layer is younger than those below it.

As explained in answer 60, rock layer *A* formed during the Ordovician Period. Based on superposition, rock layer *D* is older than rock layer *A* but younger than rock layer *E*. Find the Geologic History of New York State chart in the *Reference Tables for Physical Setting/Earth Science*.

Now locate the index fossil shown at location II in rock layer *H* in the diagrams of index fossils along the bottom of the chart. Note that the index fossil in rock layer *H* most closely resembles "D–*Valcouroceras.*" In the column labeled "Time Distribution of Fossils," locate the circled letter "**D**". From the center of the circled letter "**D**" representing *Valcouroceras*, trace left to the column labeled "Period." Note that *Valcouroceras* lived during the Ordovician Period. Therefore, rock layer *H* formed during the Ordovician Period. Thus, rock layer *H* and rock layer *A* were both formed during the Ordovician Period, or are roughly the same age, and allow location I to be correlated with location II.

By superposition, rock layer *G* is younger than rock layer *H*, and rock layer *F* is younger than rock layer *G*. Rock layer *D* is older than rock layer *A* and, by correlation, older than rock layers *F* and *G*. Finally, by superposition, rock layer *E* is older than rock layer *D*. Thus from oldest to youngest, the relative ages of rock layers *D*, *E*, *F*, and *G* are: *E-D-G-F*.

One credit is allowed for **the correct order shown below**.

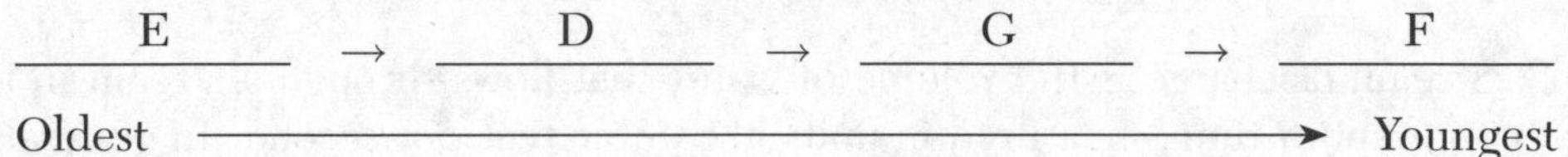

62. Layers of rock are generally deposited in an unbroken sequence. However, if forces within Earth cause rocks to be uplifted, deposition ceases. Weathering and erosion may wear away layers of rock before the land surface is low enough for another layer to be deposited. The uplift may cause rock layers to tilt or fold. The result is an unconformity, which is a break or gap in the sequence of a series of rock layers. Thus, the rocks above an unconformity are quite a bit younger than those below it. Thus, it is reasonable to infer that the rock units beneath the lower of the two unconformities were uplifted until they emerged from their depositional environment. Then their surface was weathered and eroded to form the unconformity. After that, the rock units subsided or were worn down until they were low enough for deposition to begin once again. Then several layers were deposited atop the unconformity, each burying the layers below it. Finally, the whole process occurred again, forming the upper unconformity and uppermost layers. Thus, the formation of the unconformities involved many processes, including uplift, emergence, weathering, erosion, subsidence, deposition, and burial. Any two of these processes should be listed as your answer.

One credit is allowed for *two* acceptable processes. Acceptable responses include but are not limited to:

- **uplift/emergence**
- **weathering**
- **erosion**

- **submergence/subsidence/sinking**
- **deposition/sedimentation/precipitation**
- **burial**

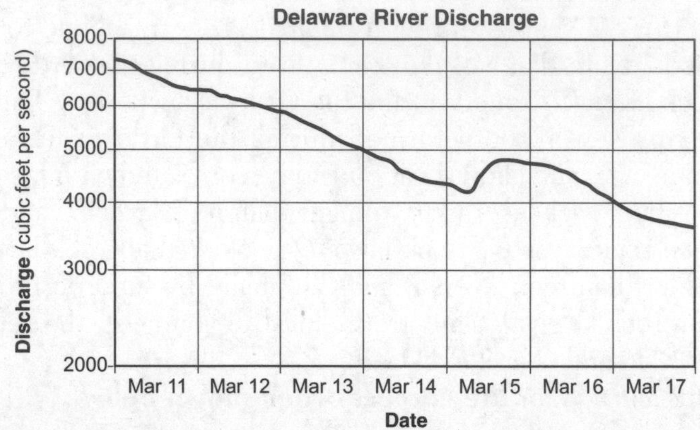

63. Stream discharge is the volume of water that flows through a stream in a given amount of time, measured in units like cubic feet per second. In general, the faster water is moving, the greater its carrying power and the more sediment the water can carry. All other things being equal, velocity is highest in streams with the highest discharge (rate of water flow). Thus, the greatest amount of sediment will be transported by the Delaware River when its discharge is the greatest. On the graph labeled "Delaware River Discharge," locate the highest point on the line. From that point, trace vertically downward to the horizontal axis labeled "Date" and note that the greatest discharge occurred on March 11.

Thus, the date when the river was most likely carrying the greatest amount of sediment was March 11.

One credit is allowed for **March 11**.

64. During a heavy rainfall or during a sudden snowmelt in the spring, the rate at which water builds up on the surface of the land exceeds the rate at which it can seep into the ground. Water that accumulates at the surface but does not seep into the ground moves downhill as runoff and adds to the volume of water in a stream. Therefore, the increase in stream discharge on March 15 was most likely due to rain or snowmelt that increased runoff into the stream, causing an increase in the volume of water in the stream.

One credit is allowed. Acceptable responses include but are not limited to:

- **precipitation**
- **rain shower**
- **snowmelt**

- **increased runoff**
- **flooding**
- **increase in the volume of water in the stream**

Note: Credit is *not* allowed for "velocity of water" or "increase in water velocity" because water velocity is a result of an increase in water volume or discharge.

65. It is given that the Delaware River flows out of the Catskills and into the Atlantic Ocean. First, find the Generalized Landscape Regions of New York State map in the *Reference Tables for Physical Setting/Earth Science,* and locate "The Catskills." Next, find the Generalized Bedrock Geology of New York State map in the *Reference Tables for Physical Setting/Earth Science.* Locate the region corresponding to the Catskills and the Delaware River along the Pennsylvania–New York State border. Note the location of the Atlantic Ocean to the southeast of the Catskills. Thus, the general compass direction toward which the Delaware River flows *along the New York State border* is toward the southeast. (Note that the Delaware River turns where the Pennsylvania–New York State border meets the New Jersey border and flows toward the south along the New Jersey border.)

One credit is allowed for **southeast/SE *or* south/S *or* south southeast/SSE**.

PART C

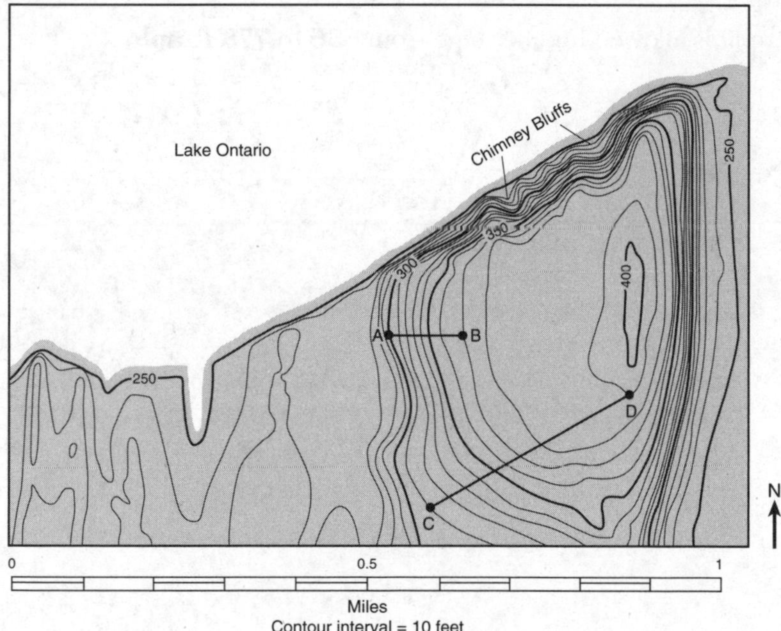

66. Find the Equations section in the *Reference Tables for Physical Setting/Earth Science,* and note the equation for gradient. To determine the gradient, proceed as follows:

a. Write the equation for determining the gradient:

$$\text{gradient} = \frac{\text{change in field value}}{\text{distance}}$$

b. Note that point *A* lies on the 300-foot contour line and point *B* lies on the 370-foot contour line. Align the edge of a piece of scrap paper with points *A* and *B* on the map, and mark off the distance from point *A* to point *B*. Compare this to the map scale below the map, and note that points *A* and *B* are 0.1 miles apart. Substitute the elevations of points *A* and *B* and the distance between them in the equation and solve:

$$\text{gradient} = \frac{370\text{ ft} - 300\text{ ft}}{0.1\text{ mile}} = \frac{70\text{ ft}}{0.1\text{ mile}} = 700\text{ ft/mile}$$

c. Label the answer with the correct units.

$$\text{Gradient} = 700\text{ ft/mile}$$

One credit is allowed for any value from **636** to **778 ft/mile**.

67. To construct a topographic profile along line *CD*, proceed as follows. Place the straight edge of a piece of scrap paper along the solid line connecting point *C* to point *D*. Mark the edge of the paper at points *C* and *D* and wherever the paper intersects a contour line. Wherever the paper intersects a contour line, label the mark with the elevation of that contour line as shown below.

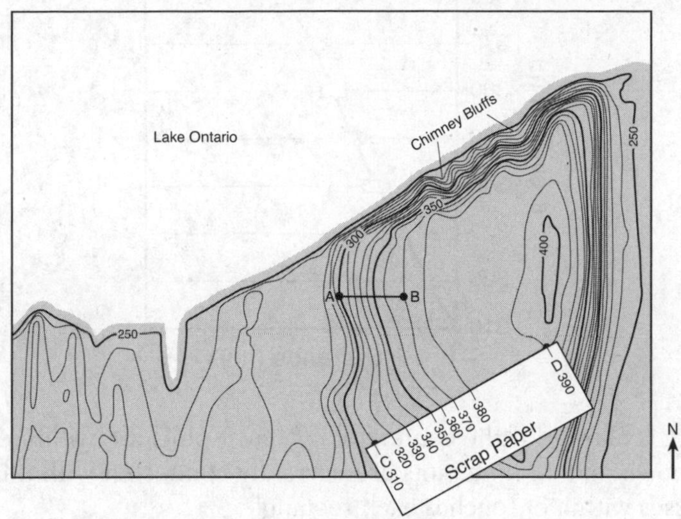

Now place this scrap paper along the lower edge of the grid provided, so points *C* and *D* on the paper align with points *C* and *D* on the grid. Then at each point where a contour line crossed the edge of the paper, draw a dot on the grid at the appropriate elevation.

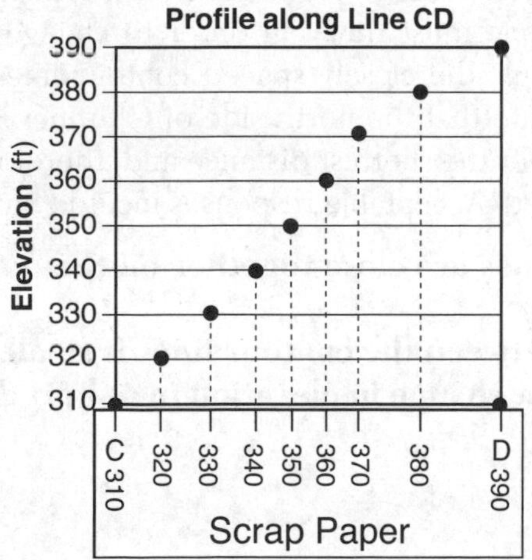

Finally, connect all of the dots in a smooth curve to form the finished profile, as shown below.

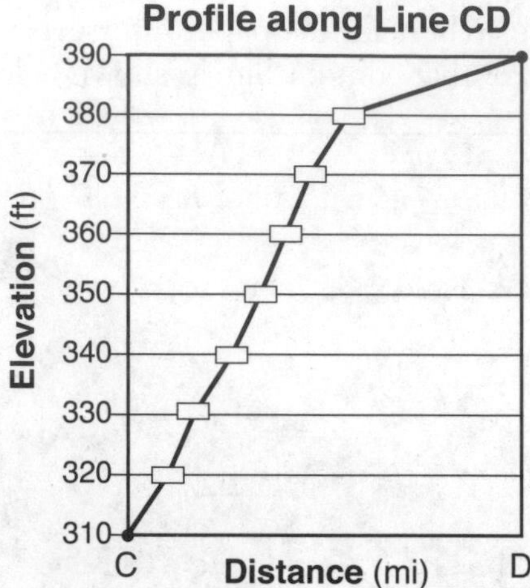

One credit is allowed if the centers of *all seven* plots are within or touch the rectangles shown and *all nine* plots are correctly connected with a line, from *C* to *D* that passes within or touches each rectangle.

Note: One credit is allowed if the line misses a plot, but is still within or touches the rectangle.

68. On a topographic map, contour lines connect points of equal elevation. The distance between one contour line and the next indicates the distance over which the elevation changes. The closer together the contour lines are on a map, the shorter the distance one must travel in order to change elevation, that is, the steeper the slope. Thus, the closely spaced contour lines on the north side of Chimney Bluffs indicate that the north side of Chimney Bluffs has the greatest change in elevation over the shortest distance and, therefore, the steepest slope.

One credit is allowed. Acceptable responses include but are not limited to:

- **The contour lines are close together on the north side of Chimney Bluffs.**
- **The spacing between the contour lines is small.**
- **There is a great change in elevation in a short distance on the north side.**

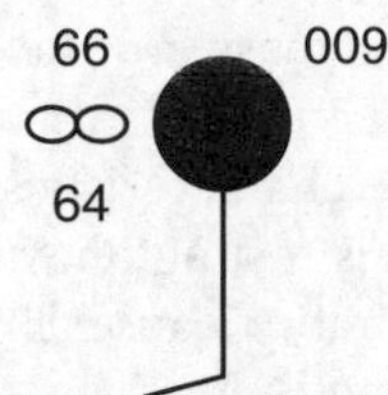

69. Find the Key to Weather Map Symbols in the *Reference Tables for Physical Setting/Earth Science* and locate the section labeled "Station Model Explanation." Note the positions in which the values for dewpoint, amount of cloud cover, barometric pressure, and present weather are placed on a station model.

Note that dewpoint (°F) is shown to the lower left of the station model. Thus, the dewpoint in Albany, New York, is 64°F.

Note that the amount of cloud cover is represented by how much of the circle at the center of the station model is shaded. The entire circle at the center of the station model is shaded. Thus, the cloud cover at Albany, New York, is 100%.

Note that barometric pressure is shown at the upper right of the station model and that the three digits listed represent the last three digits of the air pressure in millibars. Since air pressure rarely rises above 1050.0 mb and rarely drops below 950.0 mb, the general rule is to place a 9 in front of the three digits on the station model if they are greater than 500 and place a 10 in front of the three digits if they are less than 500. Then place a decimal point between the last two digits. According to the station model, the barometric pressure is represented by the digits 009. Since the three digits are less than 500, place a 10 in front of the digits and a decimal point between the last two digits. Therefore, the barometric pressure at the weather station is 1000.9 mb.

Finally, note that present weather is indicated with a symbol. According to the "Present Weather" key, the symbol shown on the station model represents haze. Therefore, the present weather at Albany, New York, is haze. Record these values in the table provided, as shown below.

One credit is allowed if ***all four* weather conditions are correctly recorded as shown below.**

Weather in Albany, New York	
Dewpoint	64°F
Amount of cloud cover	100%
Barometric pressure	1000.9 mb
Present weather	Haze

70. The characteristics of an air mass are the result of the geographic region over which it formed, or its *source region*. Air resting on or moving very slowly over a region tends to take on the characteristics of that region. For example, air that sits over the Gulf of Mexico is resting on warm water. The air is warmed by contact with the water, and evaporation causes much humidity to enter the air. As a result, the air becomes warm and moist, just like the Gulf of Mexico. An air mass that forms over water is called a maritime air mass. A warm air mass is called a tropical air mass. Therefore, the air mass that normally forms over the Gulf of Mexico is a maritime tropical air mass.

Find the Key to Weather Map Symbols in the *Reference Tables for Physical Setting/Earth Science* and locate the section labeled "Air Masses." Note that mT is the symbol for a maritime tropical air mass.

One credit is allowed for **mT**.

Notes: Credit is allowed for using either uppercase or lowercase letters. However, credit is *not* allowed if the air mass letters are reversed, such as Tm.

For students who used the Spanish edition, either exclusively or in conjunction with the English edition of the exam, credit is allowed for the correct two-letter air mass symbol as it appears in either the English or Spanish *Reference Tables for Physical Setting/Earth Science*.

71. Low-pressure centers are usually associated with warm, moist air that is rising along frontal boundaries. As warm, moist air rises, it expands, which causes adiabatic cooling. As the air cools, the air temperature approaches the dewpoint temperature. When the air is cooled to its dewpoint or below, the water vapor in the air condenses. Clouds form when water vapor in the atmosphere changes back into tiny droplets of liquid water by the process of condensation or forms ice crystals by the process of deposition (sublimation).

One credit is allowed. Acceptable responses include but are not limited to:

- **condensation**
- **cooling**
- **expansion**
- **deposition/phase change directly from water vapor to ice**

Month	Jan	Feb	Mar	Apr	May	Jun	Jul	Aug	Sep	Oct	Nov	Dec
Average High Air Temperature (°F)	14	15	21	30	41	50	54	53	47	36	28	18
Average Snowfall (in)	44	40	45	36	12	1	Trace	0.1	2	18	38	46

72. To construct a graph of the average high air temperatures for each month, do the following. In the data table, note that during the month of January, the "Average High Air Temperature (°F)" is 14. On the grid provided, locate the intersection of January on the horizontal axis labeled "Months" and 14 on the vertical axis labeled "Air Temperature (°F)." Draw an **X** at this intersection. Repeat this process for each of the remaining months and their corresponding average high air temperatures listed in the table given in the question. When all of the **X**s have been plotted, connect them with a smooth line. A graph with all points correctly plotted and connected with a smooth line is shown below.

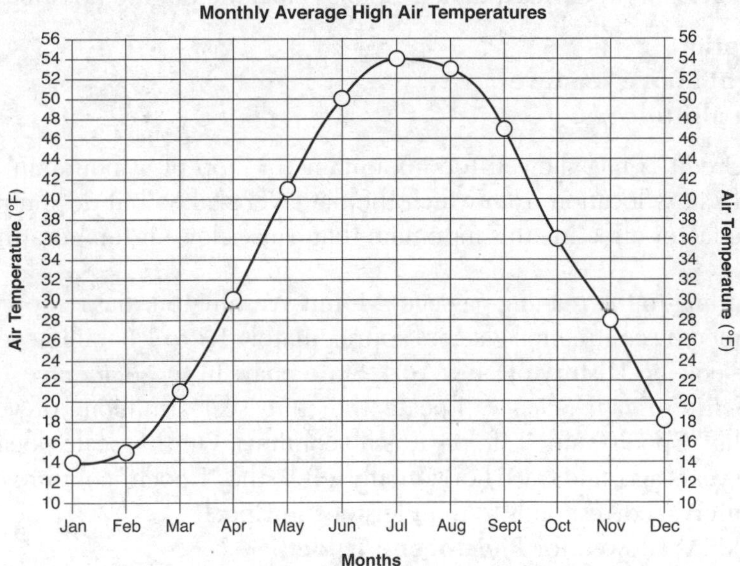

One credit is allowed if **the centers of *all twelve* plots are within or touch the circles shown and are correctly connected with a line that passes within or touches each circle**.

Note: Credit is allowed if the student-drawn line does not pass through the student plot points but is still within or touches the circles. Credit is *not* allowed if the student makes any attempt to graph average snowfall on the Monthly Average High Temperatures graph because the vertical axis is in degrees Fahrenheit, not inches.

73. It is given that Mount Washington is one of the highest mountains east of the Mississippi. Temperature is a major climate factor. Mountains are, by definition, regions of high elevation. Find the Selected Properties of Earth's Atmosphere chart in the *Reference Tables for Physical Setting/Earth Science.* Locate the graph labeled "Temperature (°C)." Note that temperature decreases rapidly with increasing altitude (elevation) in the troposphere. Elevation is a climate factor because there is a decrease in temperature with an increase in elevation. The fact that temperature decreases about 1°C for every 100-meter rise in elevation explains why high mountains may have tropical vegetation at their bases but permanent ice and snow at their peaks. Thus, the climate factor that best explains why Mount Washington has low air temperatures throughout the year is elevation (altitude above sea level).

One credit is allowed. Acceptable responses include but are not limited to:

- **elevation**
- **height above sea level**
- **high altitude**

Note: Credit is *not* allowed for "mountain" or "top of a mountain" because this identifies the location from which the data were taken but does not identify the climate factor affecting this mountain that causes low air temperatures.

74. According to the reading passage, Mount Washington "has deep-cut glacial valleys formed by continental glaciers approximately 1.7 to 1.8 million years ago." Find the Geologic History of New York State chart in the *Reference Tables for Physical Setting/Earth Science*. Locate the time scale (millions of years ago) between the "Epoch" and "Life on Earth" columns. On this scale, locate 1.7 to 1.8 million years ago, and trace horizontally left to the "Epoch" column. Note that this time interval corresponds to the Pleistocene Epoch.

One credit is allowed for **Pleistocene Epoch**.

Velocities of Galaxies Moving Away From Earth	
Galaxy	**Velocity (km/s)**
A	61,000
B	15,000
C	1200
D	39,000
E	22,000

75. It is given that the farther away a galaxy is, the faster it is moving away. Therefore, to list the galaxies in order from closest to Earth to farthest from Earth, they should be listed in order from lowest to highest velocities at which they are moving away from Earth. According to the table, this is: C-1200, B-15,000, E-22,000, D-39,000, and A-61,000.

One credit is allowed for **a correct list as shown below**.

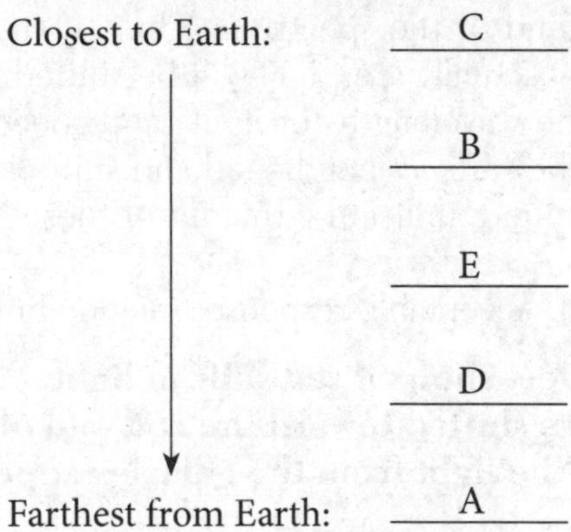

Note: One credit is allowed if the correct velocities are substituted for the letters.

76. The best explanation for the observation that all galaxies are moving away from each other is that the universe is expanding in all directions. If the motion of galaxies is then traced back in time, there is a point about 13 to 14 billion years ago at which the galaxies were very close to each other. Thus, we have a model in which the universe started out with all its matter and energy in a very small volume and then expanded outward in all directions. This motion is similar to an explosion, hence the name for this model—the Big Bang theory.

One credit is allowed for **the Big Bang theory** ***or*** **the Big Bang**.

77. When light emitted by the atoms in a star's core passes through the star's cooler outer layers, certain wavelengths of light are absorbed. The result is a pattern of dark spectral lines in the spectrum of light emitted by the star. Spectral lines are highly atom specific and can be used to identify the chemical element that produced the spectrum. When compared with the spectral lines of elements on Earth, the spectral lines of elements in distant stars have the same pattern, but that pattern is shifted from its normal position. By observing the pattern of spectral lines in light emitted by galaxies, scientists are able to determine that a galaxy is moving toward or away from Earth in the following manner. If a source of electromagnetic waves is moving *away* from an observer at the same time as the source is emitting light of a particular wavelength, fewer wave crests will reach the eye of the observer each second. The eye will interpret this as meaning that the light has a longer wavelength than it actually has, and the spectral lines will be shifted toward the longer wavelength end of the spectrum. If the source is moving toward the observer, the opposite will occur. This is called the Doppler effect.

Find the Electromagnetic Spectrum chart in the *Reference Tables for Physical Setting/Earth Science.* Locate the section of the chart labeled "Visible light." Note that light at the red end of the spectrum has a longer wavelength than light at the blue end of the spectrum. Thus, if the wavelengths of light appear longer than they actually are, they will be shifted toward the red end of the spectrum. When the wavelengths of light (and spectral lines) from distant galaxies are observed, they are *all* lengthened and shifted toward the red end of the spectrum, or red-shifted, indicating that *all* of these galaxies are all moving away from us.

One credit is allowed. Acceptable responses include but are not limited to:

- **The Doppler effect shows a red shift in light.**
- **Spectral lines are shifted toward the red end of the spectrum.**
- **Wavelengths of the light from the galaxies appear to be longer.**
- **Doppler effect**
- **Red shift**

Note: Credit is *not* allowed for "cosmic background radiation" because even though this is evidence for the Big Bang event, it is not used to determine that a galaxy is moving away from Earth.

78. Find the Characteristics of Stars chart in the *Reference Tables for Physical Setting/Earth Science.* Locate "10" on the scale labeled "Luminosity (Rate at which a star emits energy relative to the Sun)" along the left side of the chart and "8000" on the scale labeled "Surface Temperature (K)" along the bottom of the chart. Trace right from luminosity 10 and upward from temperature 8000K, and

locate the point where these two values intersect. Note that the intersection falls within the region labeled "Main Sequence (Early stage)." Now from "8000" on the scale labeled "Surface Temperature (K)" along the bottom of the chart, trace downward to the scale labeled "Color." Note that 8000K corresponds to a star with a white color.

One credit is allowed if ***both* responses are acceptable.**

Stage: **main sequence *or* early stage**

Color: **white**

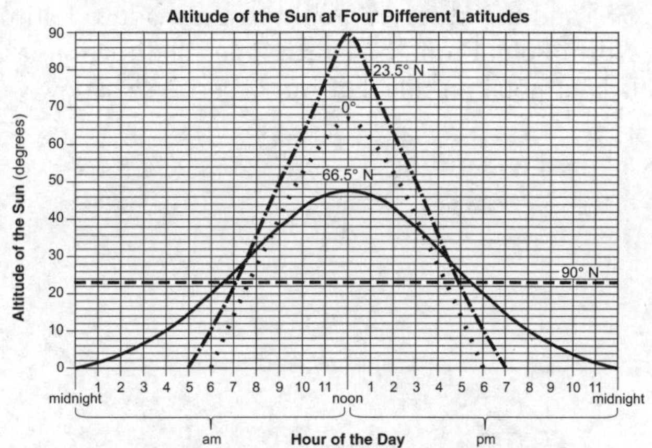

79. According to the graph, the altitude of the Sun at 23.5° N is 90°. On June 21, the summer solstice in the Northern Hemisphere, the Sun is directly overhead (altitude 90°) at the Tropic of Cancer, which is located at latitude 23.5° N. Thus, the date represented by the data shown on the graph is June 21.

One credit is allowed. Acceptable responses include but are not limited to:

- **June 20**
- **June 21**
- **June 22**
- **summer solstice**

80. As the angle of insolation (altitude of the Sun) increases from 0° toward 90°, the same amount of insolation is spread over smaller and smaller areas. Insolation is most intense when it strikes a surface at 90° (that is, perpendicular to a surface) because the insolation is concentrated in the smallest possible area. Therefore, as the noontime altitude of the Sun (angle of insolation) increases, the intensity of insolation increases. According to the graph, the altitude of the Sun is 90° at latitude 23.5° N (Tropic of Cancer). Therefore, the latitude that has the greatest intensity of insolation at approximately noon on this date is 23.5° N.

One credit is allowed for **23.5° N *or* Tropic of Cancer and acceptable evidence**. Acceptable evidence includes but is not limited to:

- **The graph shows the altitude of the Sun is greatest at 23.5° N.**
- **Sun is highest in the sky.**
- **The Sun's altitude is 90°.**
- **The Sun is directly overhead.**

81. According to the graph, if an observer is located at 90° N, the altitude of the Sun would be 23.5° and the observer would cast a shadow. Latitude 90° N corresponds to the North Pole. From the North Pole, all directions are to the south. Therefore, at 90° N at noon on this date, an observer's shadow would extend to the south.

One credit is allowed for **south**.

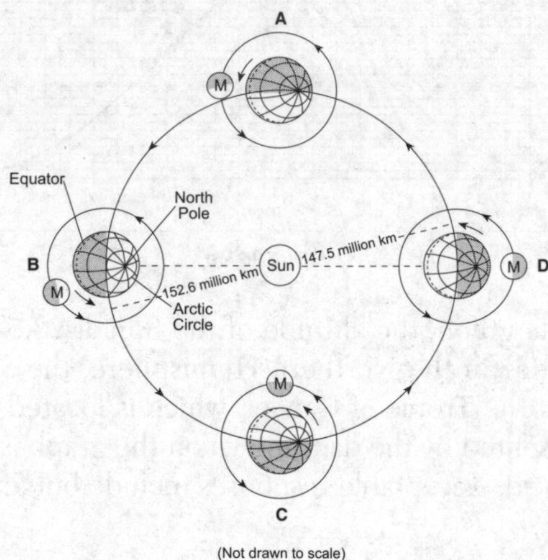

(Not drawn to scale)

82. An observer on Earth can only see that portion of the Moon that is illuminated by the Sun's rays and only from the side of Earth that is in darkness. In the diagram, the part of the Moon that is *not* shaded represents the illuminated portion of the Moon. According to the diagram, when the Moon is in position *D*, its entire illuminated portion faces Earth's nighttime side and the full moon phase would be observed.

When the moon is at position *D*, Earth is directly between the Moon and the Sun. When this occurs, the Moon can move through Earth's shadow, and a lunar eclipse can be viewed. Thus, a lunar eclipse could occur when the Moon is at position *D*.

One credit is allowed for **lettered position *D*** and **lunar eclipse**.

83. When sunlight strikes the spherical Earth, exactly one-half of Earth's sphere is illuminated and experiences daylight and exactly one-half is in darkness. The boundary between daylight and darkness forms a great circle that bisects Earth's sphere. On an equinox, the Sun's direct rays are at the equator, and the boundary between daylight and darkness passes through both the North and South Poles as it bisects Earth's spherical surface. Thus, position *A* represents an equinox. Note the arrows on Earth's orbit indicate its direction of motion. From position *A*, Earth would orbit toward position *B*. On the first day of summer (summer solstice) in the Northern Hemisphere, Earth's axis of rotation is tilted farthest toward the Sun. The position in the diagram at which Earth's axis of rotation is tilted farthest toward the Sun is position *B*. Therefore, position *B* represents the first day of summer. Therefore, position *A* represents the equinox that occurs before the first day of summer, or the spring equinox. The season between the spring equinox and the summer solstice is spring. Therefore, the Northern Hemisphere season that is beginning at position *A* is spring.

One credit is allowed for **spring**.

84. It is given that Earth is closest to the Sun near position *D*. Therefore, as Earth moves from position *C* to position *D*, the distance between Earth and the Sun decreases.

The force of gravitational attraction between any two objects is directly proportional to their masses and indirectly proportional to the square of the distance between their centers. Newton expressed this relationship in a simple mathematical expression:

$$F \propto \frac{m_1 \bullet m_2}{d^2}$$

Therefore, the force of gravitational attraction between the two objects increases as the distance between them decreases. Thus, the gravitational attraction between the Sun and Earth increases as Earth moves from position *C* to position *D* because the distance between the Sun and Earth is decreasing.

One credit is allowed. Acceptable responses include but are not limited to:

- **Earth is getting closer to the Sun.**
- **The distance between Earth and the Sun is decreasing.**
- **Earth's orbit is elliptical, and position *D* is closer to the Sun.**

85. A geocentric (*geo* = "Earth," *centric* = "centered") model of the solar system is one in which Earth is motionless at the center of the solar system and all other objects revolve around Earth. In the heliocentric (*helios* = "Sun," *centric* = "centered") model, the Sun is near the center of the solar system and all other objects in the system revolve around the Sun. Note that in the diagram, Earth orbits the Sun. Note, too, the arrows indicating that Earth rotates (is not motionless as in the geocentric model). Therefore, this diagram is a heliocentric model because it shows the Sun near the center of the diagram with a rotating Earth orbiting the Sun.

One credit is allowed. Acceptable responses include but are not limited to:

- **The Sun is near the center of the diagram.**
- **Earth orbits the Sun.**
- **The small arrows by Earth indicate that Earth rotates.**

Topic	Question Numbers (Total)	Wrong Answers (x)	Grade
Standards 1, 2, 6, and 7: Skills and Application			
Skills			
Standard 1 Analysis, Inquiry, and Design	1–5, 8, 10, 12, 14, 18, 20–25, 28, 33–36, 39–44, 46–50, 52–55, 57–62, 64–74, 76–81, 84		$\frac{100\,(60-x)}{60}=\%$
Standard 2 Information Systems			
Standard 6 Interconnectedness, Common Themes	2, 5, 7, 9, 10, 15, 17, 20, 23, 25–30, 36–39, 44–52, 54–56, 60–64, 67, 69–71, 75, 80, 82–85		$\frac{100\,(46-x)}{46}=\%$
Standard 7 Interdisciplinary Problem Solving			
Standard 4: The Physical Setting/Earth Science			
Astronomy The Solar System (MU 1.1a, b; 1.2d)	1–3, 8, 82–85		$\frac{100\,(8-x)}{8}=\%$
Earth Motions and Their Effects (MU 1.1c, d, e, f, g, h, i)	5, 7, 9, 36–39, 45, 46, 79, 81		$\frac{100\,(11-x)}{11}=\%$
Stellar Astronomy (MU 1.2b)	78		$\frac{100\,(1-x)}{1}=\%$
Origin of Earth's Atmosphere, Hydrosphere, and Lithosphere (MU 1.2e, f, h)	21		$\frac{100\,(1-x)}{1}=\%$
Theories of the Origin of the Universe and Solar System (MU 1.2a,c)	4, 6, 75–77		$\frac{100\,(5-x)}{5}=\%$

Topic	Question Numbers (Total)	Wrong Answers (x)	Grade
Meteorology Energy Sources for Earth Systems (MU 2.1a, b)			
Weather (MU 2.1c, d, e, f, g, h)	13–15, 42–44, 69–71		$\frac{100\,(9-x)}{9} = \%$
Insolation and Seasonal Changes (MU 2.1i; 2.2a,b)	15, 16, 18–20, 47, 80		$\frac{100\,(7-x)}{7} = \%$
The Water Cycle and Climates (MU 1.2g; 2.2c,d)	10–12, 17, 72, 73		$\frac{100\,(6-x)}{6} = \%$
Geology Minerals and Rocks (MU 3.1a, b, c)	33–35, 40, 41, 54		$\frac{100\,(6-x)}{6} = \%$
Weathering, Erosion, and Deposition (MU 2.1s, t, u, v, w)	29–32, 55, 56, 63–65		$\frac{100\,(9-x)}{9} = \%$
Plate Tectonics and Earth's Interior (MU 2.1j, k, l, m, n, o)	24, 25, 48–53		$\frac{100\,(8-x)}{8} = \%$
Geologic History (MU 1.2 i, j)	22, 23, 26, 27, 57–62, 74		$\frac{100\,(11-x)}{11} = \%$
Topographic Maps and Landscapes (MU 2.1p, q, r)	66–68		$\frac{100\,(3-x)}{3} = \%$
ESRT *2011 Edition Reference Tables for Physical Setting/Earth Science*	1, 3, 4, 14, 18, 20–25, 28, 33–36, 40–44, 47–50, 52, 54, 55, 58–60, 64–66, 69–71, 74, 78		$\frac{100\,(39-x)}{39} = \%$

To further pinpoint your weak areas, use the Topic Outline in the front of the book.
MU = Major Understanding (see Topic Outline)

NOTES

NOTES

NOTES

NOTES

NOTES

NOTES

NOTES